SPOTLIGHT-MODE SYNTHETIC APERTURE RADAR: A SIGNAL PROCESSING APPROACH

SPOTLIGHT-MODE SYNTHETIC APERTURE RADAR: A SIGNAL PROCESSING APPROACH

Charles V. Jakowatz, Jr.
Daniel E. Wahl
Paul H. Eichel
Dennis C. Ghiglia
Paul A. Thompson
Sandia National Laboratories
Albuquerque, New Mexico

SPOTLIGHT-MODE SYNTHETIC APERTURE RADAR:
A SIGNAL PROCESSING APPROACH

By: Charles V. Jakowatz, Jr., Daniel E. Wahl, Paul H. Eichel, Dennis C. Ghiglia, Paul A. Thompson

Library of Congress Control Number: 9547486

Printed in the United States of America.

9 8 7

springer.com

CONTENTS

FOREWORD

Historically, image formation has been accomplished by analog means. Over the past twenty-five years, however, the availability of high-speed digital computers has enabled the creation of a new field, called computed imaging or digital imaging. Computed imaging refers to the synthesis or computation of imagery using data collected from an object, a material, or a scene. Application areas include medicine, biology, material science, surveillance, navigation, and astronomy. Imaging modalities include computer tomography (CT), magnetic resonance imaging (MRI), ultrasound and acoustic imaging, x-ray crystallography, synthetic aperture radar (SAR), and radio astronomy. Two groups of researchers have driven the development of the field of computed imaging. Researchers in the first group are characterized by their tie to specific imaging applications and modalities. Typically, these persons possess intimate knowledge of the physical science and mathematics describing the imaging process within their own discipline. Researchers from this group have pioneered the development of new imaging modalities. With this beginning, each application area has tended to develop its own independent language and its own approach to modeling and data inversion. This has led to our current situation where researchers in SAR are unfamiliar with radio astronomy, researchers in radio astronomy know little about CT, and researchers in CT have no knowledge of SAR. And yet, these and many other computed imaging systems utilize similar signal processing principles. This fact has not escaped notice from a second group of researchers, who are drawn from the signal and image processing community. Researchers in this second group have interests in applications, but are experts on Fourier transforms, convolution, linear algebra, and the implementation of the associated numerical operations. Signal processing researchers who have worked on more than one imaging application are in a particularly good position to identify commonalities between different subfields within computed imaging, and to thereby help provide unification to the field.

This book on SAR adopts such an integrated, signal processing view. SAR is a means of constructing microwave images of extremely high resolution, using antennas of reasonable size. Applications abound in military, scientific, and commercial all-weather surveillance. This book considers a high-resolution type of SAR, called spotlight mode, where the radar antenna is steered to illuminate a scene using pulses transmitted from many different viewing angles. This type of SAR was originated by Jack Walker and his colleagues at the Environmental Research Institute of Michigan (ERIM), where it was originally described in range-Doppler terms. It was later discovered that spotlight-mode SAR works on the same principle as CT. Two colleagues and I, from the signal processing community, are credited with making this connection. Our study was limited to two-dimensional models for both the radar scene and the data-collection geometry. In this book, Charles Jakowatz and his colleagues from Sandia National Laboratories extend our thinking in a very major way, showing how the tomographic model can be used to describe true three-dimensional imaging. This leads to a very natural setting for the description of image layover, image formation from data collected on curved flight paths, and interferometric SAR for computing 3-D imagery and performing fine-scale change detection. All of these topics are covered, and much more. This book offers a comprehensive treatment of spotlight-mode SAR that can be understood by radar and signal processing engineers alike. It contains an excellent introduction to the basic range-resolving mechanism in radar, a complete development of the tomographic paradigm, a thorough description of the standard spotlight-mode SAR image-formation procedure, a superb discussion of autofocus algorithms, and a highly detailed and informative chapter on interferometry. Throughout, the book is well-illustrated, with many concepts demonstrated via actual SAR imagery. The writing of this book represents a major achievement, and I believe it will have a lasting impact on the SAR community. I am honored to have the privilege of introducing such a fine contribution to the technical community.

David C. Munson, Jr.
University of Illinois

PREFACE

More than two years ago, we decided to write a book about synthetic aperture radar imaging for at least four reasons. First, as researchers interested primarily in digital signal and image processing, we concluded that some of the most challenging and interesting research problems we had encountered were in the area of SAR. Second, we observed that signal processing solutions to problems in SAR often have useful application to real-world problems. For example, algorithms for automatic phase-error correction (autofocus) really do turn defocused imagery into useful products in a completely automated way, and techniques for SAR interferometry really do produce accurate terrain elevation maps of the earth's surface. Third, we noticed that there were very few books published on the particular mode of SAR imaging that this book addresses, namely that of spotlight-mode SAR. (The other major mode of SAR imaging, commonly known as strip-map SAR, has been widely written about since the early development of radar imaging by aperture synthesis some forty years ago.) Finally, we have come to believe that the principles of tomography and signal processing are the best way for many people new to spotlight-mode SAR to gain an understanding of this subject. Since David Munson and his colleagues at the University of Illinois introduced the analogy between spotlight-mode SAR and medical tomography in 1983, we have found this approach to be extremely valuable in our research and development efforts. This book is built upon those principles.

The book is intended for a variety of audiences. Hopefully, by using it, engineers and scientists working in the field of remote sensing but who do not have experience with SAR imaging will find an easy entrance into what can seem at times to be a very complicated and confusing subject. Experienced radar engineers will find that the book describes several modern areas of SAR processing that they might not have explored previously, e.g. interferometric SAR for change detection and terrain elevation mapping, or modern non-parametric approaches to SAR autofocus. Senior undergraduates and graduate students (primarily in electrical engineering) who have had courses in digital signal and image processing, but who have had no exposure to SAR could find the book useful in a one-semester course. Parts of the book are somewhat theoretical. These include Chapter 2, which develops the three-dimensional tomographic paradigm for SAR, and sections of Chapter 5 on SAR interferometry. Other portions are written from a somewhat more practical perspective. For example, Chapter 3 provides details on how to implement the polar reformatting image-formation algorithm, while Chapter 4 offers details on implementing the phase gradient autofocus (PGA) algorithm.

We owe a large debt of thanks to those who have helped us along the way. The book could not have been written without the continued support of management at the Sandia National Laboratories and the United States Department of Energy. Dr. Roger Hagengruber, vice-president at Sandia, encouraged us from the beginning to write the book and has continually supported our research. Max Koontz of the DOE sponsored much of the work that lead to our interest in SAR imaging over the last decade. Bill Childers of Sandia was invaluable to our effort in many ways. A number of Sandia staff scientists and engineers have worked with us, collected and processed SAR data for us, and taught us much about various aspects of radar imaging. A partial list of names (we undoubtedly have forgotten a few) includes: Don Lundergan (who conceived the idea of SAR interferometric change detection in 1978), Terry Calloway, Tom Flynn, Terry Bacon, Bruce Walker, Brian Burns, Bill Hensley, Doug Bickel, Thomas Cordaro, Gary Mastin, and Louis Romero. In addition, Dick Shead, Wynn Patton, Perry Gore, Kent DeGruyter, Kathy Best, and Kevin Rich did an excellent job in creating the artwork. Betty Tolman did a very thorough job of copy editing the manuscript.

We also give our thanks to Professor John Adams of California State University at Northridge for reading an early draft of the book and for encouraging us to continue. John and his colleagues at Hughes Aircraft Company also reviewed the final manuscript and provided valuable comments. We appreciate the efforts of Professor David Munson of the University of Illinois, who not only provided an extensive critique of the final manuscript and wrote the Foreword, but who also got us interested in the link between spotlight-mode SAR and tomographic imaging in the first place. Finally, the authors wish to thank their families for their patience and understanding during the very time-consuming process of book writing.

Charles V. Jakowatz, Jr. *Daniel E. Wahl* *Paul H. Eichel*
Dennis C. Ghiglia *Paul A. Thompson*

Albuquerque, NM

This work was performed in part at Sandia National Laboratories in Albuquerque, NM, supported by the U. S. Department of Energy under Contract DE-AC04-94AL85000.

1

RANGE RESOLVING TECHNIQUES

1.1 INTRODUCTION TO IMAGING RADARS

Modern airborne and spaceborne imaging radars, known as *synthetic aperture radars (SARs)*, are capable of producing high-quality pictures of the earth's surface while avoiding some of the shortcomings of certain other forms of remote imaging systems. Primarily, radar overcomes the nighttime limitations of optical cameras, and the cloud-cover limitations of both optical and infrared imagers. In addition, because imaging radars use a form of *coherent illumination*, they can be used in certain special modes such as *interferometry*, to produce some unique derivative image products that *incoherent* systems cannot. One such product is a highly accurate digital terrain elevation map (DTEM). The most recent (circa 1980) version of imaging radar, known as *spotlight-mode* SAR, can produce imagery with spatial resolution that begins to approach that of remote optical imagers. For all of these reasons, synthetic aperture radar imaging is rapidly becoming a key technology in the world of modern remote sensing.

Much of the basic "workings" of synthetic aperture radars is rooted in the concepts of *signal processing*. Starting with that premise, this book explores in depth the fundamental principles upon which the *spotlight* mode of SAR imaging is constructed, using almost exclusively the language, concepts, and major building blocks of signal processing. We contend that this approach offers a distinct advantage over the original views and formulations of the theory of spotlight-mode SAR [1], in that no highly specialized language or concepts are needed to explain any of the key ideas. In addition, throughout the text we have attempted to supplement theoretical concepts and equations with meaningful illustrative examples drawn from either computer-simulated or from real-world radar signals and images.

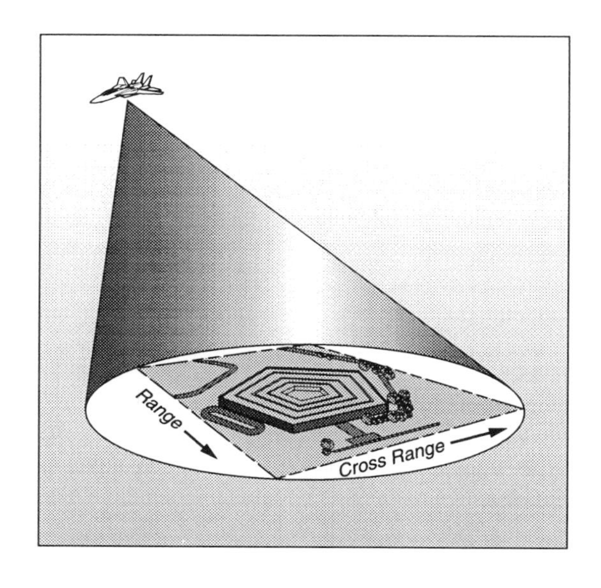

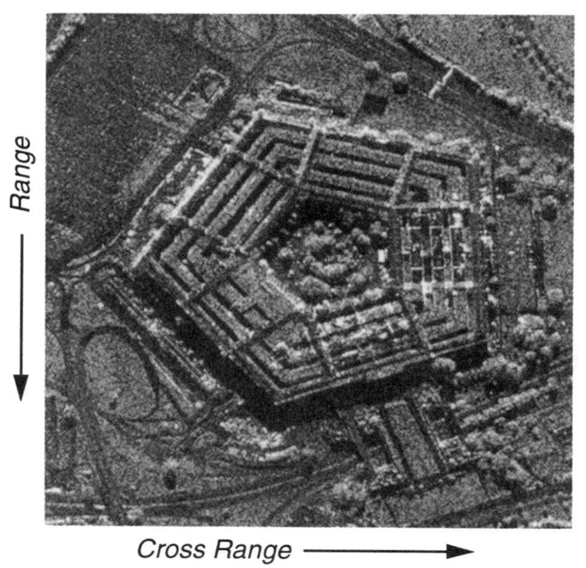

Figure 1.1 Airborne imaging radar illuminates the Pentagon. Basic SAR imaging geometry (above) and actual reconstructed SAR image (below).

We begin by introducing some basic notions that are fundamental to SAR imaging radars. The top of Figure 1.1 shows a typical geometry wherein an airborne SAR imager illuminates a patch of ground, composed in this case of the Pentagon building and its surroundings. The beam of the radar looks out to the side of the aircraft, pointing nominally in a direction orthogonal to the flight path. This direction of radiation propagation is referred to as the *range* direction, while the one nominally parallel to the flight path is called the *cross-range*, or *azimuth* direction. As the aircraft moves along its flight path, it periodically transmits pulses of microwave energy that impinge on the targets contained within the illuminated patch. Each pulse is subsequently *reflected* back to and received by the radar, where a demodulation procedure is performed. The assemblage of data collected and pre-processed in this manner is called a *phase history* and is passed on to an *image-formation processor*, which typically is located on the ground, but which could be onboard the flying platform. The image-formation processor produces as output a reconstruction of the *electromagnetic reflectivity* of the illuminated ground patch. The reflectivity is treated as a two-dimensional function that exists in the dimensions of range and cross range. The bottom of Figure 1.1 shows an actual SAR image, where the magnitude of reflected energy can be seen. Note that although the SAR image looks distinctly different from an optical photograph, the key features of the building are easily recognized. In this case the SAR was operating at a range of 6 km from the scene and produced an image with a resolution of approximately 1 meter in both dimensions. The details of why and how the SAR processor can reconstruct raw phase-history data into this kind of image and other related derivative products are provided in this book. In this chapter, we begin with a discussion of how multiple targets are resolved in the range direction. Chapter 2 deals with the cross-range resolution problem.

1.2 RANGE RESOLUTION IN IMAGING RADARS USING CW PULSES

Anyone who has shouted across a canyon and listened for the returning echo of their own "hello" has used the basic principle from which imaging radars achieve resolution in the *range direction*. The concept of echo-ranging simply states that to know an echo signal's round trip flight time and its speed of propagation is to know the *range* from the signal source to the target.[1] Figure 1.2 illustrates an airborne radar system that is illuminating several targets (reflectors) that are separated in range by some distance. In order to concentrate on the range-resolving problem, we will assume that all targets lie along a single line in the ground-range direction,

[1] Recall that "radar" is an acronym for radio detection and ranging.

i.e., at a given range, there are *not* multiple targets placed at various cross-range positions. Chapter 2 addresses the more general problem where targets lie in both range *and* in cross range.

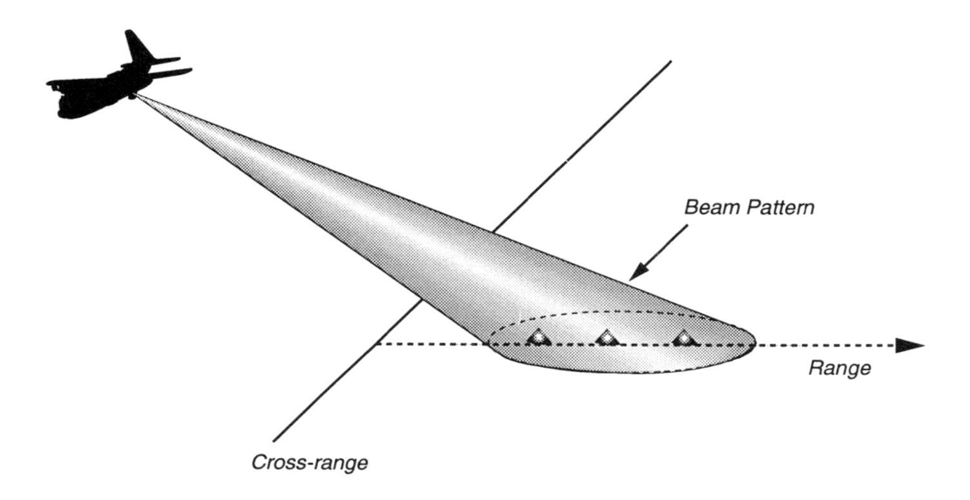

Figure 1.2 Illumination of multiple targets in the ground-range dimension.

Suppose the radar launches a continuous wave (CW) *burst* of microwave energy that is described by

$$s(t) = b(t)\ \cos(\omega_0 t)\ . \tag{1.1}$$

Here, $b(t)$ is an *envelope* function of duration τ_b seconds that amplitude modulates the carrier wave of frequency ω_0. That is, $b(t) = 0$ outside the interval $-\tau_b/2 \leq t \leq \tau_b/2$. A typical burst waveform is seen in Figure 1.3. In this case, the envelope function is generated as a raised cosine pulse

$$b(t) = \frac{1}{2}\left[1 + \cos\left(\frac{2\pi}{\tau_b}t\right)\right]\ . \tag{1.2}$$

Now consider a single point scatterer that is illuminated by the radar beam pattern. This target will reflect some of the incident energy back toward the radar. The diagram in Figure 1.4 shows the geometry used to describe the situation. The y dimension is the *ground range*, where the center of the illuminated patch is located at $y = 0$ and is offset from the platform's y position by amount y_0. The distance from the platform to the patch center is u_0, while the distance to the scatterer located at position y is given by $u_0 + u$, where u represents the *slant range*. With these

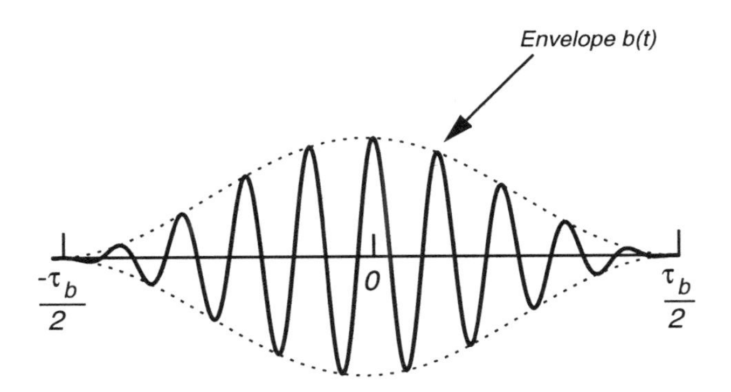

Figure 1.3 Continuous wave (CW) pulse with envelope $b(t)$.

definitions, then, the slant range and ground range are related by $u = y \cos \psi$, where ψ is the *depression angle*[2] of the collection. (This assumes that $y \ll u_0$, i.e., that the radar operates from a standoff distance that is large compared to the size of the ground patch that is illuminated.) For the present, we will develop an expression for the *range resolution* of a radar in terms of the slant range u. At the chapter's end, we will show how resolution and related parameters are easily modified if it is desirable to express them in terms of the ground range y instead. In Chapter 3, we extend this discussion significantly by examining the differences in computing and displaying a SAR image in terms of *ground-plane* vs. *slant-plane* coordinates.

The single scatterer results in the following signal received by the radar:

$$r_b(t) = A|g(u)| \cos(\omega_0 (t - \tau_0 - \tau(u)) + \angle\, g(u))\, b(t - \tau_0 - \tau(u)) \tag{1.3}$$

where $g(u)$ is the microwave reflectivity of the scatterer located at slant-range position u, and A is a scale factor that accounts for propagation attenuation and other effects. Note that the returned signal from this target is simply a *scaled and delayed version* of the signal that was transmitted. The complex-valued reflectivity of the scatterer is $g(u)$, i.e., its magnitude is $|g(u)|$ and its phase is $\angle\, g(u)$. The magnitude of this complex quantity determines the amount of the incident energy that is reflected back as the return signal, while $\angle\, g(u)$ describes a change in the phase of the waveform upon reception. In general, the value of $\angle g(u)$ is determined by the electrical properties of the target material (at frequency ω_0), as well as other factors, such as the target shape. The signal delay is simply the propagation time to and from the target (round trip) and is calculated according to the geometry of Figure 1.4

[2] We also use the term *grazing angle* interchangeably with depression angle. The grazing angle is the angle that the incident microwave radiation makes with the ground. This equivalence is strictly correct only when the scene to be imaged is flat, as is the case in Figure 1.4.

as $\tau_0 + \tau(u)$, where

$$\tau(u) = \frac{2u}{c} \tag{1.4}$$

and

$$\tau_0 = \frac{2u_0}{c} \ . \tag{1.5}$$

Equation 1.3 may be alternately expressed using the convenient complex number

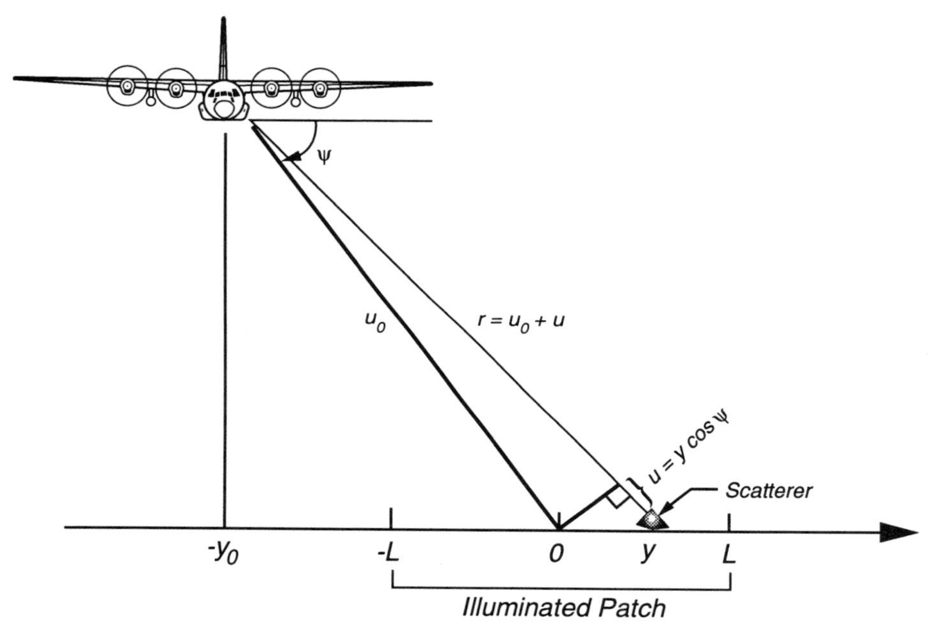

Figure 1.4 Collection geometry.

representation as

$$r_b(t) = A\,\mathbf{Re}\left\{g(u)\exp[j\omega_0(t - \tau_0 - \tau(u))]b(t - \tau_0 - \tau(u))\right\} \tag{1.6}$$

where $\mathbf{Re}\{\cdot\}$ denotes the real part of the complex quantity. A more realistic target structure consists of a *continuum* of scatterers placed along the range direction, so that the total returned radar waveform from all the scatterers within the patch is given by the integral expression

$$\begin{aligned} r_b(t) &= A\int_{-u_1}^{u_1} \mathbf{Re}\ \left\{g(u)\exp[j\omega_0(t - \tau_0 - \tau(u))]b(t - \tau_0 - \tau(u))\right\} du \\ &= A\,\mathbf{Re}\left\{\int_{-u_1}^{u_1} g(u)\exp[j\omega_0(t - \tau_0 - \tau(u))]b(t - \tau_0 - \tau(u))du\right\} \ . \end{aligned} \tag{1.7}$$

In the above equation, we must now interpret $g(u)$ as a microwave reflectivity *density* function. Equation 1.7 states that the return signal is a *superposition* of replicas of the CW burst pulse, each of which is delayed in time by an amount given by $\tau_0 + \tau(u)$, modified in phase by $\angle g(u)$, and modified in amplitude by $|g(u)|du$. Here, we assume that the ground ranges of the various scatterers in the patch illuminated by the radar beam vary from $y = -L$ to $y = L$ (see Figure 1.4), so that the slant ranges vary from $-u_1$ to u_1, where $u_1 = L \cos \psi$. We next define the *patch propagation time*, τ_p, as the difference in two-way propagation delay between a target at the near-edge and a target at the far-edge of the illuminated patch. It is given by

$$\tau_p = 2\tau(u_1) = 2\left(\frac{2u_1}{c}\right) = \frac{4L\cos\psi}{c} . \tag{1.8}$$

The returned signal described by Equation 1.7 has support on the interval given by

$$\tau_0 - \frac{\tau_p}{2} - \frac{\tau_b}{2} \leq t \leq \tau_0 + \frac{\tau_p}{2} + \frac{\tau_b}{2} . \tag{1.9}$$

The key idea behind resolving targets in range using this type of CW burst waveform is to keep the pulse duration as *small* as possible, so that targets closely separated in range, and therefore in time in the return signal, can be isolated by the radar processor. In such a case, it follows that the duration of the received signal is approximately equal to τ_p, because τ_b is chosen deliberately to be much smaller than τ_p.

A computer simulation of the radar signals described above is instructive at this juncture. The magnitude of the microwave reflectivity density function for a hypothetical portion of terrain is plotted in Figure 1.5. The phases of the reflectivities in this case are generated as a sequence of uncorrelated random numbers, uniformly distributed on the interval $[-\pi, \pi]$. (The relevance of this assumption on phase will become clear later.) Figure 1.6 depicts the corresponding radar return waveform, $r_b(t)$, generated according to Equation 1.7. In this case, a raised cosine CW burst waveform with $\tau_b \ll \tau_p$ was employed. As would be expected from the above discussion, the graph of Figure 1.6 features readily discernible delayed versions of the launched CW burst corresponding to each of the large impulse-like targets. (The CW burst contains five cycles of the RF carrier.) Distinguishing the returns of the background *clutter*, on the other hand, is not so easy. (Note that the dB scale of Figure 1.5 indicates the relative strengths of the impulse-like targets and of the background clutter. The return waveform of Figure 1.6 is plotted using a linear scale.)

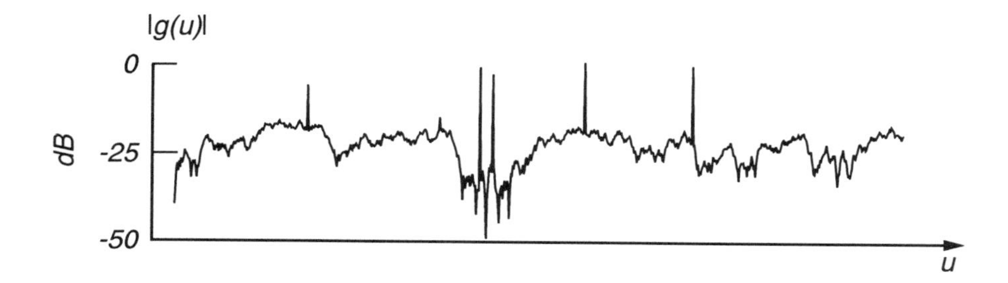

Figure 1.5 Typical terrain reflectivity density function (magnitude), as a function of u. Note that the vertical scale is in dB.

Figure 1.6 Return waveform from continuous reflectivity. Each strong reflector appears as a shifted, scaled replica of the CW burst. For the two strong targets spaced closely together, the return echos overlap.

The resolution to which $g(u)$ can be recovered from the return waveform depends on the duration of the transmitted CW burst waveform. One useful measure of effective duration that is appropriate for a pulse envelope waveform, $f(t)$, is given by

$$\tau_e = \frac{\int_{-\infty}^{\infty} f(t)dt}{f(0)} . \tag{1.10}$$

A corresponding measure of effective bandwidth is given by a similar expression, using the Fourier transform of the pulse, $F(\omega)$

$$B_e = \frac{1}{2\pi} \frac{\int_{-\infty}^{\infty} F(\omega)d\omega}{F(0)} . \tag{1.11}$$

While a rectangular envelope function would have an effective duration equal to τ_b, *shaped pulses* in general have smaller effective durations, and correspondingly *larger* effective bandwidths. From the defining equation for the Fourier transform, the above measures can be shown to have a product which is constant[3]

$$\tau_e B_e = 1 . \tag{1.12}$$

[3] See Cooper and McGillem [2], pp. 94-96 for a discussion of this.

The *time-bandwidth product* of pulse waveforms, by this definition, is always equal to unity in cycle measure, or 2π in radian measure. (As we will see in Section 1.3 when we discuss *dispersed* signals, however, the time-bandwidth product of certain waveforms cannot simply be measured from the duration and bandwidth of the pulse envelope because these waveforms encode bandwidth in other ways, a fact that is of vital importance to the design of real radar systems.) For the particular case of the raised cosine pulse of Equation 1.2, the calculation of Equation 1.10 for τ_e gives a value of $\tau_b/2$, and its corresponding measure of effective bandwidth is $2\pi B_e = 2\pi/\tau_e = 4\pi/\tau_b$. Figures 1.7 and 1.8 show the raised cosine burst and its Fourier transform, and indicate these effective duration and bandwidth measures.

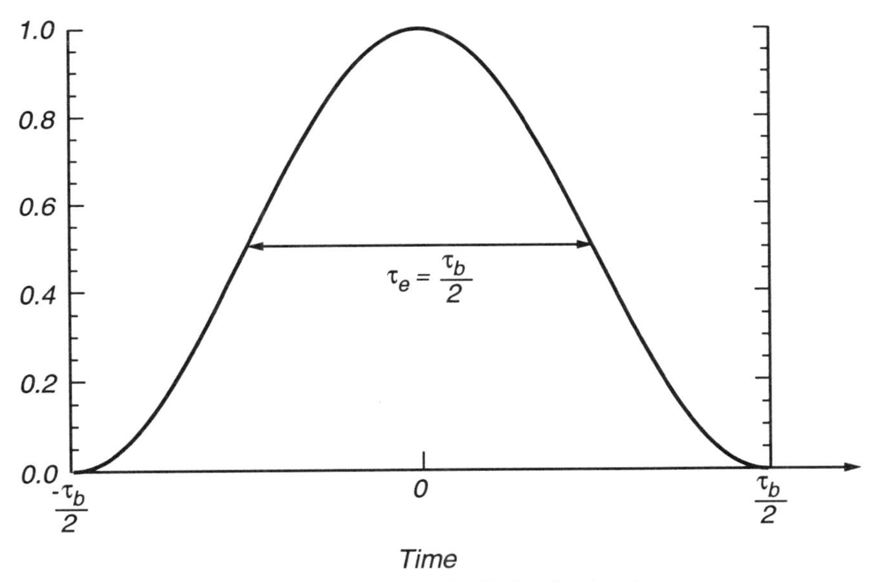

Figure 1.7 Raised cosine pulse envelope, with effective duration shown.

From the above discussions, it seems reasonable to assume that targets having return echos that are separated in time by τ_e seconds or more should be separable. In fact, the only processing required to accomplish this should be to remove the effects of the carrier frequency, so that a baseband signal of bandwidth $2\pi/\tau_e$ emerges. To this end, the returned waveform of Equation 1.7 is processed by *mixing* (multiplying) it with *in-phase* and *quadrature* (I and Q) sinusoids at frequency ω_0 and subsequently low-pass filtering the output of the mixer. The result of mixing the return signal

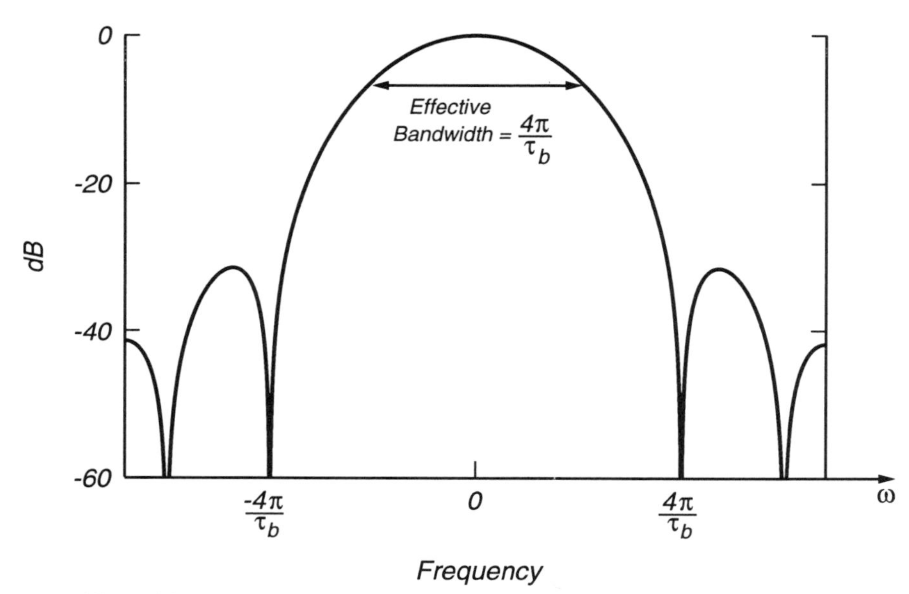

Figure 1.8 Fourier transform of pulse envelope, with effective bandwidth shown.

with the in-phase carrier term, given by $\mathbf{Re}\,\{\exp(j\omega_0\,(t-\tau_0))\}$, is:[4]

$$\tilde{r}_{bI}(t) = \frac{A}{2}\mathbf{Re}\left\{\int_{-u_1}^{u_1} g(u)\,\exp[j\omega_0\,(2t-\tau(u)-2\tau_0)]\,b(t-\tau(u)-\tau_0)\,du\right\}$$
$$+\frac{A}{2}\mathbf{Re}\left\{\int_{-u_1}^{u_1} g(u)\,\exp[-j\omega_0\,\tau(u)]b(t-\tau(u)-\tau_0)\,du\right\}. \tag{1.13}$$

The first term of the last result is centered on twice the carrier frequency, while the second term is baseband. Therefore, low-pass filtering the down-mixed signal will retain only the baseband term as

$$\bar{r}_{bI}(t) = \frac{A}{2}\mathbf{Re}\left\{\int_{-u_1}^{u_1} g(u)\,\exp[-j\omega_0\,\tau(u)]b(t-\tau(u)-\tau_0)\,du\right\}. \tag{1.14}$$

In a similar fashion, mixing the signal of Equation 1.7 with the *quadrature* term, $-\sin(\omega_0(t-\tau_0)) = -\mathbf{Im}\{\exp(j\omega_0(t-\tau_0))\}$, followed by low-pass filtering, will yield

$$\bar{r}_{bQ}(t) = \frac{A}{2}\mathbf{Im}\left\{\int_{-u_1}^{u_1} g(u)\,\exp[-j\omega_0\tau(u)]b(t-\tau(u)-\tau_0)\,du\right\}. \tag{1.15}$$

[4] The trigonometric identities relevant to mixing are: $\cos\alpha\cos\beta = \frac{1}{2}[\cos(\alpha-\beta)+\cos(\alpha+\beta)]$ and $\sin\alpha\cos\beta = \frac{1}{2}[\sin(\alpha-\beta)+\sin(\alpha+\beta)]$.

We then use a *complex signal* as a convenient mathematical construction to represent the two channels (rails) of the output of the demodulator, i.e. $\bar{r} = \bar{r}_I + j\bar{r}_Q$. Changing the variable of integration in Equations 1.14 and 1.15 to τ (see Equation 1.4), and including the nominal delay τ_0 in the left-hand side, the processed return as a complex signal becomes

$$\bar{r}_b(t) = a_0 \int_{-\frac{\tau_p}{2}}^{\frac{\tau_p}{2}} g\left(\frac{c}{2}\tau\right) \exp(-j\omega_0\tau) b(t-\tau)\, d\tau \tag{1.16}$$

where a_0 is a constant. The new limits of integration are $\pm\tau_p/2$, where τ_p is as given in Equation 1.8.

Two different interpretations of the above processed return are useful at this point. First, we employ a representation of Equation 1.16 as a convolution integral in order to demonstrate that targets separated in time by at least τ_e seconds in the processed return signal are indeed distinguishable. Specifically, $\bar{r}_b(t)$ may be viewed as the convolution of the burst pulse envelope with the function, $\bar{g}(\tau)$, which we define as

$$\bar{g}(\tau) = g\left(\frac{c}{2}\,\tau\right) \exp(-j\omega_0\tau)\ . \tag{1.17}$$

That is

$$\bar{r}_b(t) = a_0\, \bar{g}(t) \otimes b(t) \tag{1.18}$$

where the symbol $\otimes$ represents convolution. Note that the function $\bar{g}(\tau)$ is a scaled version of the reflectivity function, modified by a linear phase term that involves the radar center frequency. The output of the convolution of Equation 1.18 is effectively a *smoothed* version of $\bar{g}(\tau)$. The sliding window of the integration on $\bar{g}(\tau)$ is the pulse burst envelope function, $b(\tau)$, and its effective duration, τ_e, determines the measure to which the reflectivity function can be resolved. That is, smoothing $\bar{g}(\tau)$ over an interval length of τ_e implies that two targets will be *resolvable* in the processed return $\bar{r}_b(t)$ if they are separated in time by at least τ_e seconds, and therefore separated in slant range by

$$\rho_u = \frac{c\tau_e}{2}\ . \tag{1.19}$$

Because the effective bandwidth of the burst is the reciprocal of its effective duration, we can write the resolution equation equivalently as

$$\rho_u = \frac{c}{2B_e}\ . \tag{1.20}$$

Figure 1.9 shows the result of processing the return signal of Figure 1.6 with the mixing/low-pass filtering operation described above. The dotted plot is of the magnitude of the processed return, which estimates $|g(u)|$, while the solid curve shows the original $|g(u)|$ from Figure 1.5. Note that the strong point targets are reconstructed as replicas of the CW burst envelope, as predicted by Equation 1.18. Note also that two of the point targets are spaced sufficiently close together that the duration of the burst pulse is too large for them to be well-resolved in the processed return.

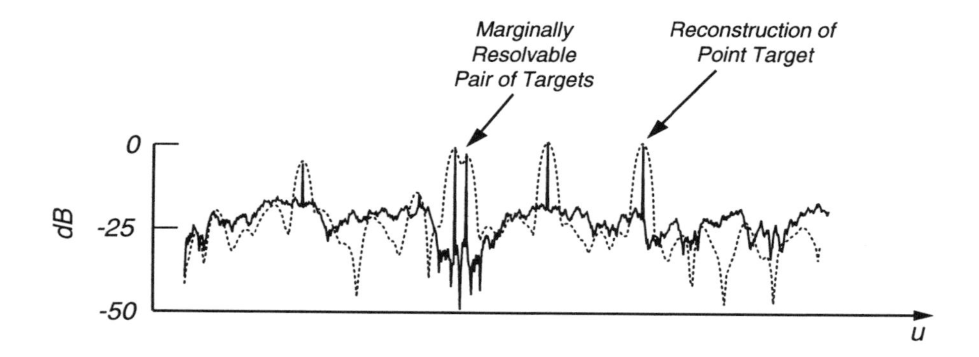

Figure 1.9 Estimate of terrain reflectivity after processing (demodulating) CW burst return waveform of Figure 1.6.

The reader at this point may be wondering why one would go to the trouble of performing the *synchronous detection* procedure described above to process the return, if all that is desired is an estimate of $|g(u)|$. That is, why should we mix the return with in-phase and quadrature sinusoids, when a simple envelope detector would produce the same result for the magnitude of $g(u)$? The answer is that the formation of a synthetic aperture radar image will necessitate the *coherent integration* of a number of such pulses transmitted and received by the radar from a series of positions along its flight path. This coherent integration requires that phase as well as magnitude information (as represented mathematically by the complex signal) be transduced by the processed return. A second interpretation of the demodulated return $\bar{r}_b(t)$, constructed this time in terms of its Fourier transform, gives further insight into this aspect of the processed signal.

The Fourier transform of $\bar{r}_b(t)$ can be calculated as[5]

$$R_b(\omega) = a_1 \, G\left[\frac{2}{c}(\omega + \omega_0)\right] B(\omega) \tag{1.21}$$

where a_1 is a constant, $B(\omega)$ is the Fourier transform of $b(t)$, and $G(U)$ is the Fourier transform of $g(u)$, with U a *spatial-frequency* variable. The frequency variables are

[5] Use the scaling, delay, and convolution theorems of Fourier transforms to obtain this result.

related as[6]

$$U = \frac{2}{c}\,\omega\,. \tag{1.22}$$

The above expression for $R_b(\omega)$ has an interesting interpretation. It states that values of the Fourier transform of the processed return, $\bar{r}_b(t)$, are equal to values of *a certain portion of* the Fourier transform of the terrain reflectivity function, $g(u)$. The portion of the $G(U)$ spectrum so determined is dictated by the support of $B(\omega)$ and by ω_0. If we consider the support of $B(\omega)$ equal to $2\pi B_e$ (see Figure 1.8), it follows that $G(U)$ is known for a range of *offset spatial frequencies* given by

$$\frac{2}{c}(\omega_0 - \pi B_e) \leq U \leq \frac{2}{c}(\omega_0 + \pi B_e)\,. \tag{1.23}$$

Equations 1.21 through 1.23 imply that the processed radar return waveform represents a *narrowband reconstruction* of the terrain reflectivity function, $g(u)$. As shown in Figure 1.10, the spatial bandwidth of this reconstruction is proportional to the effective bandwidth of the transmitted pulse, B_e. The center spatial frequency is proportional to the radar center frequency, ω_0. Finally, we note that this model assumes that the reflectivity g is independent of the frequency of the incident energy.

This description of the processed return in terms of offset spatial frequencies will be important when, in Section 1.3, we examine how dispersed waveforms can capture the same information regarding $g(u)$, by using the pulse compression technique. More importantly, this Fourier description becomes vital to explaining the cross-range resolving mechanism of a spotlight-mode SAR, via coherent integration of a large number of processed returns. To that purpose, we will return to a much expanded discussion of these ideas in Chapter 2.

Several key points regarding the above discussion of the range resolving mechanism should be emphasized: 1) the resolution (distance between separable targets) in the range dimension is inversely proportional to the radar bandwidth. For a typical imaging radar system, a ρ_u of 3 meters might be desired. This requires a bandwidth of 50 MHz, or equivalently, a CW burst pulse with duration of $2 \cdot 10^{-8}$ seconds (20 nanoseconds). If a resolution improved by ten times to 0.3 m were desired, then an increase of the bandwidth to 500 MHz would be necessary; 2) the range resolving power of the radar is *not* a function of the radar center frequency; 3) the range resolving power of the radar is *not* a function of the range from platform to scene center; 4) the processed return, $\bar{r}_b(t)$, reconstructs a narrowband version of $g(u)$, where the spatial frequencies transduced are such that the center of the spatial frequency interval is proportional to ω_0 and the width of the interval is proportional to the bandwidth of the transmitted waveform; 5) both magnitude and

[6] The units of radians/sec for ω become radians/m for U.

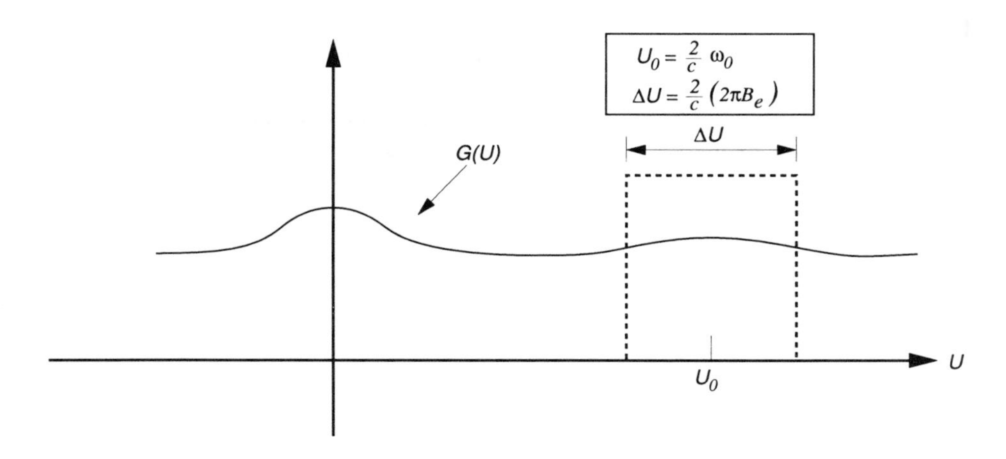

Figure 1.10 Offset spatial frequencies transduced by the radar.

phase components of the complex-valued reflectivity function, $g(u)$, are transduced by the process of quadrature demodulation.

Typical airborne imaging radar systems are used at ranges of several tens of kilometers, while the nominal range of spaceborne SARs is another order of magnitude greater. Unfortunately, the concept of the short-duration CW pulse described above quickly runs into practical difficulties. Specifically, the power level required by the microwave tube used to transmit the very short bursts of CW over such large distances becomes prohibitively high. That is, the *energy per pulse* that can be delivered in this case is low, because of the relatively low duty cycle factor. As a result, the operational standoff range of a system using CW bursts can be quite limited. There is, however, an alternative scheme for transmitting microwave energy that allows high bandwidths, and therefore good range resolution, but that uses long-duration waveforms, so that the energy delivered per pulse is relatively high (for a fixed level of tube power). This would appear at first to violate the condition of unity time-bandwidth product as prescribed by Equation 1.12. However, that requirement was derived strictly for pulse-like waveforms, and in fact does not apply to signals that encode bandwidth in certain other ways. These issues are the subjects of Section 1.3.

Summary of Critical Ideas from Section 1.2

- **The range resolving mechanism in imaging radars employs the well-known principle of echoing-ranging.**
- **A short CW burst is a natural choice for the echo waveform, although its short duty cycle makes it undesirable because of relatively small levels of energy per pulse that can be transmitted.**
- **The return signal from a continuum of reflectors in the range direction is described by a superposition integral, where each scatterer is represented by a delayed and scaled replica of the launched waveform. The complex reflectivity of each target determines the magnitude and phase of its representation in the return.**
- **An estimate of the complex reflectivity as a function of range is obtained by demodulating the carrier with in-phase and quadrature sinusoids, followed by low-pass filtering.**
- **A Fourier analysis of the range processor output shows that what the radar transduces is a narrowband estimate of the reflectivity function, i.e., the estimate is effectively derived from offset Fourier data, wherein the center spatial frequency is proportional to the radar center frequency, and the spatial bandwidth is proportional to the radar bandwidth.**
- **The range resolution depends on the radar bandwidth, but not on the range to the platform nor on the center frequency. It is given by:**

$$\rho_u = \frac{c}{2B_e} \, .$$

1.3 RANGE RESOLUTION USING DISPERSED WAVEFORMS

In this section we show that there are alternative ways for encoding radar signals that can be effectively used to achieve high resolution in range, but that do not suffer the same consequences of low average power levels that short CW bursts do. Such schemes involve the use of *coded waveforms* and of their associated signal processing techniques. We will discuss one kind of coded signal known as a *stretched* or *dispersed* waveform and an associated signal processing technique called *pulse compression*. The result will be to demonstrate that a radar's range resolution is not necessarily determined by the duration of the transmitted waveform.

To this end, consider a radar system that launches instead of the CW burst pulse of Figure 1.3, the *linear FM chirp* waveform described by $\mathbf{Re}\{s(t)\}$, with

$$s(t) = \exp[j(\omega_0 t + \alpha t^2)] \tag{1.24}$$

where $s(t)$ is 0 everywhere outside of the interval $-\tau_c/2 \leq t \leq \tau_c/2$. This waveform has the same linear phase term, $\omega_0 t$, as the CW burst, but, in addition, the new function has a *quadratic* phase term given by αt^2. Because frequency may be generally interpreted as the first derivative of phase, this term corresponds to a linearly increasing frequency, hence the name *linear FM chirp*. The quantity 2α, which has the units of $radian^2\ sec^{-2}$, is the so-called *chirp rate*, while ω_0 is the center frequency. The frequencies encoded by the chirp extend from $\omega_0 - \alpha\tau_c$ to $\omega_0 + \alpha\tau_c$, so that the bandwidth of the signal (in Hertz) is given by

$$B_c = \frac{\alpha\tau_c}{\pi} \,. \tag{1.25}$$

An example of such a chirp waveform (for $\alpha > 0$) is shown in Figure 1.11.

The utility of the FM chirp waveform in imaging radars comes about because the duration of this signal can be made long compared to that of the CW burst pulse, and yet the result is the *same effective bandwidth*. This is contrary to the basic notion developed in Section 1.2 for pulse waveforms, that longer durations imply smaller bandwidths, or equivalently, that the time-bandwidth product is always constant and equal to one cycle. The linear FM chirp, as well as certain other signals, exhibit the interesting property of possessing *very large time-bandwidth products*. The linear FM chirp is perhaps the most utilized of large time-bandwidth waveforms for imaging radars [3],[4],[5].

To demonstrate the large time-bandwidth product aspect of dispersed waveforms, we utilize a computer simulation of a linear FM chirp waveform and a CW burst

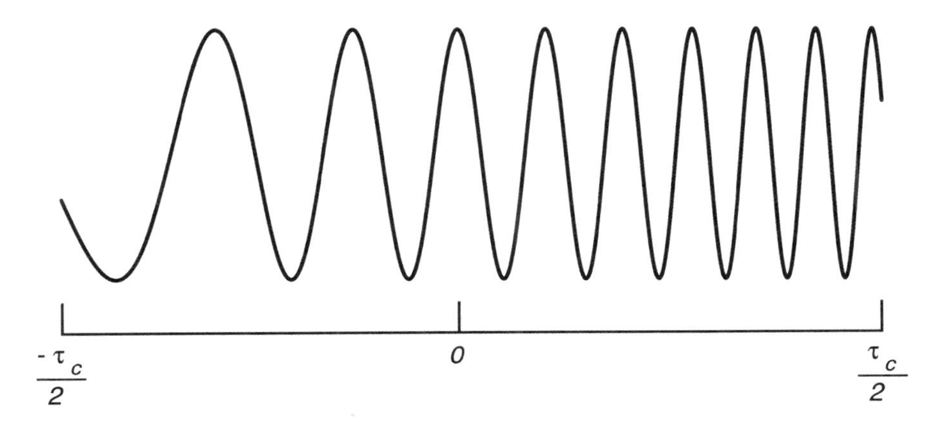

Figure 1.11 Linear FM chirp waveform.

that have the same bandwidth. The duration of the FM chirp, however, is 500 times greater than that of the effective duration of the CW burst. A plot of the discrete Fourier transform of the two waveforms, as shown in Figure 1.12, demonstrates the assertion that the effective bandwidths are equal. Note that the transform of the chirp is essentially flat over its range of frequencies.

We saw in Section 1.2 that the range resolution of a radar is determined by the bandwidth of its transmitted waveform. As a result, it should be possible to use the linear FM chirp signal in place of the CW burst to achieve the same resolution, but with much greater energy per pulse, assuming a fixed level of microwave tube power. In fact, the increase in delivered energy per pulse is in the ratio of the durations of the two waveforms, τ_c/τ_b. This, in turn, leads to a corresponding increase in signal-to-noise ratio in a radar image, which has beneficial effects on such image quality parameters as contrast.

What we will now demonstrate with a more careful mathematical analysis is that although the FM chirp in fact has a duration equal to τ_c, it can *behave* like a pulse with duration equivalent to the inverse of its bandwidth, i.e., $\tau_{eq} = 1/B_c$. The signal processing that allows this to happen is known as *pulse compression.* The amount of this compression is given by $\tau_c/\tau_{eq} = \tau_c B_c = \alpha\tau_c^2/\pi$, which is the *time-bandwidth product* (in cycles) of the waveform. In Section 1.2 we used as a measure of the bandwidth of a CW burst waveform the bandwidth of the pulse envelope. This was appropriate because the burst pulse rides on a constant carrier

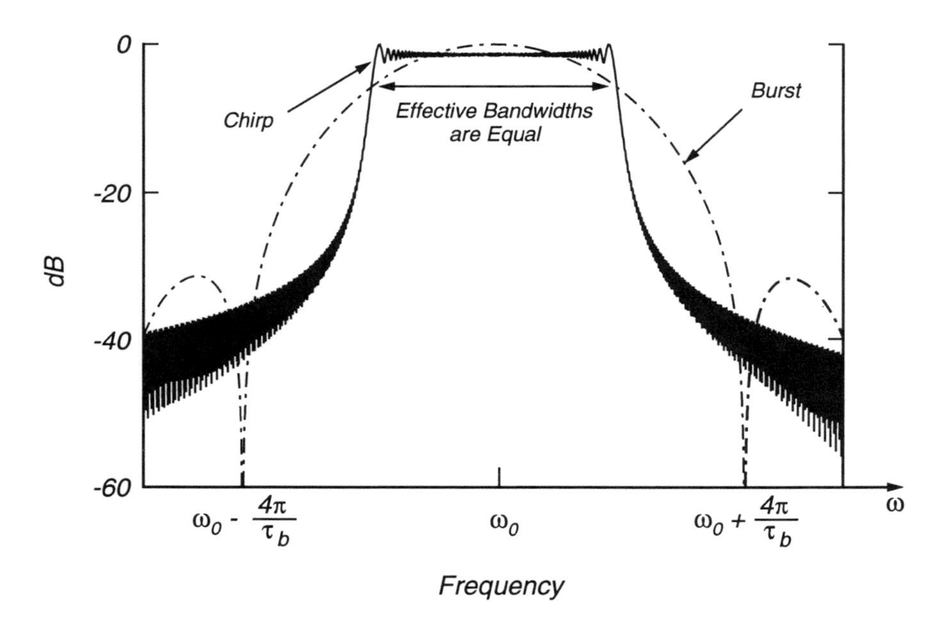

Figure 1.12 Fourier transform magnitudes of CW burst and FM chirp waveforms.

frequency, which merely translates the spectrum up by ω_0. In the case of the chirp waveform, however, because the signal is *frequency modulated*, its bandwidth is not just equal to the bandwidth of the envelope (a rectangular pulse of duration τ_c). On the contrary, as the duration of the envelope becomes larger, the bandwidth of the FM chirp becomes proportionately larger. The time-bandwidth products of typical FM chirps used in real imaging radars can be as large as 10^4. It is precisely these large values of pulse compression ratios and corresponding increases in transmitted energy levels that allow modern imaging radars to operate at useful standoff distances.

Analogous to the case of the CW pulse, the mathematical expression for the chirp return echo signal may be written as a superposition of multiple scaled and delayed versions of the chirp waveform, where the scaling is prescribed by the complex reflectivity value of each scatterer and the time delays are determined by the distance between a particular scatterer and the patch center (see Equation 1.7). If the radar beam illuminates those and only those targets lying at slant ranges constrained to $-u_1 \leq u \leq u_1$, the equation becomes

$$r_c(t) = A\mathbf{Re}\left\{\int_{-u_1}^{u_1} g(u)\,\exp[j[\omega_0\,(t-\tau_0-\tau(u)) + \alpha\,(t-\tau_0-\tau(u))^2]]du\right\}. \tag{1.26}$$

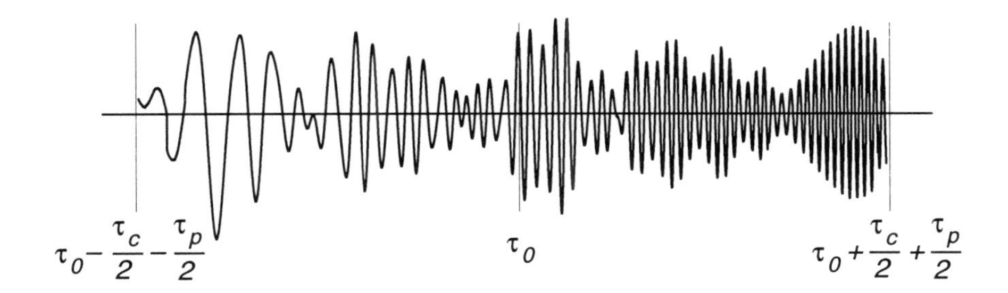

Figure 1.13 Returned waveform from continuous reflectivity using linear FM chirp.

(The interval of times for which the above equation is valid will be examined later in this section when we analyze a technique for processing (demodulating) the return.)

Computer simulations are again useful at this point to demonstrate some of the important aspects of the signals described above. The signal shown in Figure 1.13 is the simulated radar return from the terrain reflectivity function of Figure 1.5, when a linear FM chirp waveform is launched instead of a CW burst. Figure 1.13 is therefore analogous to Figure 1.6, although there is a major difference in these two returns. While the signal returned from the CW burst has a duration of approximately the patch propagation time τ_p (because $\tau_p \gg \tau_b$), the duration of the chirp return signal is essentially τ_c. This is because the duration of the chirp is deliberately chosen to be several times greater than τ_p for a reason that will become apparent later. Figure 1.14 is a timing chart showing the relationship between the chirp replicas returned from targets at the near-edge, center, and far-edge of the ground patch. Because of the reversal of situations regarding pulse length relative to patch propagation time for the burst and chirp waveforms, their return signals exhibit a marked difference in appearance. Whereas copies of the CW pulse from each of the large impulse-like targets are easily recognized in the signal of Figure 1.6, the waveform of Figure 1.13 does not possess this property. In fact, the return more resembles a single copy of the chirp waveform.

As may be expected from the above discussion of the differences in the nature of the returned signals of Figures 1.6 and 1.13, the processing required to extract an estimate of $g(u)$ in the linear FM chirp case is somewhat more complicated than simply removing the carrier frequency term via demodulation, as was done for the CW burst case. This is because the chirp waveform must be *deconvolved* from the convolution integral represented by Equation 1.26.

To demonstrate why Equation 1.26 does in fact represent a convolution integral, we again use the change of variable prescribed by Equation 1.4 and define a new

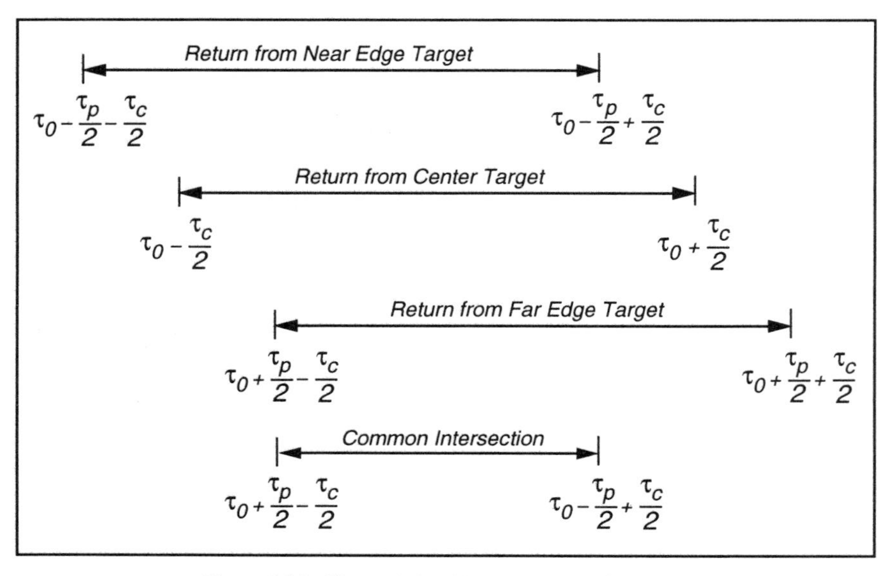

Figure 1.14 Time relationships of returned chirps.

function, $g_1(\tau)$ as

$$g_1(\tau) = g\left(\frac{c}{2}\tau\right) = g(u) . \tag{1.27}$$

Then, with $s(t)$ to denote the chirp (Equation 1.24), the equation for $r_c(t)$ can be interpreted as the convolution of g_1 and s, with the output evaluated at $t + \tau_0$. That is

$$r_c(t + \tau_0) = A_1 \mathbf{Re} \left\{ \int_{-\tau_1}^{\tau_1} g_1(\tau) s(t - \tau) d\tau \right\} . \tag{1.28}$$

Although there is more than one method for deconvolving $s(t)$ from $r_c(t)$ to obtain an estimate for $g_1(t)$, there is a particularly attractive technique for the case in which $\tau_c \gg \tau_p$, commonly referred to as *deramp* processing. This method is used in spotlight-mode synthetic aperture radars. Deramp processing is a chirp deconvolution procedure accomplished by using the following three steps: 1) *mixing* the returned signal with delayed in-phase and quadrature versions of the transmitted FM chirp; 2) *low-pass filtering* the mixer output; and 3) *Fourier transforming* the low-passed signal. Deramp processing also represents one particular form of the generic procedure called *pulse compression.* At the outset, it is certainly not obvious that these steps will accomplish the desired deconvolution of the chirp waveform. The mathematical analysis that follows demonstrates this interesting compression property of linear FM chirp waveforms.

To begin our analysis of deramp processing, we make one observation and one assumption. First, we observe that Equation 1.26 is valid only for times that are prescribed by

$$\tau_0 - \frac{\tau_c}{2} + \frac{\tau_p}{2} \leq t \leq \tau_0 + \frac{\tau_c}{2} - \frac{\tau_p}{2} . \tag{1.29}$$

The above inequalities restrict the processing window to only that common time segment for which chirp returns from all targets in the ground patch exist simultaneously, as shown in the timing diagram of Figure 1.14. This, in turn, implies that the arguments of each of the delayed chirp pulses in Equation 1.26 will be valid for all times considered. Second, we assume that

$$\tau_c \gg \tau_p . \tag{1.30}$$

This assumption was alluded to earlier in the description of the chirp return waveform of Figure 1.13. As will be seen at the conclusion of our deramp analysis, this constraint is chosen for a very important design consideration in actual spotlight-mode SAR imaging systems.

At the start of the deramp procedure, the return signal of Equation 1.26 is mixed with *in-phase* and *quadrature* versions of the FM chirp that are delayed by an amount precisely equal to the round-trip propagation time τ_0 to the center of the ground patch. This, of course, assumes that τ_0 is known for a given pulse. In real SAR imaging systems, this has major consequences not only for the design of the onboard motion compensation system, but for the signal reconstruction algorithms as well. The fact that τ_0 is only known imperfectly (even with the best of onboard electronic navigation systems) makes it necessary to have an important post-processing technique in SARs known as *autofocus*, or *automatic phase-error correction*. This subject is treated extensively in Chapter 4.

The deramp mixing terms are given by

$$c_I(t) = \cos(\omega_0(t - \tau_0) + \alpha(t - \tau_0)^2) \tag{1.31}$$

and

$$c_Q(t) = -\sin(\omega_0(t - \tau_0) + \alpha(t - \tau_0)^2) . \tag{1.32}$$

Use of the same trigonometric identities employed in the mathematics of demodulation of the carrier in the CW pulse case of Section 1.2 yields the following expression for the in-phase term of the mixer output (see reference [6]):

$$\tilde{r}_{cI}(t) = \frac{A}{2}\mathbf{Re}\left\{ \int_{-u_1}^{u_1} g(u) \exp[j[\omega_0(2t - \tau(u) - 2\tau_0) + \alpha((t - \tau_0)^2 + (t - \tau(u) - \tau_0)^2)\,]]\, du \right\}$$

$$+\frac{A}{2}\mathbf{Re}\left\{\int_{-u_1}^{u_1} g(u)\ \exp[j[\alpha\tau^2(u)-\tau(u)(\omega_0+2\alpha(t-\tau_0))]]\,du\right\}. \quad (1.33)$$

The first term of Equation 1.33 is removed by a low-pass filter, as it is centered on the frequency $2\omega_0$. The output of this filter, which is a baseband signal, is therefore

$$\bar{r}_{cI}(t) = \frac{A}{2}\mathbf{Re}\left\{\int_{-u_1}^{u_1} g(u)\ \exp\left[j\left[\alpha\tau^2(u)-\tau(u)(\omega_0+2\alpha(t-\tau_0))\right]\right]\,du\right\}. \quad (1.34)$$

In a similar fashion, the filtered quadrature component is obtained as

$$\bar{r}_{cQ}(t) = \frac{A}{2}\mathbf{Im}\left\{\int_{-u_1}^{u_1} g(u)\ \exp\left[j\left[\alpha\tau^2(u)-\tau(u)(\omega_0+2\alpha(t-\tau_0))\right]\right]\,du\right\}. \quad (1.35)$$

This process of mixing with I and Q versions of the chirp, followed by low-pass filtering, is often referred to as *quadrature demodulation*. The I and Q output signals from the quadrature demodulator can then be represented mathematically as a complex signal

$$\bar{r}_c(t) = \frac{A}{2}\int_{-u_1}^{u_1} g(u)\ \exp\left[j\left[\alpha\tau^2(u)-\tau(u)(\omega_0+2\alpha(t-\tau_0))\right]\right]\,du\,. \quad (1.36)$$

If we ignore for the moment[7] the quadratic phase term

$$\Phi_{skew} = \alpha\tau^2(u) = \alpha\left(\frac{2u}{c}\right)^2 = \frac{4\alpha u^2}{c^2} \quad (1.37)$$

we can write Equation 1.36 as

$$\bar{r}_c(t) = \frac{A}{2}\int_{-u_1}^{u_1} g(u)\ \exp\left[j\left[-\frac{2u}{c}(\omega_0+2\alpha(t-\tau_0))\right]\right]\,du\,. \quad (1.38)$$

Notice that the integrand of Equation 1.38 involves the general form of the Fourier transform kernel, $\exp(-juU)$, where (see Equation 1.22)

$$U = \frac{2}{c}\,\omega = \frac{2}{c}(\omega_0+2\alpha(t-\tau_0))\,. \quad (1.39)$$

As a result, we see that the processed return chirp signal is precisely equal to the Fourier transform of the reflectivity function, *evaluated over a specific limited range of frequencies* that are determined by the time-support of $\bar{r}_c(t)$. Because the

[7] We will return to this phase term in Chapter 2 (Section 2.6), when we discuss how it limits the size of the image patch reconstructed in a spotlight-mode SAR image.

interval on which we are processing this signal is given by Equation 1.29, the Fourier transform of $g(u)$ is determined over the range of *spatial frequencies* given by

$$\frac{2}{c}(\omega_0 - \alpha(\tau_c - \tau_p)) \leq U \leq \frac{2}{c}(\omega_0 + \alpha(\tau_c - \tau_p)) . \tag{1.40}$$

Given that we have assumed that $\tau_c \gg \tau_p$, and because the bandwidth of the chirp (in radians) is given by $2\pi B_c = 2\alpha\tau_c$, Equation 1.40 can be approximated by

$$\frac{2}{c}(\omega_0 - \pi B_c) \leq U \leq \frac{2}{c}(\omega_0 + \pi B_c) . \tag{1.41}$$

When Equation 1.41 is compared to Equation 1.23 of Section 1.2, it is concluded that the *same narrowband offset spatial frequencies of* $g(u)$ are transduced by using either a CW burst or a linear FM chirp waveform (assuming both have center frequency ω_0). This is true *only if the bandwidths are equal*, i.e., $B_e = B_c$. In this case, the spatial bandwidth transduced by the chirp waveform matches that of Figure 1.10 for the CW burst case, with B_e replaced by B_c:

$$\Delta U = \frac{2}{c}(2\pi B_c) . \tag{1.42}$$

The two situations are of course different in that a final Fourier transformation (also called range compression) of $\bar{r}_c(t)$ must be made in the chirp case to obtain the estimate of the reflectivity function. Stated another way, use of a linear FM chirp and subsequent quadrature demodulation results in an intermediate signal that measures a portion of the Fourier spectrum of the reflectivity. In this sense, range has been "converted" to spatial frequency. For the case of the CW burst on the other hand, the Fourier-domain data never appear as an intermediate step of the processing. The output of the demodulator/low-pass filter directly estimates the terrain reflectivity.

Because the spatial frequencies transduced by the chirp waveform match those obtained using a CW burst when the bandwidths of the two waveforms are the same, the resulting range resolution must be the same. That is, the expression for the resolution using the chirp must be Equation 1.20 with B_e is replaced by B_c:

$$\rho_u = \frac{c}{2B_c} . \tag{1.43}$$

Additional insight into the range resolution expression is found when analyzing the processed return from a single isolated point target that is mathematically described by a Dirac delta function, $g(u) = \delta(u)$ (arbitrarily located at $u = 0$). The result is commonly referred to as the *impulse response function (IPR)* of the radar. In this case, the Fourier data are constant over the interval ΔU given by Equation 1.42, because the transform of $\delta(u)$ is constant everywhere. The resulting reconstruction of the target following Fourier inversion (range compression) is then described by

$$|\hat{g}(u)| = \frac{\Delta U}{2\pi}\frac{\sin[u\Delta U/2]}{[u\Delta U/2]} = \frac{\Delta U}{2\pi}\,\text{sinc}\left[\frac{u\Delta U}{2\pi}\right] \tag{1.44}$$

where the *sinc* function is defined as

$$\text{sinc}(x) = \frac{\sin[\pi x]}{\pi x} . \tag{1.45}$$

The first zero crossing of this reconstruction occurs at a distance of $u = 2\pi/\Delta U$ from the origin, as shown in Figure 1.15. As a result, if we say that two point reflectors of equal size must be separated by at least this amount in order to be distinguishable, then Equation 1.43 easily follows from Equation 1.42. Without question, the *relative phases* of two targets that we wish to resolve are very important in this regard. The dotted plot of Figure 1.16 shows the net response from a pair of point targets separated by $u = 2\pi/\Delta U$, when one target phase differs from the other by 135 degrees. Note that the "dip" in this response is consistent with our usual notion of resolvability. On the other hand, the solid curve shows the response when the targets are separated by the same distance but when both have the same phase. In this case, there is no dip to indicate that two targets are present. Instead, a much-widened single mainlobe appears. It must be concluded that while the expression of Equation 1.43 does provide a certain useful and practical measure of resolution, the issue of resolvability is really more complicated than this simple formula might suggest. In Chapter 3, certain techniques for modifying the nature of the point target response are discussed.

The next series of computer simulations demonstrates that deramp processing of the chirp return radar signal does indeed perform as predicted by the mathematics described above. The same terrain reflectivity density function that was employed for the simulation of the CW burst case (Figure 1.5) was again used to generate the return waveform $r_c(t)$ as described by Equation 1.26. The chirp was chosen to have the same center frequency and bandwidth as the CW burst data which resulted in Figures 1.6 and 1.9, with the time-bandwidth product of the chirp chosen to be 500. The τ_c/τ_p ratio was approximately 9. Figure 1.17 shows the result of processing the simulated $r_c(t)$ via the deramp method.[8] (The solid line is the actual reflectivity magnitude of Figure 1.5, while the dotted line is the result of processing.) Note the similarity in this reconstruction to that of Figure 1.9. As implied in the discussion above, the reconstructions from the CW burst and the FM chirp have the same

[8] Just prior to range compression, a certain Fourier-domain window is applied to the data. This reduces the sidelobes of the response, at the cost of slightly widening the mainlobe. Chapter 3 discusses windowing in more detail.

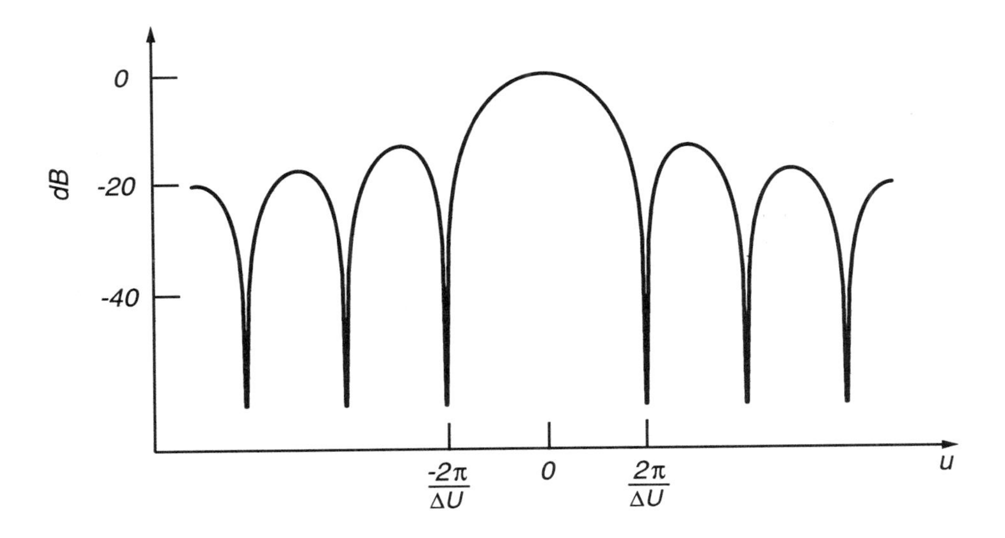

Figure 1.15 Sinc function response for point target with spatial bandwidth ΔU.

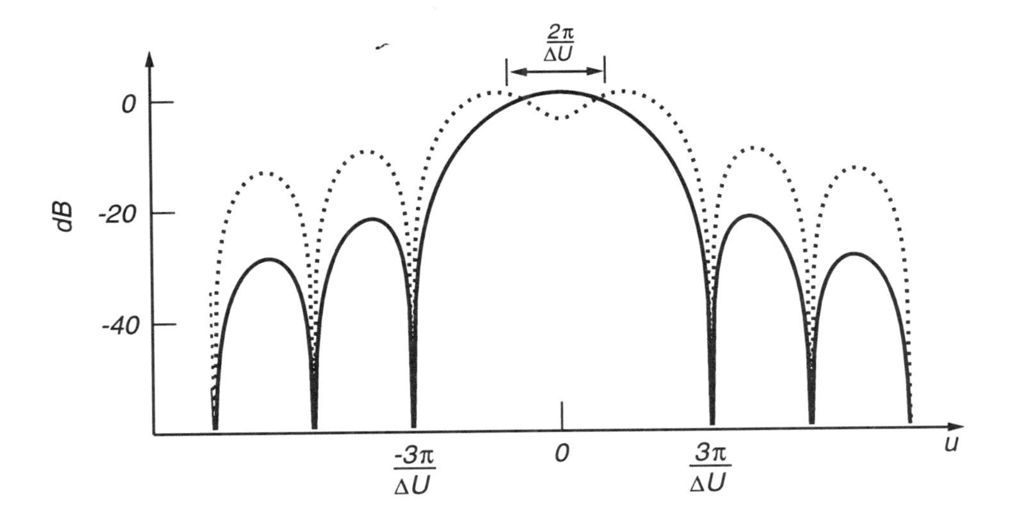

Figure 1.16 Reconstructions for pair of point targets separated by $2\pi/\Delta U$, with target phase differences of 0° (solid line) and 135° (dotted line).

inherent resolution because the bandwidths of the two transmitted waveforms are the same. Use of the chirp in this case, however, allows a transmitted duration which is 500 times greater than that of the burst. Figure 1.18 shows the result of constructing and processing a different simulated return from the same terrain reflectivity function, where the bandwidth of the transmitted chirp was doubled from that of the previous case. This was accomplished by doubling the chirp rate for the same chirp duration. Note that the reconstructed responses of point targets now appear to have approximately half the width as before, and that this is sufficient to resolve the two point targets that are not fully separable in the reconstruction of Figure 1.17.

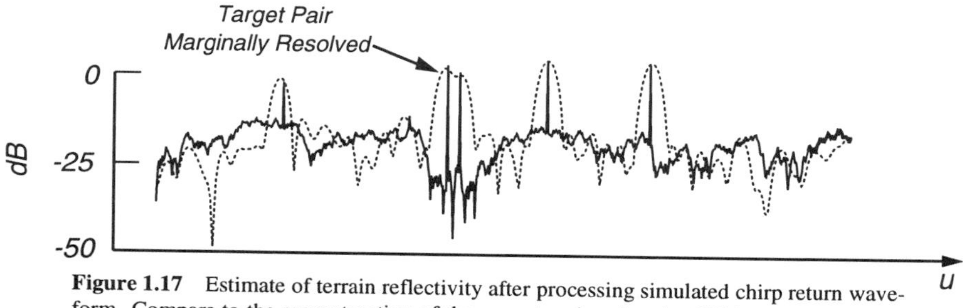

Figure 1.17 Estimate of terrain reflectivity after processing simulated chirp return waveform. Compare to the reconstruction of the same terrain using the CW burst (Figure 1.9).

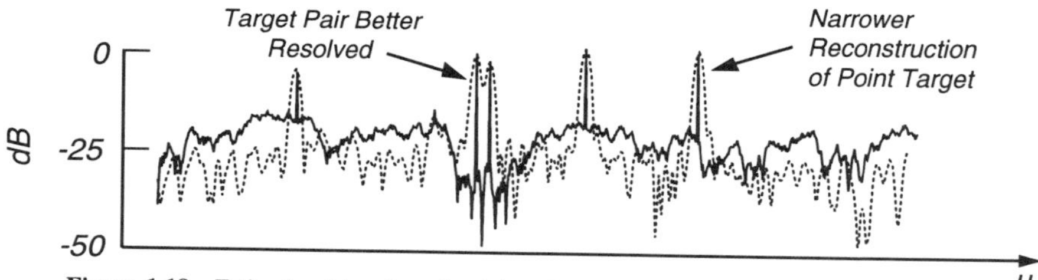

Figure 1.18 Estimate of terrain reflectivity after processing chirp return waveform. In this case, the transmitted chirp has twice the bandwidth employed in the data processed to obtain the result of Figure 1.17. Note the narrower reconstruction of single point targets as well as the better resolved closely spaced pair.

At this point, we return to the notion of slant range vs. ground range that we raised earlier in the chapter. All of the equations we have derived thus far involve an expression for the terrain reflectivity function in terms of the slant range u (Figure 1.4). Alternatively, we could have chosen to use the ground range y instead, and defined the reflectivity function as $g_0(y)$. Use of ground range in this manner ties the reflectivity function to earth coordinates. The analog of Equation 1.38 rewritten in terms of y and $g_0(y) = g(u)$ then becomes

$$\bar{r}_c(t) = A_2 \int_{-L}^{L} g_0(y) \exp\left[j\left[-\frac{2y\cos\psi}{c}(\omega_0 + 2\alpha(t-\tau_0))\right]\right] dy . \qquad (1.46)$$

The spatial frequency U of Equation 1.39 is replaced by a new spatial frequency given by

$$Y = \frac{2\cos\psi}{c}\omega = \frac{2\cos\psi}{c}(\omega_0 + 2\alpha(t-\tau_0)) . \qquad (1.47)$$

The Fourier transform $G_0(Y)$ is determined over the interval of Y values (see Equation 1.41) given by

$$\frac{2\cos\psi}{c}(\omega_0 - \pi B_c) \leq Y \leq \frac{2\cos\psi}{c}(\omega_0 + \pi B_c) . \qquad (1.48)$$

Similarly, the range resolution expression of Equation 1.43 becomes

$$\rho_y = \frac{c}{2\cos\psi B_c} . \qquad (1.49)$$

Note that the size of ρ_y is greater than that of ρ_u by a factor of $1/\cos\psi$. Does this suggest that better resolution can be achieved by using the slant-range as opposed to ground-range reconstruction of the terrain reflectivity profile? Of course the answer to this question is no, because two targets spaced apart in slant range by an amount Δu are also spaced further apart in ground range by the same factor, i.e. $\Delta y = \Delta u/\cos\psi$. Therefore, the *effective* range resolution is the same in both coordinate systems. Chapters 2 and 3 discuss the issue of ground range vs. slant range as it impacts radar imagery.

The mixing operation in the deramp procedure is generally performed with analog circuitry. This is because it is performed on signals centered on the carrier frequency, which is typically on the order of Gigahertz. Because the sampling rates required to do this digitally are prohibitive, the deramping is performed onboard the radar platform, and this output is digitized. (In a SAR imaging system, these data are typically telemetered to a ground station for image-formation processing, although in some systems all of the processing is accomplished onboard the platform.) The required sampling rate is determined by the bandwidth of the demodulated signal, which is at baseband. Its bandwidth, however, is *not* simply the bandwidth of the transmitted chirp, B_c, as might be expected. Analysis of the coefficient of t in Equation 1.38 reveals that the highest frequency of the demodulated signal (in Hertz) is

$$f_{max} = \frac{4\alpha u_1}{2\pi c} = \frac{\alpha\tau_p}{2\pi} . \qquad (1.50)$$

Because the bandwidth of the transmitted chirp is $B_c = \alpha\tau_c/\pi$, it is easily shown that the required sampling rate for the demodulated signal (twice its highest frequency) is equal to

$$f_s = \frac{\tau_p}{\tau_c} B_c \,. \tag{1.51}$$

It now becomes clear why choosing τ_c to be much greater than τ_p is advantageous. Under this condition, the reduced bandwidth of the output of the quadrature demodulator allows $\bar{r}_c(t)$ to be sampled at a rate much lower than the chirp bandwidth B_c. This can be very important in practical SAR system design, as high sampling rates of the demodulator output may be undesirable from a hardware viewpoint. For those situations in which it is desirable for other reasons to have $\tau_c \leq \tau_p$, there is an alternate way to process chirp return data that is commonly known as *matched filtering*. This version of processing involves demodulation with in-phase and quadrature carrier sinusoids, followed by filtering with a baseband chirp kernel that matches the transmitted chirp (except for the carrier frequency term); hence the name *matched filter*. This results in output data that have bandwidth equal to B_c. Because this type of processing is typically employed in a conventional strip-map SAR , it is worthy of some discussion,

The reason that τ_c is chosen to be less than τ_p in a conventional strip-map SAR is that these systems are generally designed to cover a very large ground swath in the range dimension, using relatively *coarse* range resolution. (This is in opposition to the situation for the spotlight mode, where very fine range resolution is obtained over smaller patch sizes.) When the swath size becomes large, τ_p is correspondingly large. If you attempt to choose τ_c to exceed τ_p, you quickly run into trouble because the repetition rate at which successive pulses from the radar can be transmitted (pulse repetition frequency, or PRF) becomes limited. The limit arises because the terrain echo from targets located on the far side of the ground swath on one pulse must not overlap in time with the echo from targets on the near side of the swath on the next pulse. In turn, the constraint on PRF limits the achievable cross-range resolution. This follows because a wider beam illumination pattern (in the cross-range dimension), required for higher cross-range resolution in a strip-mapping SAR, requires a higher sampling rate to support it to avoid spatially aliasing.

Summary of Critical Ideas from Section 1.3

- **Dispersed (stretch) waveforms may be used in place of CW bursts to obtain high range resolution in imaging radars.**

- **A linear FM chirp is an example of a dispersed waveform commonly employed in radars, due to its desirable pulse compression properties.**

- **The bandwidth of the FM chirp is given by $B_c = \alpha\tau_c/\pi$, and the resulting slant-range resolution is given by: $\rho_u = c/(2B_c)$. This is not a function of either the radar center frequency or of the range from the platform to patch center.**

- **The ground-range resolution is given by:**

$$\rho_y = \frac{c}{2\cos\psi B_c}\ .$$

  A reconstruction computed on the basis of slant range does not resolve two targets any better than does one computed on a ground-range basis, because they are spaced closer together in slant range than in ground range by the same $\cos\psi$ factor.

- **For equal bandwidths, an FM chirp pulse is dispersed in time by a factor equal to its time-bandwidth product, $B_c\tau_c$, compared to a CW burst pulse.**

- **Using the FM chirp allows greater transmitted energy per pulse (for a fixed level of microwave tube power) compared to that of the CW burst, because the signal is "on" for a larger portion of the time, i.e., the chirp has a higher duty cycle factor.**

- **The steps of deramp processing are: 1) demodulation with in-phase and quadrature versions of the FM chirp, delayed appropriately by the round-trip time to the patch center; 2) low-pass filtering; and 3) range compression (Fourier transformation). This processing sequence effectively executes the deconvolution of the chirp waveform from the return signal, leaving an estimate of the terrain reflectivity function.**

Summary of Critical Ideas from Section 1.3 (cont'd)

- **Deramp processing is advantageous when $\tau_c > \tau_p$, because this results in the demodulator output having bandwidth that is less than B_c. This, in turn, reduces the rate at which the signal must be digitized. The bandwidth reduction ratio (radar bandwidth to demodulator output bandwidth) is given by the ratio of the chirp length, τ_c, to the patch propagation time, τ_p.**

- **For the case in which $\tau_c < \tau_p$ (typical for conventional strip-mapping SARs), another form of deconvolution called matched filter processing is utilized.**

REFERENCES

[1] J. Walker, "Range-Doppler Imaging of Rotating Objects," *IEEE Transactions on Aerospace and Electronic Systems*, Vol. AES-16, No. 1, pp. 23-51, January 1980.

[2] G. R. Cooper and C. D. McGillem, *Methods of Signal and System Analysis*, Holt, Rinehart, and Winston, Inc., New York, 1967.

[3] J. R. Klauder, A. C. Price, S. Darlington, and W. J. Alberscheim, "The Theory and Design of Chirp Radars," *Bell Systems Technical Journal*, Vol. 39, pp. 745-808, 1960.

[4] A. W. Rihaczek, *Principles of High-Resolution Radar*, Peninsula Publishing, Los Altos, CA, 1985.

[5] J. L. Eaves and E. K. Reedy, *Principles of Modern Radar*, Van Nostrand Reinhold, New York, 1987.

[6] D. C. Munson, J. D. O'Brien, and W. K. Jenkins, "A Tomographic Formulation of Spotlight-Mode Synthetic Aperture Radar," *Proceedings of the IEEE*, Vol. 71, No. 8, pp. 917-925, August 1983.

2

A TOMOGRAPHIC FOUNDATION FOR SPOTLIGHT-MODE SAR IMAGING

2.1 THE CROSS-RANGE RESOLUTION PROBLEM

In this chapter we develop a rigorous mathematical foundation that describes how a spotlight-mode SAR can achieve resolution simultaneously in the range and cross-range dimensions. As will be seen, the solution to the range resolution problem via the classic notion of echo separation using high-bandwidth pulses appears at first to be irrelevant to the issue of separability of targets in cross range. However, when we employ the interpretation of deramp-processed linear FM chirp returns as direct transductions of certain *spatial frequencies* of the terrain reflectivity function, a methodology for separation of targets in both dimensions emerges.[1] In fact, this Fourier-domain description of processed returns makes the cross-range resolvability dilemma appear mathematically very much like a problem from the field of medical science, the solution to which led to an entirely new mode of x-ray diagnostic imaging technology during the early 1970s. This is the now well-known and standardized process of *x-ray computerized axial tomography*, known commonly as CAT scanning, or CT, for *computed tomography*.

We will spend the first part of this chapter developing the critical ideas and associated mathematics that make medical tomography possible. We will then show how spotlight-mode SAR imaging can be formulated, with a few modifications, using the major tenets of this very same framework. Finally, we will employ this paradigm to derive all the important mathematical expressions used to describe the performance of a spotlight-mode SAR imaging system. The beauty of the approach lies in its ability to explain this important mode of SAR almost entirely in terms of well-known

[1] Although our analysis will assume use of chirp waveforms, the mathematical development does not require this. As we saw in Chapter 1, for example, the same spatial-frequency information is transduced by CW pulse waveforms having equivalent bandwidth.

concepts from the discipline of *signal processing*. David C. Munson and his colleagues at the University of Illinois were the first to suggest this tomography/SAR analogy and to present it formally in the literature [1]. Their development made the critical ideas of spotlight-mode SAR accessible to a person who has some background in the fundamentals of signal processing, but who is not necessarily schooled in the specialized language of *range-Doppler* imaging that accompanied the early developments of this subject [2]. A number of important points were not covered in Munson's original paper. In particular, those points that involve imaging of three-dimensional (elevated) targets were not discussed. Jakowatz and Thompson [3] later presented a general three-dimensional treatment of the SAR/tomography analogy. In this chapter, we start with the simpler two-dimensional model to motivate the cross-range resolving problem, and subsequently develop a complete three-dimensional tomographic paradigm for spotlight-mode SAR imaging.

To introduce the fundamental cross-range resolving problem, we first examine Figure 2.1, which shows an imaging scenario where a radar beam illuminates a large ground patch of radius L. A nominally-sized aircraft radar antenna with a length of 1.0 meter would, for example, form a beam pattern with a cross-range dimension equal to 1500 m when operating from a range of 50 kilometers and using a wavelength of 3 cm. (This wavelength corresponds to a frequency of 10 GHz, which is in the X-band portion of the microwave spectrum.) For simplicity, we assume that the target structure across the scene is described by a two-dimensional microwave reflectivity density function, $g(x, y)$, where x and y are spatial coordinates in the ground plane. That is, for the purposes of this introductory argument, we imagine a scene where all the targets lie on a flat surface.

The goal of the SAR imaging system is to produce an estimate of the two-dimensional function, $|g(x, y)|$, which will then become a *radar image* of the scene illuminated. (Note that in our geometry the y axis defines the range direction.) Although appropriate processing of a single returned pulse can resolve targets in the range dimension, as we saw in Chapter 1, all targets lying along the *same constant-range contour* will be received by the radar at precisely the same time, and therefore cannot be distinguished. Because the radar launches spherical wavefronts, the constant-range contours that intersect the ground patch are described by circular arcs. For the case of the radar platform that operates at standoff distances that are large compared to the scene diameter, however, these curves are well approximated with straight lines.

Therefore, what is transduced by a single received pulse at a particular time t_1 *cannot* simply be related to the value of the complex reflectivity function at any particular cross-range/range position, (x, y). What is measured instead is the *integration* of the reflectivity values from *all* targets that lie along the corresponding constant (ground) range line, $y = y_1$, in the scene patch (see Figure 2.1). That is, for the case in which

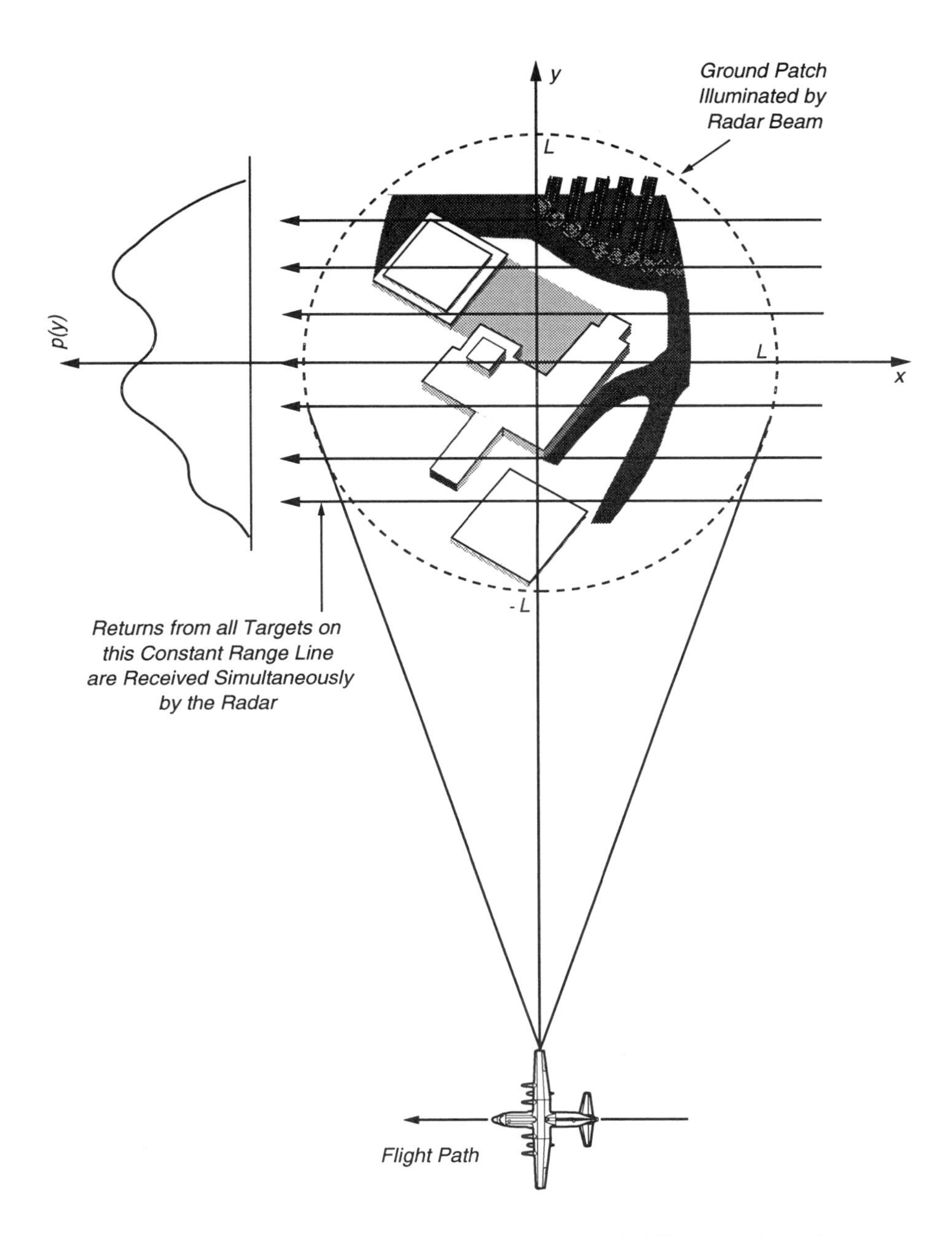

Figure 2.1 Imaging scenario depicting the cross-range resolvability issue. Targets lying on the same constant-range line but separated in cross range cannot be distinguished in the return of a single pulse, when a broad beam is used to illuminate the entire ground patch.

a linear FM chirp is employed as the transmitted waveform, the expression for the return radar signal at the receiver (prior to deramp processing) is given by

$$r_c(t) = A\,\mathbf{Re}\left\{\int_{-L}^{L} p(y)\,\exp[j[\omega_0\,(t - \tau_0 - \tau(y)) + \alpha(t - \tau_0 - \tau(y))^2]]\,dy\right\} \tag{2.1}$$

with

$$\tau(y) = \frac{2\,y\,cos\psi}{c} \tag{2.2}$$

and

$$p(y) = \int_{-L}^{L} g(x,y)dx\;. \tag{2.3}$$

(The scene reflectivity $g(x,y)$ is taken to be zero everywhere outside a circle of radius L, because we assume that the antenna beam pattern only illuminates targets within the circle.) If this return is processed via the deramp technique, the output of the quadrature demodulator (prior to range compression) is then analogous to Equation 1.46, with $g(y)$ replaced by $p(y)$:

$$\bar{r}_c(t) = A_1 \int_{-L}^{L} p(y)\,\exp\left[j\left[-\frac{2y\cos\psi}{c}(\omega_0 + 2\alpha(t-\tau_0))\right]\right]\,dy\;. \tag{2.4}$$

Equations 2.3 and 2.4 indicate that no pointwise estimate of $g(x,y)$ across the entire scene can be made from this single processed return. Instead, what one could obtain by performing the range compression step on $\bar{r}_c(t)$ is an estimate of certain *line integrals* of $g(x,y)$ taken along the cross-range direction, as represented by values of $p(y)$. This suggests that what is required for separation of targets in cross range is an illuminating radar beam pattern for which the projection onto the ground is confined to a very narrow strip of cross-range space, as is shown in Figure 2.2. If an aircraft carrying such an antenna were to launch and receive pulses in a sequence of positions as it progressed along its flight path as indicated in Figure 2.3, then a two-dimensional image could be constructed by placing the range-compressed pulses into columns of a two-dimensional array. The spacing at which the aircraft would send pulses would be chosen equal to (or slightly less than) the cross-range width of the beam pattern on the ground, so that contiguous coverage would be achieved. Such a system is termed a *real-aperture* imaging radar.

Unfortunately, a certain practical limitation arises in a real-aperture system due to the fact that narrower beam patterns require using correspondingly larger physical antennas. To demonstrate this, we again consider a radar operating at a nominal range to the ground patch of 50 kilometers. Suppose that we desire a radar image for which the resolution in both range and cross range is 1 meter. This level of resolution would be required, for example, in order to count the number of vehicles

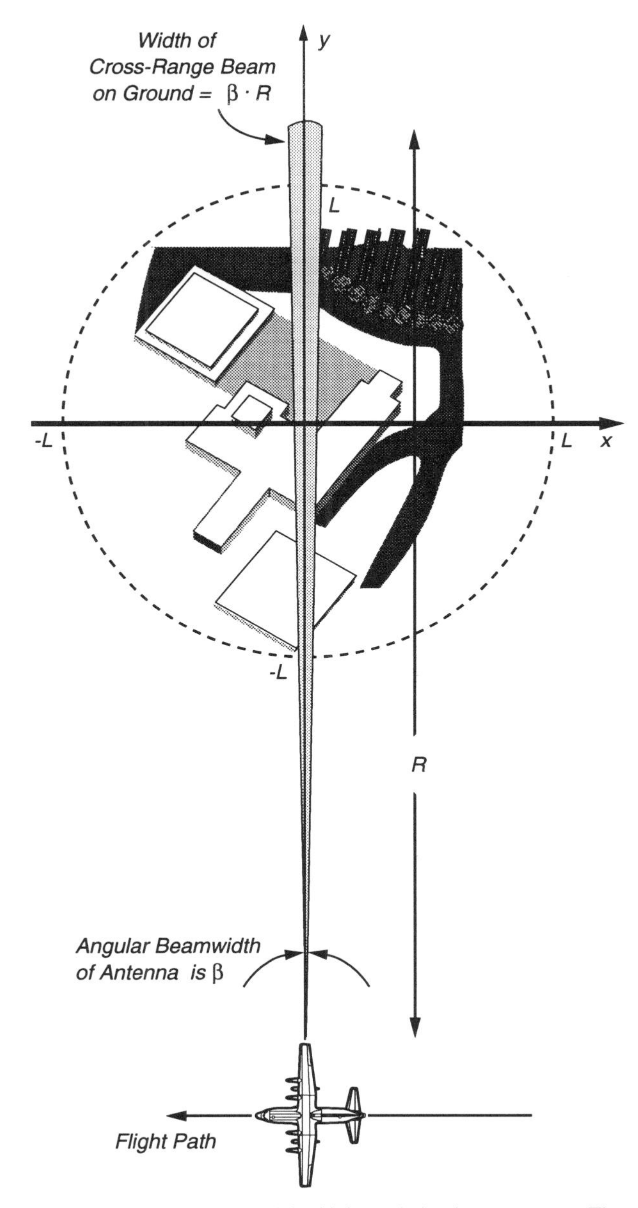

Figure 2.2 Narrow beam pattern required for high resolution in cross range. The cross-range resolution is determined by the width of the beam on the ground patch. This is nominally given by the product of the range R and the angular beamwidth β of the antenna. This angular beamwidth is determined by the ratio of the wavelength λ to the diameter of the physical aperture D.

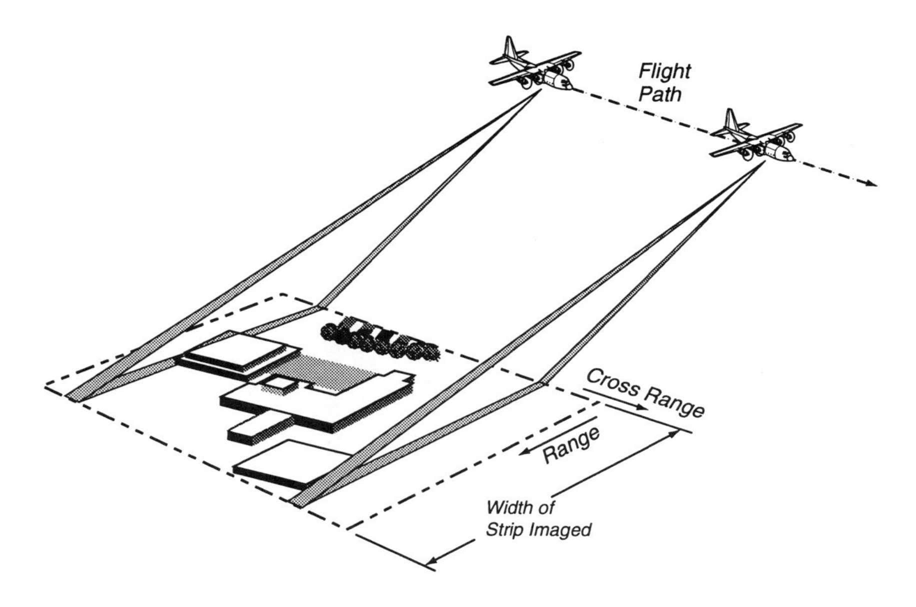

Figure 2.3 Collection geometry for real-aperture side-looking imaging radar. Cross-range resolution is provided by confining the beam in the along-track (cross-range) direction to a narrow strip on the ground. The motion of the aircraft then "paints" the image in this direction.

(trucks, tanks, aircraft, etc.) present in a scene. From our results in the last chapter, we know that a range resolution of 1 meter requires a bandwidth of approximately 150 MHz, as calculated from Equation 1.43. For a real-aperture imaging radar system capable of achieving the same resolution in cross range, however, the size of antenna required to confine the beam sufficiently in azimuth turns out to be rather large. The nominal angular width of a beam produced by a radiating aperture is given by

$$\beta = \frac{\lambda}{D} \tag{2.5}$$

where β is the angular beamwidth (measured in radians), D is the width of the aperture, and λ is the radar wavelength. In this case, we are looking for a narrow beamwidth in cross range, so that the critical dimension of the aperture is the horizontal one. The cross-range width of the beam on the ground is then given by

$$W_{cr} = R\beta = R\frac{\lambda}{D} \tag{2.6}$$

where R is the range from the ground location to the radar platform (see Figure 2.2). For our hypothetical case in which we desire $W_{cr} = 1\ m$, let us again assume that $\lambda = 0.03\ m$. The calculation of the required width of the physical aperture onboard the aircraft is

$$D = \frac{\lambda R}{W_{cr}} = \frac{0.03 \cdot (50 \cdot 10^3)}{1} = 1500\ m\ . \tag{2.7}$$

The width of the physical antenna required is 1500 meters. Because such a structure is clearly impractical, we see that a real-aperture imaging radar in this particular scenario is incapable of attaining the desired cross-range resolution of 1 meter.

On the other hand, a realistic physical antenna with a length of 10 meters mounted along the belly of an aircraft yields a cross-range resolution (for the same operational parameters assumed above) of 150 meters. Images formed with this level of resolution have utility in mapping certain natural features (mountains, rivers, forests, etc.), but they are insufficient for the task of identifying many man-made structures. Note that because the long dimension of the antenna must be placed along the length of aircraft fuselage in this case, the radar becomes by necessity a *side-looking* imager. In fact, these radars are usually referred to as Side-Looking Airborne Radar (SLAR) systems. By contrast, a nose-mounted system looking forward and down using a 1.0 meter diameter antenna scanned either mechanically or electrically in azimuth (see Figure 2.4) can only resolve objects to 1500 meters in cross range from 50 kilometers away. This results in a very coarse image of the ground. In earlier days these images were used as navigation aids, and of course the same radar is useful for detecting other aircraft and/or weather patterns in its forward vicinity by looking straight ahead.

Apparently, the only way to improve cross-range resolution for the real aperture SLAR is to either reduce the platform standoff range or to reduce the angular beamwidth by reducing the wavelength. Decreasing the standoff range is undesirable for many military scenarios, and is clearly contrary to the notion of space-based systems. Reduction of the radar wavelength is limited primarily by electromagnetic propagation effects. In order to make the radar impervious to effects of cloud cover and precipitation with virtually no absorption loss, it is generally required to employ those portions of the microwave spectrum corresponding to frequencies no greater than those of K_u-band, i.e., to frequencies lower than 15 GHz, or wavelengths greater than 2 cm. For frequencies above K_u-band, there are several windows for which atmospheric absorption of the microwave energy is not crippling, although it is appreciably worse than at 15 GHz. These are at 35, 90, and 135 GHz. As a result, the propagation-induced lower limit on λ is at best on the order of a few millimeters. In addition, there is a practical lower bound on the ratio λ/D that can be built into real antennas, imposed by manufacturing tolerances in surface precision. These limits effectively restrict real antennas to angular beamwidths of no smaller than about 10^{-4} radians. The result of the above considerations is that the best we could hope for in a real-aperture imaging radar operating at our nominal standoff range of 50 km would be a 2 mm wavelength system that employs an antenna with

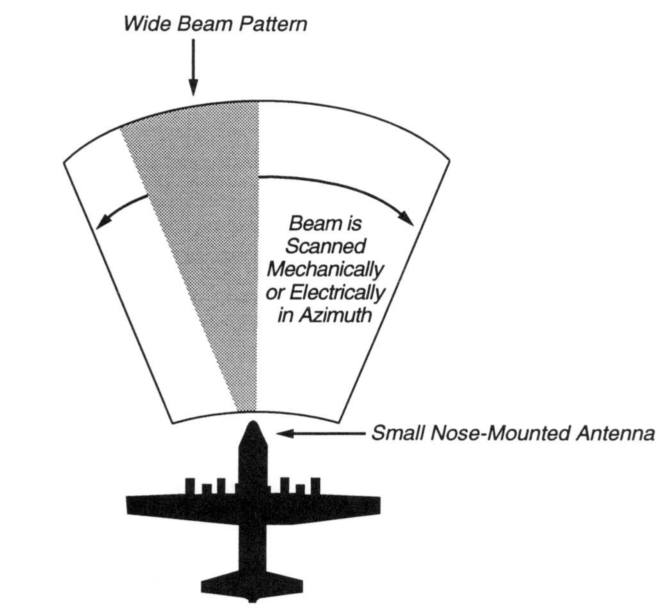

Forward-Looking Radar Using Nose-Mounted Antenna

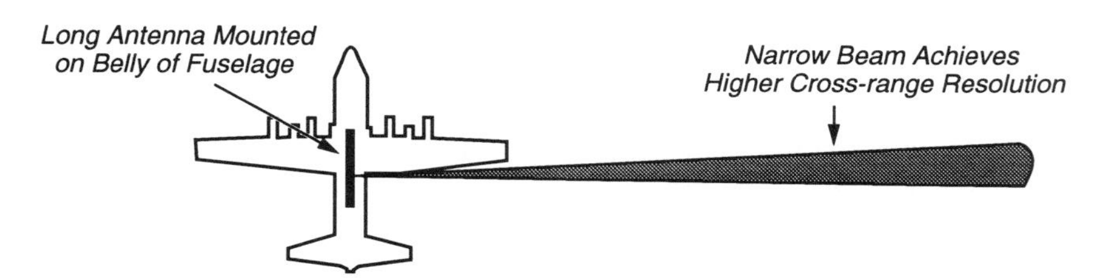

Side-Looking Airborne Radar (SLAR) with Longer Belly-Mounted Antenna

Figure 2.4 Comparison of forward-looking (top figure) and side-looking (bottom figure) aircraft imaging radars. For the forward-look mode, the antenna is mechanically or electrically scanned in azimuth to "paint" the image in cross range. The relatively small size of the nose-mounted antenna dictates a wider beamwidth and therefore coarser cross-range resolution. In the side-looking (SLAR) mode, the antenna can be made larger by placing it along the length of the belly of the aircraft fuselage. This leads to a narrower beamwidth and therefore finer cross-range resolution. The forward motion of the aircraft provides self-scanning of the image in the cross-range dimension.

a length of 20 m (this produces $\beta = 10^{-4}$), yielding a cross-range resolution of 5 m. If finer resolutions and/or longer operating ranges are desired, an alternative to real-aperture systems must be sought.

The calculations above are typical of those which motivated the early developers of imaging radars to seek alternative schemes that could achieve high cross-range resolution, and yet simultaneously employ a reasonably small size physical antenna, use frequencies of K_u-band or lower, and operate at very long standoff ranges. The solution for meeting all these demands, namely that of synthetic aperture radars, eventually produced a type of imaging system that avoids the range-dependent cross-range resolution dictated by Equation 2.6. This system has a distinct advantage over real-aperture systems, and ultimately has allowed modern imaging radars to be built that operate at frequencies essentially free from weather-induced propagation losses (i.e., K_u-band and longer wavelengths) while producing very fine cross-range resolution, even from considerable standoff ranges. Section 2.2 introduces basic notions that are critical to the synthetic aperture solution.

Summary of Critical Ideas from Section 2.1

- **Real-aperture radars have cross-range resolutions that are range-dependent.**
- **Fine cross-range resolution (e.g., 1 meter) would require unrealistically large physical antenna widths for practical standoff ranges and microwave center frequencies.**

2.2 THE CONCEPT OF APERTURE SYNTHESIS AS A SOLUTION TO THE CROSS-RANGE RESOLVABILITY PROBLEM

The solution to the cross-range resolution problem starts with the careful reconsideration of Figure 2.1. We have already seen that a single processed pulse provides information only about the line integrals of the reflectivity function taken along a single direction (the x direction in Figure 2.1). Another pulse transmitted from viewing angle θ (with the antenna turned slightly to still aim at the ground patch center) also provides, upon deramping and range compression, information about a set of line integrals of the same scene reflectivity function, but taken along a different direction. As shown in Figure 2.5, the coordinates $(\bar{x}, \bar{y})$, representing cross-range and range in the ground plane as viewed from the direction θ, are obtained by a rotation of the (x, y) axes, with θ defined as counterclockwise from the x axis.

This linear coordinate transformation is orthonormal, and given by

$$\begin{aligned} x &= \bar{x}\cos\theta - \bar{y}\sin\theta \\ y &= \bar{x}\sin\theta + \bar{y}\cos\theta \end{aligned} \tag{2.8}$$

while the corresponding inverse transformation is

$$\begin{aligned} \bar{x} &= x\cos\theta + y\sin\theta \\ \bar{y} &= -x\sin\theta + y\cos\theta. \end{aligned} \tag{2.9}$$

We can then generalize Equation 2.3 to express $p_\theta(\bar{y})$, the integrated reflectivity function of $g(x, y)$ for angle θ, as

$$p_\theta(\bar{y}) = \int_{-L}^{L} g(x(\bar{x}, \bar{y}), y(\bar{x}, \bar{y}))d\bar{x} \tag{2.10}$$

$$= \int_{-L}^{L} g(\bar{x}\cos\theta - \bar{y}\sin\theta, \bar{x}\sin\theta + \bar{y}\cos\theta)d\bar{x}. \tag{2.11}$$

The return signal is expressed as

$$r_c(t) = A_1 \,\mathbf{Re}\left\{\int_{-L}^{L} p_\theta(\bar{y})\; s\left(t - \frac{2(R + \bar{y}cos\psi)}{c}\right)\, d\bar{y}\right\} \tag{2.12}$$

$$= A_1\mathbf{Re}\left\{\int_{-L}^{L} p_\theta(\bar{y})\; \exp[j[\omega_0\,(t - \tau_0 - \tau(\bar{y})) + \alpha\,(t - \tau_0 - \tau(\bar{y}))^2]]\, d\bar{y}\right\}. \tag{2.13}$$

The corresponding deramped return is analogous to Equation 2.4

$$\bar{r}_c(t) = A_1 \int_{-L}^{L} p_\theta(\bar{y})\; \exp\left[j\left[-\frac{2\bar{y}\cos\psi}{c}(\omega_0 + 2\alpha(t - \tau_0))\right]\right]\, d\bar{y} \tag{2.14}$$

from which $p_\theta(\bar{y})$ can be recovered by range compression.

As the *aperture synthesis* concept unfolds in this chapter, we will see that an appropriate collection of many such functions obtained over an interval of viewing angles, $\Delta\theta$, does in fact contain sufficient information from which $g(x, y)$ can be unscrambled, or *reconstructed.* As we will demonstrate, the degree of cross-range resolution in the reconstruction depends only on λ and on the size of $\Delta\theta$, that is, on

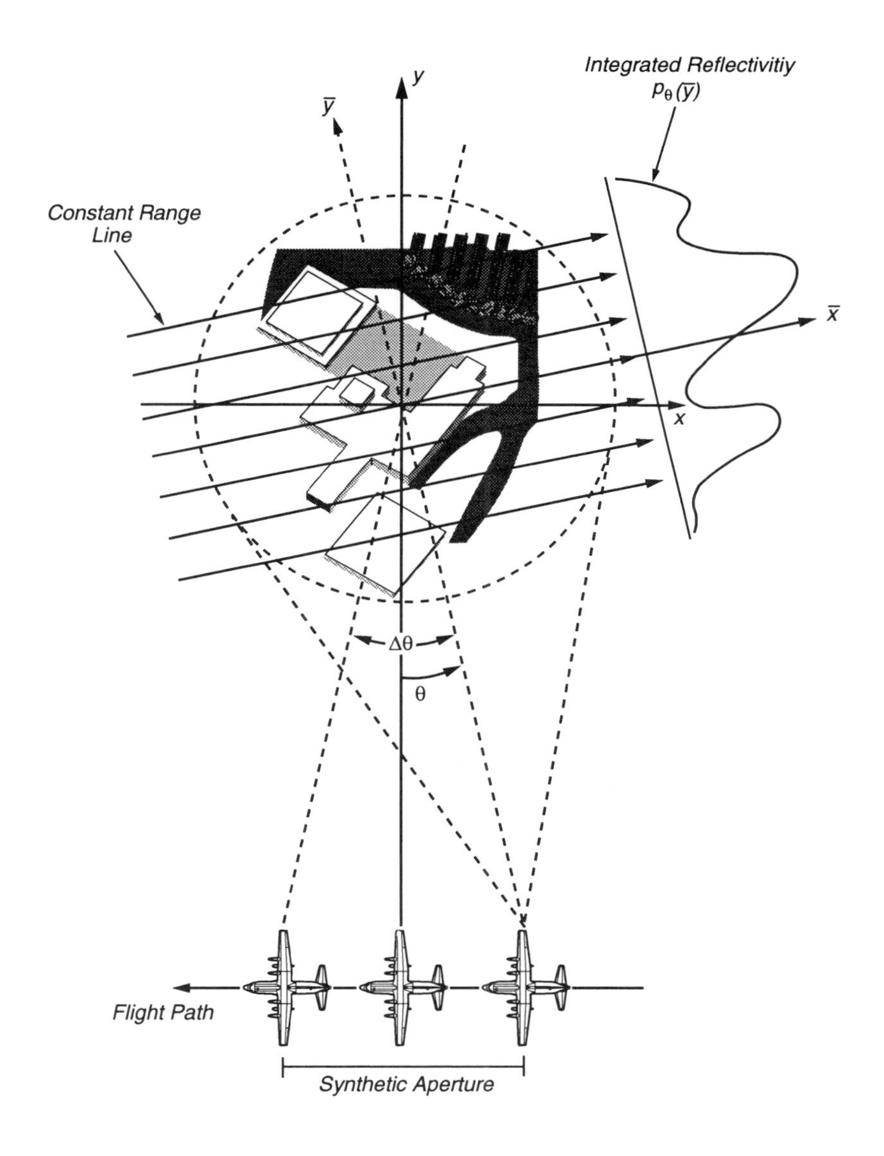

Figure 2.5 Concept of synthetic aperture radar imaging. Processing of multiple pulses launched at the ground patch as the aircraft moves along its flight path allows improved cross-range resolvability of targets. The longer the flight path, which is the synthetic aperture, the finer the achievable cross-range resolution. The effects of a long physical antenna are created synthetically via data processing, so that a nominally-sized radar antenna can be used. The antenna is steered continuously (slewed) to always aim at the patch center.

the diversity of angles from which pulses are transmitted and received. Provided that data spanning this angular interval are collected, the standoff range of the platform is unimportant. By this process, the beneficial effects of a large antenna can be *synthesized* via data processing, thus allowing fine cross-range resolution to be achieved *independent of the operating range*, from data gathered with a nominally-sized antenna. The path over which the aircraft flies to transmit and receive the collection of pulses becomes the *synthetic aperture* (see Figure 2.5). Before constructing a formal mathematical framework that will lead to an algorithm for combining a set of functions $p_\theta(\bar{y})$ into a reconstructed SAR image, we will briefly review the historical development of imaging radars that employ the aperture synthesis concept.

During the early 1950s an engineer named Carl Wiley, while employed at the Goodyear Aircraft Corp., developed the initial concepts of what we now know as SAR imaging [4]. The patent for the invention was not issued until 1965 [5]. In that patent the terminology of synthetic aperture radar was not used. Instead, Wiley discussed *Doppler beam sharpening*. Through developments performed during the ensuing years at a number of American aerospace companies and universities, Wiley's ideas evolved into what is now commonly referred to as *strip-mapping* SAR. The basic collection geometry for such a system is shown in Figure 2.6.

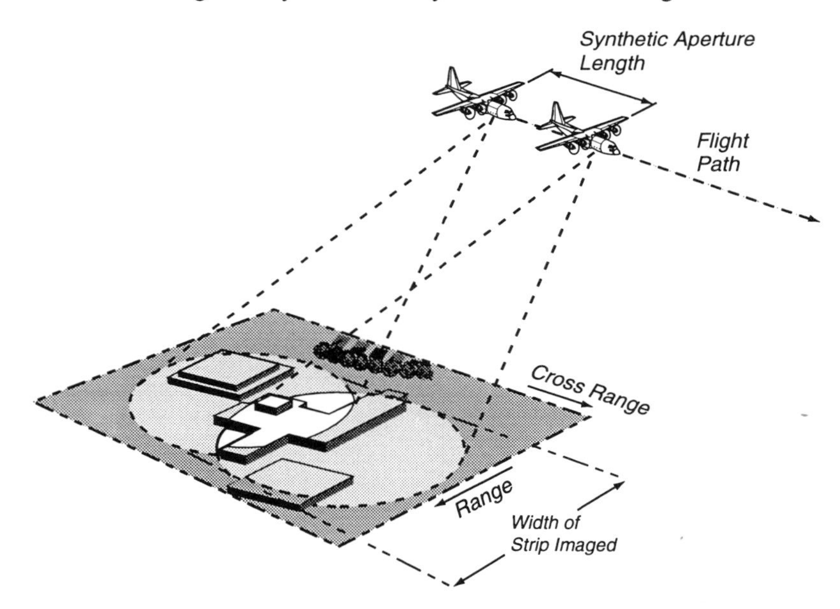

Figure 2.6 Collection geometry for a strip-mapping SAR.

In the simplest version of this type of SAR collection, the antenna is aimed orthogonal to the flight path and sweeps out a large beamwidth on the ground. The aircraft transmits and receives radar pulses periodically as it traverses the flight path. For a given position of the aircraft, the sequence of returns obtained along a segment of the flight path centered at this position is collectively processed to form the *effective* return signal from an antenna much larger than the one actually illuminating the ground. The longer the length of space over which this integration occurs, the narrower is the synthesized beamwidth, and so the better is the achievable cross-range resolution. The length of this segment, which is the synthetic aperture, cannot exceed the width of the ground beam pattern in cross range because this is the maximum distance for which a given point on the ground appears in the illuminating beam patterns from all points along the segment. This leads to the interesting situation in which the width (along-track dimension) of the physical antenna is made *as small as possible*, so as to obtain the largest available illuminating cross-range beam pattern, and therefore the largest possible synthetic aperture length and correspondingly finest cross-range resolution. This is totally opposed to the established notion for the real-aperture radar system described earlier, wherein the *largest* possible width of physical antenna was desired in order to produce the narrowest illuminating beam pattern. Cutrona [6] was the first to show that the achievable resolution in a strip-mapping SAR is equal to one-half of the width (along-track dimension) of the antenna. An important condition that accompanies this resolution limit, however, is that pulses must be transmitted with spacing along the flight path also equal to the resolution, i.e., equal to one-half of the physical antenna width. This is required so that the illuminated beam pattern on the ground is not *aliased* due to undersampling. (A more detailed discussion of this aperture sampling requirement is presented in Section 2.4.4.) The net result is that finer cross-range resolutions require proportionately higher pulse repetition frequencies, for a given platform velocity.

Although the above discussion might imply that any desired level of cross-range resolution is easily attainable with a strip-mapping SAR that employs an appropriate antenna size and PRF, there are other considerations that suggest that the strip-map mode may not be optimal in all collection scenarios. For example, consider the situation where it is necessary to image a relatively small patch of the earth with high resolution. Specifically, assume that the patch to be imaged is 1 km square, that we desire 1 foot (.33 meters) resolution in both range and cross range, and that the slant range to the scene center is 70 km. This collection scenario is shown in Figure 2.7. A conventional strip-mapping SAR (with $\lambda = 3.0$ cm) that produces the desired image requires an antenna with a width of 0.66 m, which implies an angular beam of 2.60 degrees. This results in an illuminated ground footprint 3.18 km long in the cross-range dimension. As per the requirement described above, it is necessary that pulses be transmitted with spacing of 0.33 m along the synthetic aperture to avoid spatial aliasing.

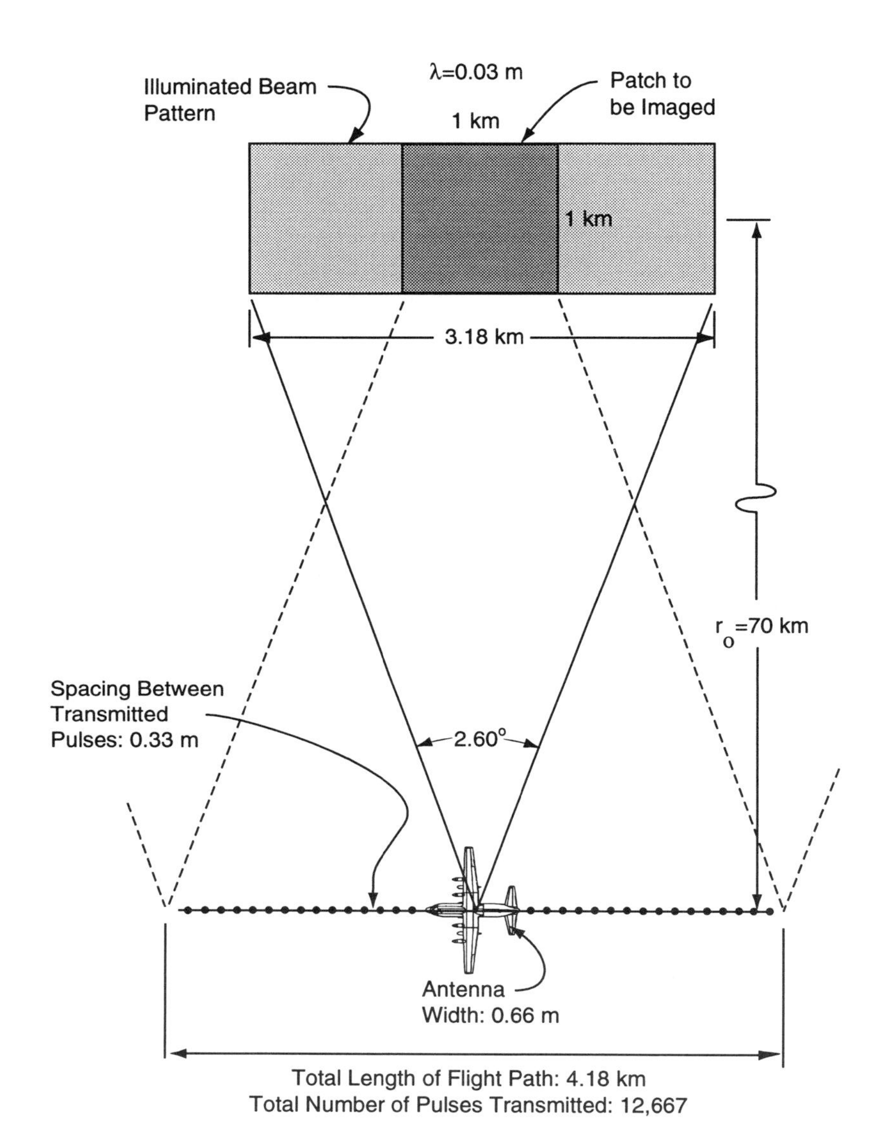

Figure 2.7 Strip-map mode used to collect data for a high-resolution image over a relatively small patch. A large number of pulses must be collected. The large illuminating beam footprint increases power demands.

Further examination of Figure 2.7 reveals that the total length of the flight path required to obtain the data necessary for imaging the 1 km square patch is the sum of the beamwidth and the width of the patch, i.e., 3.18 km + 1 km = 4.18 km. A 0.33 m spacing between transmission points then requires a total of 12,667 pulses. It is possible to reduce this rather large number of required pulses *if we are only concerned with imaging the 1 km square patch*, i.e., if we do not desire to image any other portion of the strip. This is accomplished by using a completely different mode of collecting the radar pulse data. The new collection methodology is termed *spotlight mode*, which is the chief focus of this text. The invention of this form of radar imaging is generally credited to Jack Walker [2]. Actually, without stating it as such, we have already introduced the notion of spotlight mode (see Figure 2.5). The chief difference between this and a strip-mapping SAR is that the radar beam is now continually steered, or *slewed*, so as to constantly illuminate the same ground patch from all positions of the flight path. Figure 2.8 depicts the basic flight geometry for the spotlight-mode collection. The name for this mode derives from the fact that the radar is actually "spotlighting" the ground patch for the entire time of flight across the synthetic aperture, much as one might do with a searchlight beam.

Figure 2.9 suggests how the spotlight mode can lead to a reduced number of collected radar pulses compared to the conventional strip-map SAR for the imaging of the 1 km square section of ground in our example. By using a beam that only illuminates the patch to be imaged, the width of the required physical antenna is three times greater than the strip-map antenna (2.1 m versus 0.66 m). The required spacing between transmitted pulses is also greater by the same proportion (1 m versus 0.33 m). The net result is that only 3,000 pulses are required by the spotlight-mode system, instead of the 12,667 pulses needed for the strip-map modality. In addition, the spotlight-mode collection offers an advantageous power situation, where the higher gain represented by the wider physical antenna reduces the microwave transmitter power required for imagery having the same SNR.

The remainder of this book describes various signal processing aspects and applications of spotlight-mode SAR. As it turns out, the paradigm and processing procedures used for strip-mapping systems commonly differ substantially from the ones we develop here for spotlight mode. (In some situations, however, spotlight-mode algorithms are used to form images from data collected in a strip-map mode.) The reader who is interested in understanding how strip-mapping SARs are designed and how the associated data processing is performed should select any of a wide variety of texts written on this subject. The book by Curlander and McDonough, for example, provides a good treatment [7].

We are now ready to pose formally the critical question for spotlight-mode SAR image reconstruction. Given that a set of deramped, range-compressed pulses col-

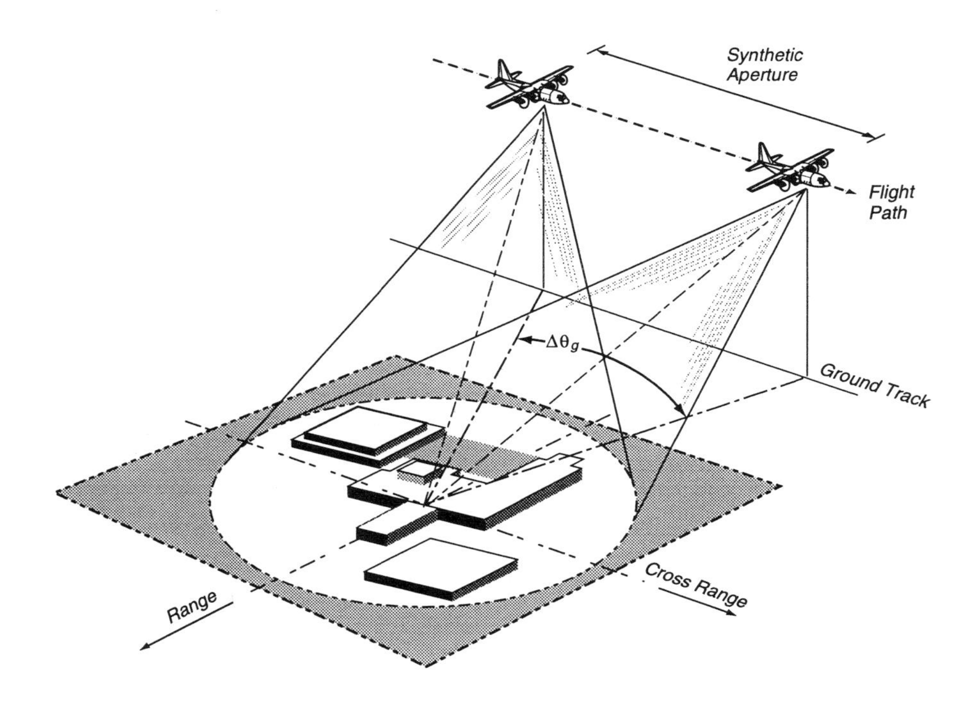

Figure 2.8 Spotlight-mode collection geometry. In this imaging modality the illuminating radar beam is steered continually so as to "spotlight" the scene. Unlike the case of conventional strip-mapping SARs, the synthetic aperture in spotlight mode can be larger than the size of the illuminating beam pattern in the along-track dimension.

lected over some range of viewing angles constitute a certain set of line integrals of the scene reflectivity function $g(x, y)$, how can an estimate $\hat{g}(x, y)$ be produced from this collected data? As we suggested in the introduction to this chapter, the solution to this same mathematical problem was implemented circa 1970 in a different arena. It was manifested as a methodology that revolutionized both the science and business of medical x-ray imaging. The technique, called *Computerized Axial Tomography* (CAT), is now used extensively for medical diagnostics, and has been expanded for use beyond x-rays to include ultrasound tomography, positron emission tomography (PET), and single photon emission tomography (SPECT) [8]. In Section 2.3 we explain the mathematical foundations of medical tomography. Section 2.4 shows how, with certain modifications, the same framework can be used to derive an elegant formulation of the collection and reconstruction of spotlight-mode SAR data. We then address a number of important ancillary questions that include: What is an expression for the resulting cross-range resolution as a function of the angular viewing diversity? On what spatial interval along the synthetic aperture must

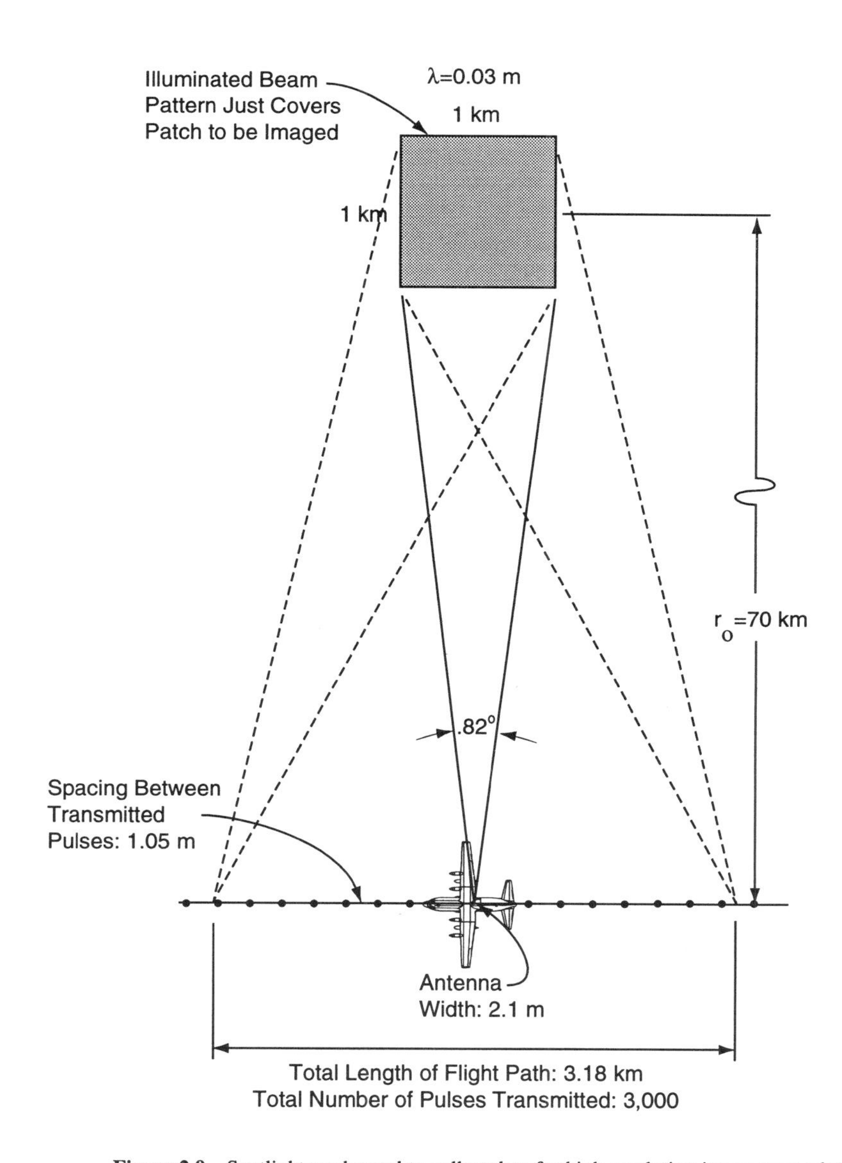

Figure 2.9 Spotlight mode used to collect data for high-resolution image over relatively small patch. Fewer pulses need to be collected in comparison with the strip-map case. In addition, the power demands are eased for the spotlight-mode case because the illuminated beam footprint is reduced.

pulses be launched? Are there conditions on scene size and/or standoff range that eventually limit the tomographic paradigm?

Summary of Critical Ideas from Section 2.2

- **Aperture synthesis is a way to achieve fine cross-range resolution by coherent integration of a series of radar returns transmitted from a variety of positions along a flight path. The effects of a large physical antenna are thereby synthesized via data processing.**
- **SAR can be executed in two fundamental modes: a) strip-map and b) spotlight.**
- **Strip-map SARs are generally used to image large area strips of the earth at coarser resolutions.**
- **Spotlight-mode SARs are generally advantageous where high-resolution imagery is desired over small earth patches.**
- **In a spotlight-mode collection, the radar antenna is continuously slewed to keep the beam on the target patch to be imaged.**

2.3 MATHEMATICS OF COMPUTERIZED AXIAL TOMOGRAPHY (CAT)

2.3.1 Introduction to Medical X-Ray CAT

From the time that x-rays were first used in medical imaging in 1896[2] until the 1970s, the mode of radiographic imaging that became standardized is shown in Figure 2.10 An x-ray source illuminates the patient's head (or other body part) and produces an image which is a two-dimensional projection of the patient's internal three-dimensional structure. What is actually transduced is the x-ray attenuation coefficient. Because the image obtained on the x-ray film is a projection, each point in the image represents the integration of the three-dimensional x-ray attenuation coefficient profile along one *ray*, or line, passing through the patient. While this form of image proves to be very useful for diagnosing many medical abnormalities and injuries, there are also many situations when it provides no benefit. For example, small cancerous tumors may go completely undetected in this kind of image, because the difference in the x-ray attenuation coefficient between healthy and cancerous tissue can be quite small. Therefore, a measurement of the integration of the attenuations along an entire ray may not reveal a small amount of abnormal tissue in the middle of healthy tissue.

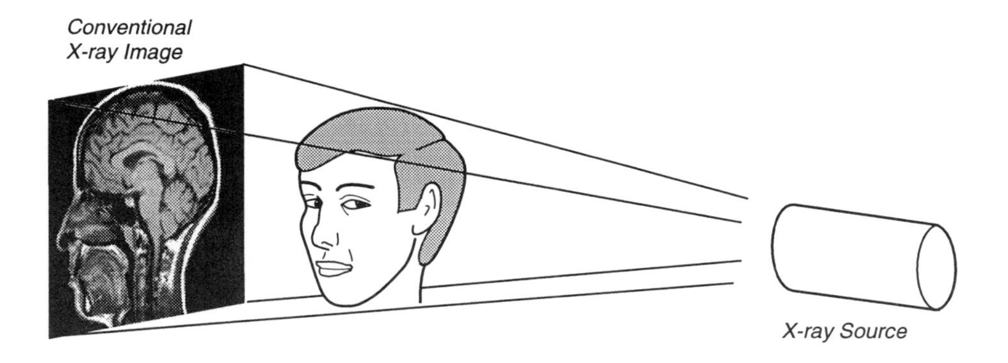

Figure 2.10 Modality of conventional x-ray imaging.

[2] The initial discovery of x-rays was by the German scientist Wilhelm Roentgen in 1895. The medical community immediately recognized the value of his discovery and within a year began using x-ray images for diagnosing bone fractures and other maladies.

What is needed to overcome the problem outlined above is a system that produces estimates of the patient's three-dimensional x-ray attenuation profile $g(x, y, z)$ on a point-by-point basis, so that the resulting image values do not represent integrations across a multitude of spatial positions. In this way, subtle differences in tissue type could be detected. A methodology and design of associated equipment for creating this type of three-dimensional pointwise estimate of the x-ray attenuation coefficient was first formally suggested in 1972 by G. N. Hounsfield, who obtained a British patent for his proposed imaging apparatus [9], [10]. Hounsfield's invention represented the birth of modern x-ray computed axial tomographic imaging.[3] The acronyms *CAT* (computed axial tomography) and *CT* (computed tomography) are both commonly used to describe these medical imaging systems. The basic geometry for collecting the x-ray data that allows computed axial tomographic reconstructions is shown in Figure 2.11. A highly collimated beam of x-rays is directed at the patient's head so that only a thin *slice* of the skull is irradiated. In fact, *tomography* is derived from the Greek word $\tau o \mu o \varsigma$ (*tomos*), meaning *section* or *slice*. The beam is then translated and rotated, collecting data that can be used to reconstruct an image of that thin cross-section. As the source and detector are translated in synchrony, the received x-ray intensity after transmission through the patient is measured at a series of positions.

The result of the source and detector translation[4] is to sweep out a single *projection* function, $p_\theta(u)$. This is shown in the upper portion of Figure 2.12. Note that the (u, v) coordinate system is related to that of (x, y) by counterclockwise rotation from the x axis through angle θ. This orthonormal linear transformation is the same as that of Equation 2.8, with $(\bar{x}, \bar{y})$ replaced by (u, v):

$$\begin{aligned} x &= u \cos\theta - v \sin\theta \\ y &= u \sin\theta + v \cos\theta. \end{aligned} \tag{2.15}$$

The corresponding inverse transformation is

$$\begin{aligned} u &= x \cos\theta + y \sin\theta \\ v &= -x \sin\theta + y \cos\theta. \end{aligned} \tag{2.16}$$

Each value of the projection function is the result of integrating the two-dimensional

[3] Actually, at least two relevant papers significantly preceded Hounsfield's patent, one by Oldendorf [11] and the other by Cormack [12]. Both Hounsfield and Cormack were later recognized for their pioneering contributions to medical tomography when they jointly received the Nobel Prize for Medicine in 1979.

[4] This type of scanning source/detector combination was only used in early versions of CAT devices. Modern machines employ a fan (divergent) beam and a linear array of detectors, so that the linear scanning motion is not required. This speeds up the collection time considerably. However, the parallel beam geometry is the one that provides the direct analogy to spotlight-mode SAR. Thus, we describe it here.

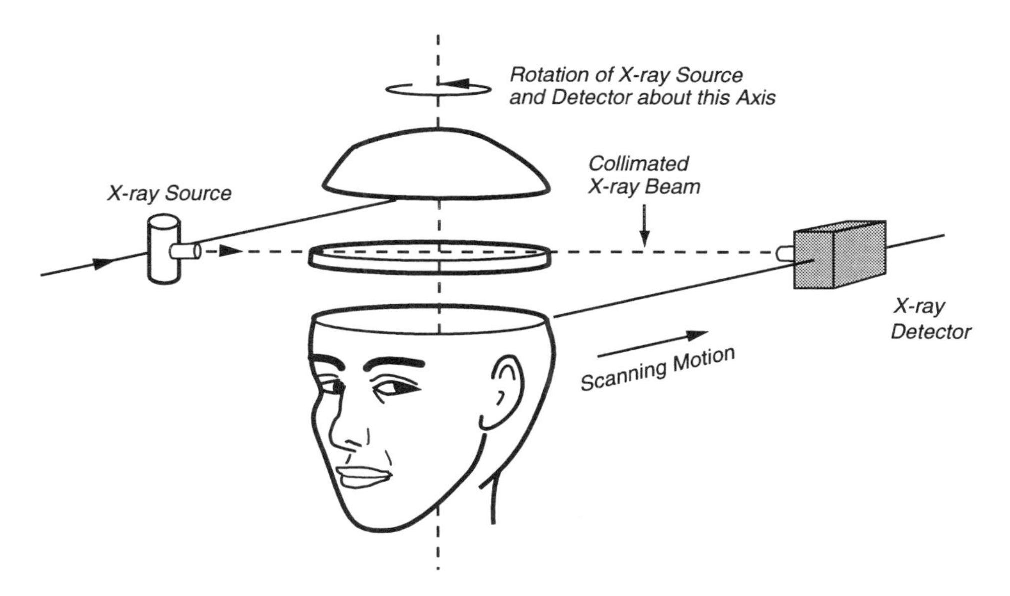

Figure 2.11 Modality of x-ray computerized axial tomography (CAT) imaging . At each angular orientation of the x-ray system relative to the patient, a series of sampled projection functions is obtained via linear translation (synchronous scanning) of the x-ray source and detector. The projection functions represent values of the line integrals of the two-dimensional x-ray attenuation coefficient profile for the thin slice of the patient's head irradiated by the highly collimated beam. The estimate of this profile from the projection data via tomographic reconstruction becomes the CAT image.

x-ray attenuation coefficient profile $g(x,y)$ for the given slice along a single ray. That is, for a given ray the transmitted x-ray intensity I_0, the received x-ray intensity I_r measured by the detector, and the attenuation coefficient function $g(x,y)$ are related by the exponential attenuation relationship

$$I_r = I_0 \exp\left\{-\int_{-L}^{L} g(x(u,v), y(u,v))dv\right\} . \tag{2.17}$$

Rearranging Equation 2.17 and using Equation 2.15 gives

$$ln\left[\frac{I_0}{I_r}\right] = p_\theta(u) = \int_{-L}^{L} g(u\cos\theta - v\sin\theta, u\sin\theta + v\cos\theta)dv . \tag{2.18}$$

Here, we assume that $g(x,y) = 0$ everywhere outside of the circle of radius L, centered at the origin, i.e., the patient's skull fits entirely within this circle. As a result, the projection functions are all zero outside the support defined by $-L \leq u \leq L$.

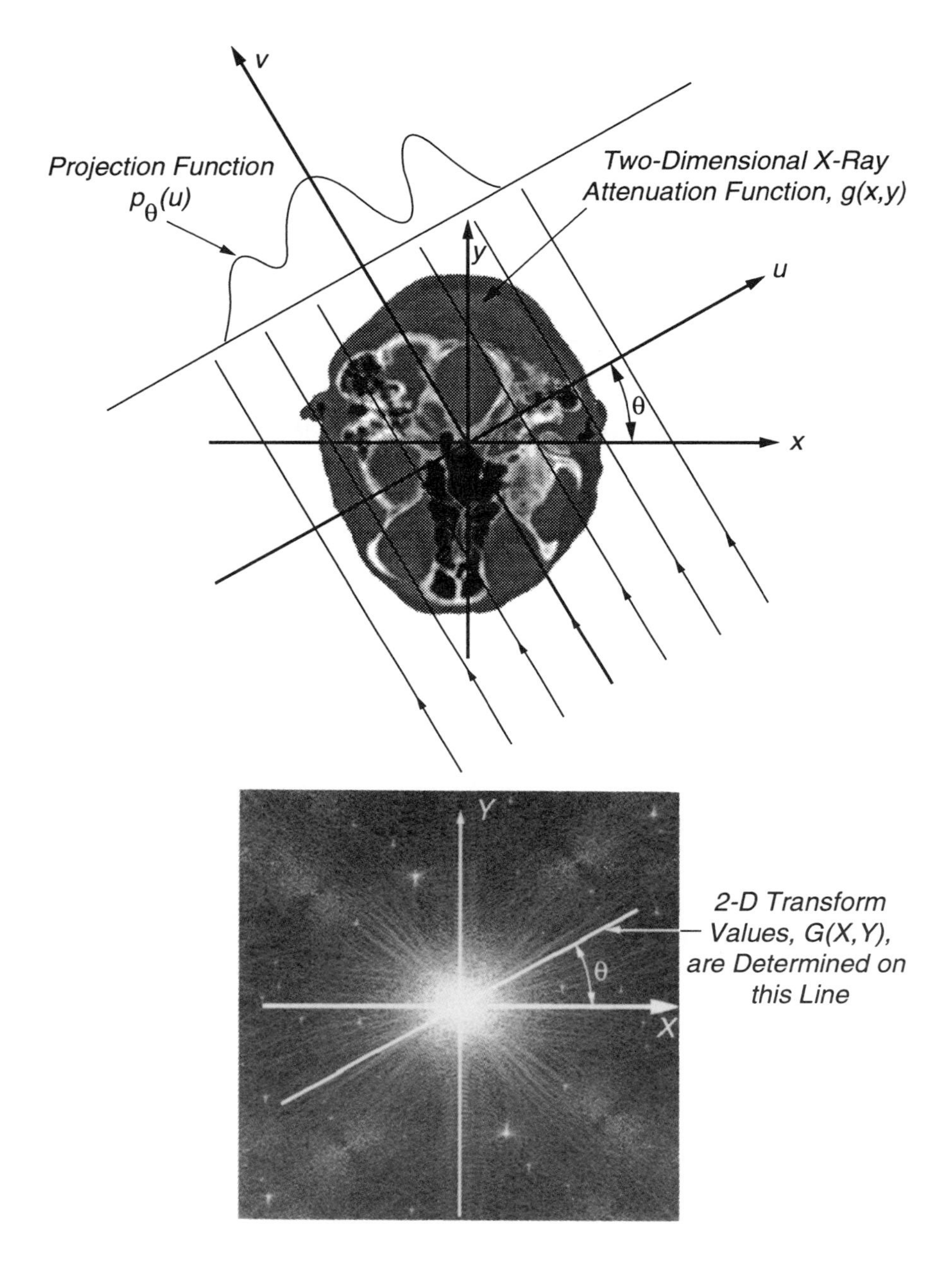

Figure 2.12 Projection-slice theorem as applied to x-ray tomography. Values of the one-dimensional Fourier transform of a projection function taken at angle θ are equal to values of the two-dimensional Fourier transform of $g(x, y)$ along a line in the Fourier plane that lies at angular orientation θ with respect to the X axis.

The x-ray source and detector are rotated about an imaginary vertical axis through the center of the patient's skull, as shown in Figure 2.11. At each of a series of angular orientations θ, spanning $180°$, a projection function is obtained. The set of projection functions is then processed by computer according to a tomographic reconstruction algorithm to produce an estimate $\hat{g}(x,y)$ of the attenuation profile for that particular slice. The digital display of this computed estimate becomes the tomographic image. The required sampling rates in both the u and θ dimensions are determined by the spatial bandwidth of the image to be reconstructed, as will be seen shortly. The three-dimensional skull structure is determined by translating the entire scanning apparatus vertically, and sequentially obtaining data for slices at a number of different levels. The *stacking* of this series of reconstructed two-dimensional slices constitutes the desired three-dimensional estimate $\hat{g}(x,y,z)$ of an entire body section.

2.3.2 The Projection-Slice Theorem in Tomography

The lower portion of Figure 2.12 shows the key concept used in the reconstruction of tomograms from a set of projection data. The foundational relationship is that the one-dimensional Fourier transform of any projection function $p_\theta(u)$ is equal to the two-dimensional Fourier transform $G(X,Y)$ of the image to be reconstructed, evaluated along a line in the Fourier plane that lies at the same angle θ measured from the X axis. This important result is known as the *projection-slice theorem* [13], and may be stated more precisely as

$$G(U\cos\theta, U\sin\theta) = P_\theta(U) \tag{2.19}$$

where

$$P_\theta(U) = \int_{-\infty}^{\infty} p_\theta(u)e^{-juU}\,du \tag{2.20}$$

and

$$G(X,Y) = \int_{-\infty}^{\infty}\int_{-\infty}^{\infty} g(x,y)e^{-j(xX+yY)}\,dx\,dy\,. \tag{2.21}$$

The power of the projection-slice theorem should now be evident. As projections are taken over a range of θ, the one-dimensional Fourier transform values of the projection data determine values of the two-dimensional Fourier transform $G(X,Y)$ along lines of the same angular orientations. If projections are taken in angular increments of δ_θ and span $180°$ of viewing directions, a circular region of the two-dimensional Fourier plane, centered at the origin, will be swept out. Because the Fourier transforms of the projection functions are computed as discrete transforms of sampled functions, a finite set of samples on each radial line will be determined.

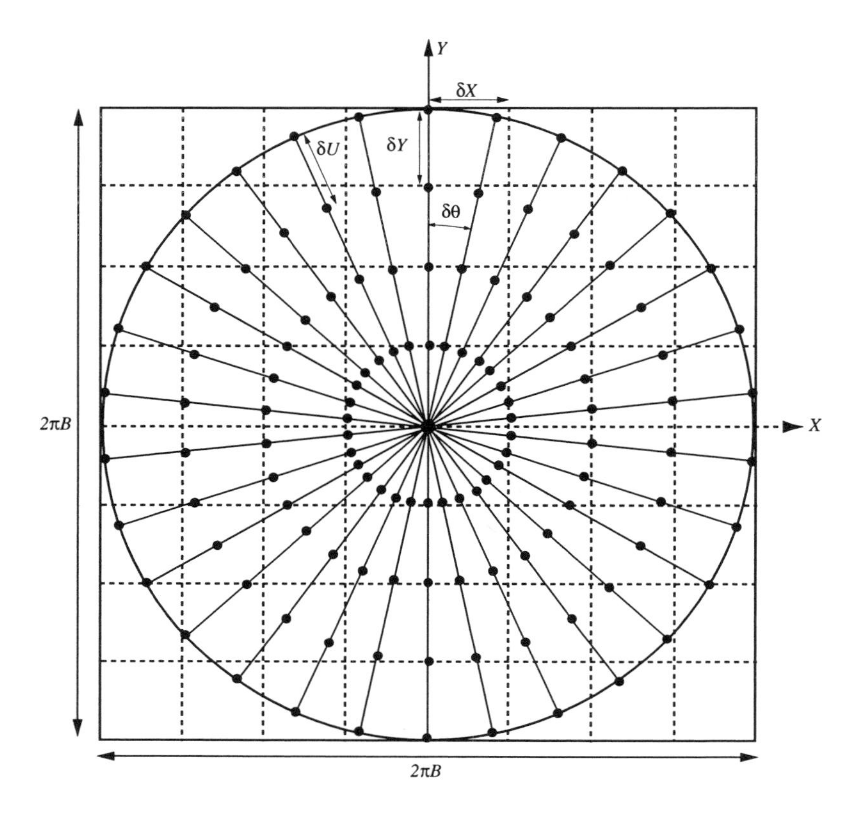

Figure 2.13 Region of 2-D Fourier space determined from transformed projections, showing polar raster on which data are sampled. Each radial line of Fourier data is provided from the transformed projection data taken at the same angular orientation. By scanning over the range of angles from 0 to 180 degrees, a circular region of Fourier space is swept out.

Figure 2.13 shows the resulting *polar raster* of sampled data that is obtained in this way. The prescription for obtaining the reconstructed image $\hat{g}(x, y)$ is to perform an inverse discrete Fourier transform of the data on the polar raster. We will describe two methods for accomplishing this inversion. First, however, we should prove the projection-slice theorem, because it is the cornerstone for the derivation of these algorithms.

The projection-slice theorem is easily proved using a basic theorem from Fourier transforms. The proof proceeds as follows. Consider the projection function obtained for the special case of $\theta = 0$:

$$p_0(x) = \int_{-L}^{L} g(x, y) dy \, . \tag{2.22}$$

Its Fourier transform is

$$P_0(X) = \int_{-L}^{L} \left\{ \int_{-L}^{L} g(x,y)dy \right\} e^{-jxX} dx = \int_{-L}^{L} \int_{-L}^{L} g(x,y)e^{-jxX} dxdy . \tag{2.23}$$

We use the finite limits of $(-L, L)$ on these integrals because we assume $g(x, y)$ is zero everywhere outside of the circle centered at the origin with radius L. The Fourier transform of $g(x, y)$, evaluated along the line corresponding to $\theta = 0$ is

$$G(X,0) = \int_{-L}^{L} \int_{-L}^{L} g(x,y)e^{-j(xX+y\cdot 0)} dxdy = \int_{-L}^{L} \int_{-L}^{L} g(x,y)e^{-jxX} dxdy . \tag{2.24}$$

Because Equations 2.23 and 2.24 are the same, the projection-slice theorem is proven for the special case of $\theta = 0$. To extend the result to arbitrary values of θ, we invoke a theorem from Fourier transforms that states that the Fourier transform of a θ-rotated version of an arbitrary function, g, is equal to the θ-rotated version of the Fourier transform of g. (This theorem applies to Fourier transforms of arbitrary dimensions and is proven in Appendix A.) We then note that the projection of $g(x, y)$ at angle θ is equal to the projection at angle $0°$ of a θ-rotated version of $g(x, y)$. Combining this result with the projection-slice theorem for $\theta = 0$ completes the proof.

2.3.3 Algorithms for Tomographic Reconstruction in Medical CAT

The Polar Reformatting Algorithm

We now proceed to describe an early algorithm used for tomographic image reconstruction from projection data in medical x-ray CAT systems.[5] This involves two-dimensional Fourier inversion of the polar raster data (see Figure 2.13), after first interpolating the data to a Cartesian raster. Therefore, we refer to it either as the Fourier inversion algorithm or the polar reformatting algorithm. The interpolation to the Cartesian array is performed because a fast algorithm for finite Fourier transformation, namely the FFT routine, is available for such data, while no rapid Fourier transform algorithm exists for data sampled on a polar raster.

At this point, some considerations on sampling rates for the tomographic data collection are in order. Figure 2.13 assumes that the two-dimensional transform of $g(x, y)$

[5] The first x-ray tomograms were computed using solvers of large, sparse systems of linear equations. These methods were generally known as algebraic reconstruction techniques [14].

has finite support on a circle of diameter $2\pi B$, where B has spatial-frequency units of cycles/meter.[6] We would then expect to reconstruct the image to a nominal resolution of B^{-1} meters in both x and y. We would like to answer the question of what the sampling rates for the projection data in the u and θ dimensions should be to support such a reconstructed image.

To this end, we note that by the projection-slice theorem, each projection function must also be bandlimited to B cycles/meter. The sampling criterion would therefore require that each projection function $p_\theta(u)$ be sampled at a minimum rate of $\delta u = B^{-1}$. It is also required that δX and δY on the interpolated sample grid be no greater than π/L, if the entire scene of diameter $2L$ is to be reconstructed alias-free. From Figure 2.13, we see that this gives the upper limit on $\delta\theta$ (in radian measure) as

$$\delta\theta \leq \frac{\delta X}{\pi B} = \frac{\delta u}{L} \,. \tag{2.25}$$

If the projection data are sampled such that N samples span the patch diameter of $2L$ with sample spacing δu, then $N = 2L/\delta u$. As a result, the (minimum) number of projections, M, required across the total π radians of viewing angles must be

$$M = \frac{\pi}{\delta\theta} = \pi\frac{L}{\delta u} = \frac{\pi}{2}N \,. \tag{2.26}$$

It should be noted that in areas close to the center of the Fourier domain, the polar samples are denser. In general, the result is that the higher spatial frequencies will be interpolated less accurately than the lower spatial frequencies. We will return to an in-depth study of polar reformatting in Chapter 3.

The Convolution/Back-Projection (CBP) Algorithm

The other tomographic reconstruction method that we will explore is known as the *convolution/back-projection (CBP)* algorithm. It is also sometimes called the *filtered back-projection* algorithm. Its derivation begins with the expression for the inverse Fourier transform of the frequency-domain data written in polar coordinates. Using polar coordinates in both domains gives

$$g(\rho\cos\phi, \rho\sin\phi) = \frac{1}{4\pi^2}\int_{-\pi/2}^{\pi/2} d\theta \int_{-\infty}^{\infty} G(r\cos\theta, r\sin\theta)|r|e^{jr\rho\cos(\phi-\theta)}dr \,. \tag{2.27}$$

[6] Because we assume that $g(x, y)$ has finite support on the circle of radius L, we recognize that its Fourier transform theoretically cannot also have finite support. Therefore, we are really only assuming that $g(x, y)$ is *essentially* bandlimited, in the sense that *most of the energy* of its transform lies in the circle of radius $2\pi B$.

The projection-slice theorem (Equation 2.19) allows the above inversion equation to then be rewritten as

$$g(\rho\cos\phi, \rho\sin\phi) = \frac{1}{4\pi^2}\int_{-\pi/2}^{\pi/2} d\theta \int_{-\infty}^{\infty} P_\theta(r)|r|e^{jr\rho\cos(\phi-\theta)}dr \;. \tag{2.28}$$

Equation 2.28 defines the convolution/back-projection tomographic reconstruction algorithm. The term *convolution* is explained in the following way. Because multiplication of Fourier transforms is equivalent to the convolution of the corresponding space-domain functions, the inner integral in Equation 2.28 can be interpreted as the convolution of the projection function, $p_\theta(\rho)$, with a filtering kernel, $h(\rho)$, the Fourier transform of which is equal to $|r|$. Because the convolution is evaluated at $\rho\cos(\phi-\theta)$, the formula may be expressed as

$$g(\rho\cos\phi, \rho\sin\phi) = \frac{1}{4\pi^2}\int_{-\pi/2}^{\pi/2} Q(\rho\cos(\phi-\theta))d\theta \tag{2.29}$$

where Q is the filtered (convolved with h) version of the projection function:

$$Q = p_\theta \otimes h \;. \tag{2.30}$$

Here, $\otimes$ denotes the convolution operator. The *back-projection* aspect of the algorithm becomes clear when we rewrite the reconstruction equation in terms of Cartesian image-domain coordinates as

$$g(x,y) = \frac{1}{4\pi^2}\int_{-\pi/2}^{\pi/2} Q(x\cos\theta + y\sin\theta)d\theta \;. \tag{2.31}$$

The argument of Q in Equation 2.31 is obtained easily as

$$\rho\cos(\phi-\theta) = \rho\cos\phi\cos\theta + \rho\sin\phi\sin\theta = x\cos\theta + y\sin\theta \;. \tag{2.32}$$

The evaluation of Q at $x\cos\theta + y\sin\theta$ implies that this value of the filtered projection function is *back-projected* along a line in the same direction in which the projection function was obtained, i.e., orthogonal to the u axis (Equation 2.16), as shown in Figure 2.14. The reconstruction of a given pixel at location (x, y) can therefore be viewed as the summation of values back-projected from each of the filtered projection functions. One-dimensional interpolation of the filtered projection functions is required, as opposed to two-dimensional interpolation of Fourier data in the polar reformatting algorithm.

The CBP algorithm is the one employed in virtually all modern medical CAT scanners. The reasons why CBP has supplanted the polar reformatting technique for medical tomography are several and are discussed in references [14] and [15].[7]

[7] More will be said about the computational aspects of CBP versus polar reformatting in the context of SAR image reconstruction in Section 2.4.4.

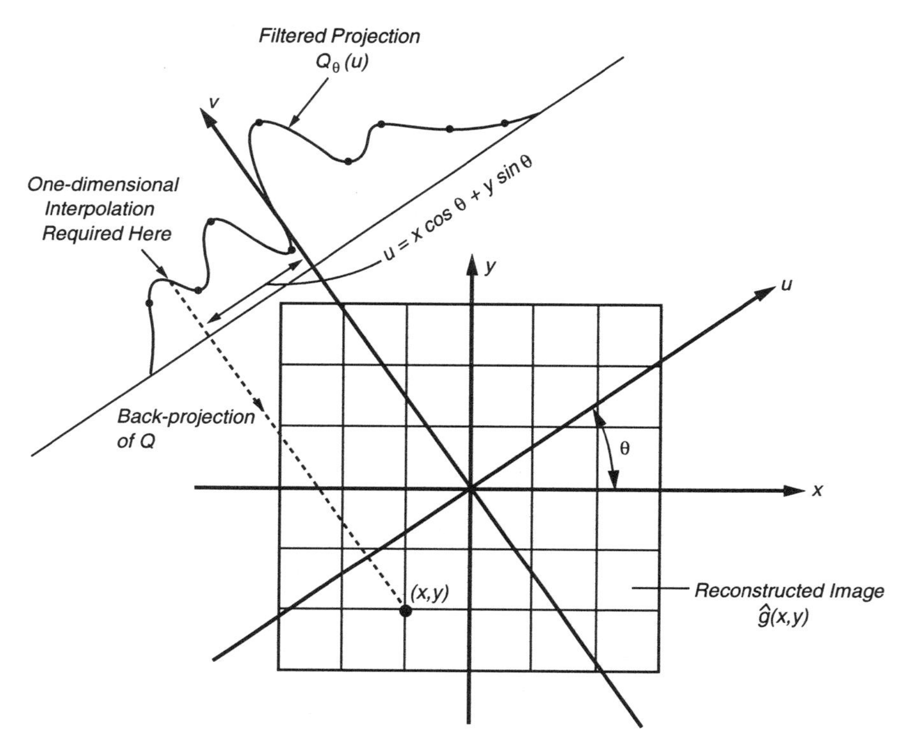

Figure 2.14 Convolution/back-projection (CBP) algorithm. Each point in the reconstructed image is obtained by integration of values back-projected from the filtered (convolved) projection functions. This algorithm involves only one-dimensional interpolations of the filtered projections, $Q_\theta(u)$.

The tomographic reconstruction formula shown in Equations 2.29 and 2.30 has an interesting history. This inversion of a function's line integrals to recover the function itself is attributable to J. Radon, who published its derivation in 1917 [16]. Presumably, Radon never realized the practical implications of the abstract mathematics he had discovered. The Radon transform of an image is the mapping of the image into its complete set of projection functions, which we denote as $p(\theta, u)$ (Equation 2.18), i.e.,

$$\mathbf{R}[g(x,y)] = p(\theta, u) \tag{2.33}$$

while the *inverse Radon transform* reconstructs the image from its complete projection set (Equation 2.28):

$$g(x,y) = \mathbf{R}^{-1}[p(\theta, u)] \, . \tag{2.34}$$

The Radon transform can be extended to higher dimensions. In fact, we will employ this result in three dimensions in our most general treatment of spotlight-mode SAR.

The remainder of this chapter, together with Chapter 3, will describe how spotlight-mode SAR can be treated as a tomographic reconstruction problem, and how the details of Fourier inversion for SAR image formation are actually implemented.

Summary of Critical Ideas from Section 2.3

- **An x-ray tomographic imaging system collects a series of projection functions from a thin slice of a patient's body.**
- **A projection function at a given viewing angle represents line integral data for the x-ray attenuation coefficient profile for the slice irradiated.**
- **The projection-slice theorem relates values of the one-dimensional Fourier transform of a projection function at a given angle to values along a line at the same angular orientation of the two-dimensional Fourier transform of the x-ray attenuation profile of the slice to be imaged.**
- **Projections taken through a full 180 degree range of viewing angles, upon Fourier transformation, sweep out data on a polar raster in two-dimensional Fourier space.**
- **Fourier inversion of the spatial-frequency data then becomes the reconstructed tomographic image.**
- **Two algorithms for tomographic reconstruction are the polar reformatting algorithm and the convolution/back-projection (CBP) algorithm.**
- **Modern medical CAT systems use CBP.**

2.4 A THREE-DIMENSIONAL TOMOGRAPHIC FRAMEWORK FOR SPOTLIGHT-MODE SAR

2.4.1 Introduction

From the discussion of medical CAT scanning presented in Section 2.3 and from the introduction to the cross-range resolvability issue in imaging radars presented earlier, it should now be apparent that the two problems bear a striking resemblance. Specifically, we now recognize that each integrated microwave reflectivity function $p_\theta(\bar{y})$ (see Equations 2.10 and 2.14) that follows deramping and range compression in a spotlight-mode SAR collection, is in tomographic terms a *projection function* of the scene reflectivity $g(x, y)$. To be sure, there are several obvious differences between the two imaging modalities: 1) medical x-ray CAT uses data from a transmission mode, while SAR employs reflected signals; 2) the scene reflectivity function and its associated projection functions in SAR are complex-valued, while the analogous quantities in medical CAT are real-valued; and 3) in spotlight-mode SAR, the aircraft flight path appears to span a rather limited set of viewing angles; whereas, the medical CAT scanner obtains projections that completely encircle the patient, i.e., span the full 180 degrees of viewing angles. We will see that in spite of these and several other more subtle differences, the medical tomographic algorithms can in fact be used for reconstruction of a spotlight-mode radar image from a set of processed return pulses collected over the synthetic aperture.

At this point, we could pursue the direct analogy between medical CAT and the two-dimensional version of the spotlight-mode SAR imaging problem we have posed so far. In 1983, David Munson and his colleagues [1] did precisely that when they published the first formal work describing spotlight-mode SAR as tomography. Their development assumed the same two-dimensional radar reflectivity function that we used to motivate the SAR reconstruction problem in Sections 2.1 and 2.2. There, the earth scene to be imaged was treated as a flat surface. We made no allowance for targets being elevated above the ground plane. This simplifying assumption made the analogy with medical tomography straightforward. As we just saw in Section 2.3, x-ray CAT is described and implemented in terms of a sequence of two-dimensional slices, i.e., thin flat cross-sections of a patient's body are typically reconstructed from data collected with a highly collimated x-ray beam. In actual SAR imaging scenarios, however, radar targets always exhibit some degree of elevation change across the scene. As this has important ramifications in the reconstructed two-dimensional SAR image, it is necessary to allow for such elevation changes in the imaging model. In addition, this three-dimensional formulation provides an excellent foundation for the topic of SAR *interferometry* that will be developed in Chapter 5.

As a result of the considerations outlined above, our development of spotlight-mode SAR as tomography bypasses the two-dimensional framework; instead, it builds a paradigm wherein the reflectivity function is taken to be three-dimensional.[8] Our first result from this model will be to show that deramp-processed returns (prior to range compression) from a spotlight-mode SAR collection represent samples from a certain portion of the three-dimensional Fourier transform of the target reflectivity function. Using further tomographic analysis of the associated reconstructed two-dimensional SAR image, we then show that we can interpret the effects of elevated targets. In Section 2.4.2, we establish a pair of three-dimensional extensions of the projection-slice theorem, which was the cornerstone of the two-dimensional x-ray CAT development. With these as our major tools, we then create the spotlight-mode SAR tomographic formulation for generalized three-dimensional scenes.

2.4.2 Generalized Three-Dimensional Tomography

The foundation of Munson's two-dimensional spotlight-mode SAR formulation is the *projection-slice* theorem, which is also the cornerstone of the development of medical x-ray CAT. Extension of this result from two to three dimensions produces two alternate versions of the theorem, depending on how the third dimension is utilized. Both these relationships are relevant to the three-dimensional tomographic paradigm for spotlight SAR and are detailed below.

We begin by stating the two versions of the projection-slice theorem in their simplest and most intuitive form, and then generalize the results. Consider a complex-valued function in three dimensions denoted by $g(x, y, x)$, with its three-dimensional Fourier transform given by

$$G(X, Y, Z) = \iiint g(x, y, z) e^{-j(xX + yY + zZ)}\, dx\, dy\, dz\ . \tag{2.35}$$

The first theorem involves a one-dimensional projection function formed by integrating $g(x, y, z)$ over two of the three spatial dimensions:

$$p_1(x) = \iint g(x, y, z)\, dy\, dz\ . \tag{2.36}$$

The projection-slice theorem states that the one-dimensional Fourier transform of $p_1(x)$, denoted $P_1(X)$, is identical to the one-dimensional function obtained by

[8] Because our development will in several ways parallel Munson's two-dimensional formulation, the reader is encouraged to read reference [1].

evaluating $G(X, Y, Z)$ along the X axis:

$$P_1(X) = \int p_1(x)e^{-jxX}\, dx = G(X, 0, 0) . \tag{2.37}$$

The process of evaluating $G(X, Y, Z)$ along a certain line defines what we will refer to as a *trace* (i.e., a *linear slice*) of $G(X, Y, Z)$. Accordingly, we will refer to this as the *linear trace* version of the projection-slice theorem, because it relates linear projection functions to traces of the three-dimensional Fourier transform.

The other three-dimensional version of the projection-slice theorem involves a two-dimensional projection function of the form

$$p_2(x, y) = \int g(x, y, z)dz . \tag{2.38}$$

The two-dimensional Fourier transform of $p_2(x, y)$, according to this theorem, is equal to a two-dimensional *slice* of $G(X, Y, Z)$ taken on the (X, Y) plane:

$$P_2(X, Y) = \int\int p_2(x, y)e^{-j(xX+yY)}dxdy = G(X, Y, 0) . \tag{2.39}$$

We will refer to this second projection-slice relationship as the *planar slice* theorem, because it relates planar projection functions to planar Fourier transform slices.

The two versions of the projection-slice theorem stated above easily follow from Equation 2.35 and the definitions of the respective projection functions. They specifically apply to projections in the cardinal directions. By invoking the same rotational property of Fourier transforms that we employed for the derivation of the two-dimensional form of the theorem in Section 2.3.2, both of the three-dimensional versions may be generalized to an arbitrary orientation in three-dimensional space [13]. This rotational property (see Appendix A) states that if $f(\mathbf{x})$ and $F(\mathbf{X})$ are a Fourier transform pair, the Fourier transform of a θ-rotated version of $f(\mathbf{x})$ is a θ-rotated version of $F(\mathbf{X})$. The result of applying this property to Equations 2.37 and 2.39 is that the angular orientation of the trace or plane that defines the projection function in the spatial domain is coincident with the orientation of the trace or slice in the Fourier transform domain. This fundamental result is shown in Figure 2.15 for the case of the linear trace theorem, and in Figure 2.16 for the case of the planar slice theorem.[9] In Section 2.4.3, we will first use the linear trace version to describe returns from a set of radar pulses that compose a spotlight-mode collection, and later invoke the planar slice theorem for interpreting the associated reconstructed two-dimensional SAR image.

[9] A discussion of both forms of projection-slice theorems, as well as their relationship to the Radon transform, is given in [17].

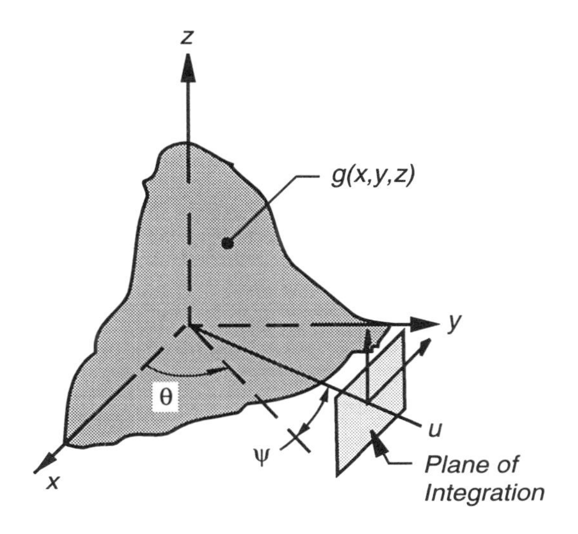

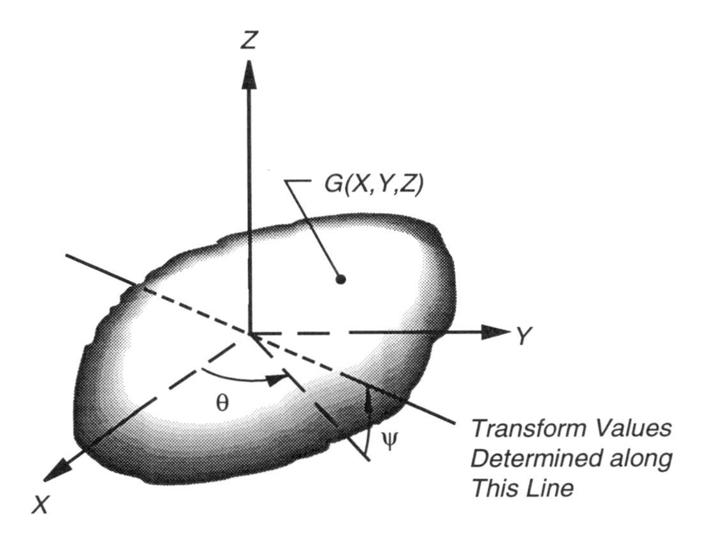

Figure 2.15 Depiction of the first of the two three-dimensional projection-slice versions used in the tomographic paradigm for spotlight-mode SAR (linear trace version). The top figure depicts the space-domain projection function obtained from integrating $g(x, y, z)$ over two spatial dimensions, used for describing the SAR data collection. Bottom figure shows the line (trace) of the three-dimensional Fourier transform of $g(x, y, z)$ which, according to this version of the projection-slice theorem, is equal to the one-dimensional Fourier transform of the linear projection function above.

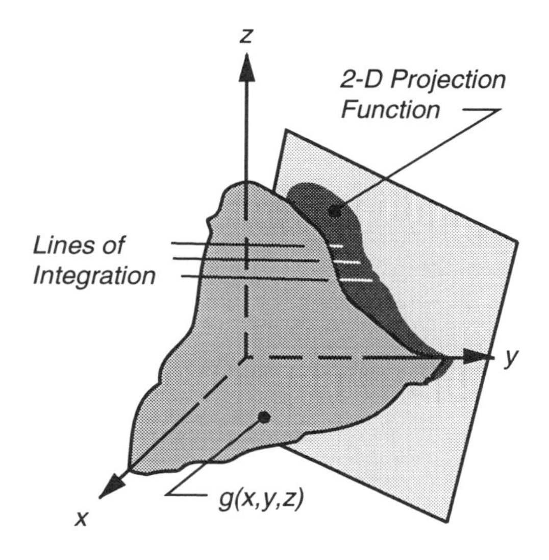

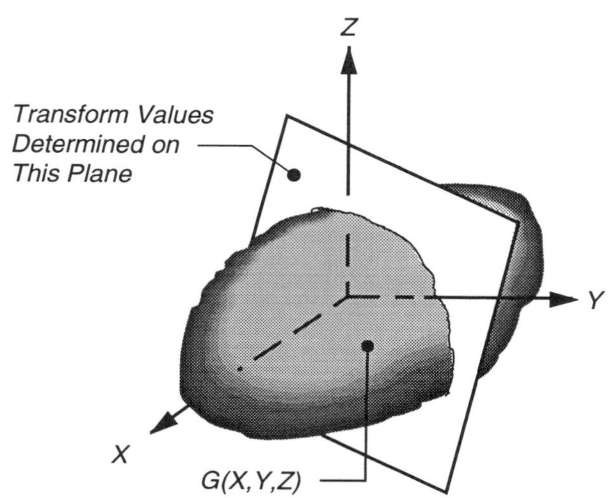

Figure 2.16 Depiction of the second of the two three-dimensional projection-slice versions used in the tomographic paradigm for spotlight-mode SAR (planar-slice version). The top figure depicts the space-domain projection function obtained from integrating $g(x, y, z)$ along a single spatial dimension, resulting in a two-dimensional projection function. This is used for describing the layover/projection effects in the two-dimensional reconstructed spotlight-mode SAR image. Bottom figure shows the planar slice of the three-dimensional Fourier transform, taken at the same angular orientation as the planar projection of $g(x, y, z)$ above. According to this version of the projection-slice theorem, it is equivalent to the two-dimensional Fourier transform of the planar projection function.

2.4.3 Spotlight-Mode SAR Imaging of a Three-Dimensional Scene

For the SAR imaging scenario considered here, we assume that the function $g(x, y, z)$ represents the three-dimensional radar reflectivity density function for a portion of the earth surface illuminated by the radar's antenna beam. The radar energy is assumed to be non-penetrating, so the function $g(x, y, z)$ is generally constrained to be zero everywhere except on a surface, and also to be zero at every point on this surface which is shadowed. We further assume that $g(x, y, z)$ does not vary over the range of viewing angles spanned by the synthetic aperture nor does it vary over the range of frequencies employed. The goal of the SAR imaging process, then, is to estimate (reconstruct) either $g(x, y, z)$, or an appropriate related function, such as a two-dimensional projection of $g(x, y, z)$ onto a plane, from the set of radar pulses gathered across the synthetic aperture.

The collection geometry of Figure 2.17 shows that an azimuthal angle θ and a grazing angle ψ together describe a particular direction from which the radar both transmits a pulse and receives the terrain echo. The same figure also defines a rotated set of coordinates (u, v, w). We now write a new form of projection function associated with viewing angles (θ, ψ) as

$$p_{\theta,\psi}(u) = \int\!\!\int g[x(u, v, w), y(u, v, w), z(u, v, w)]\, dv\, dw \tag{2.40}$$

where u is the slant range. We again assume that the waveform transmitted by the radar is the linear FM chirp pulse given by $\mathbf{Re}\{s(t)\}$ where

$$s(t) = \begin{cases} e^{j(\omega_0 t + \alpha t^2)} & \text{if } |t| \le \tau_c/2 \\ 0 & \text{otherwise} \end{cases} \tag{2.41}$$

Here, 2α is the FM chirp rate, τ_c is the pulse duration, and ω_0 is the RF center frequency. The radar return received at position (θ, ψ) follows the form of Equation 2.12 in Section 2.2 as

$$r_{\theta,\psi}(t) = A\, \mathbf{Re}\left\{ \int_{-u_1}^{u_1} p_{\theta,\psi}(u)\, s\left(t - \frac{2(R+u)}{c}\right) du \right\} \tag{2.42}$$

where R is the range from the scene center (aim point) to the radar and u_1 is the maximum slant range for any target illuminated by the beam. The difference between Equation 2.42 and Equation 2.12 is that the new projection function $p_{\theta,\psi}(u)$ involves a two-dimensional integration of the three-dimensional radar reflectivity function over plane surfaces as per the linear trace theorem, rather than the one-dimensional integration of a two-dimensional reflectivity function $g(x, y)$ along straight lines, as shown in Figure 2.5. (More precisely, the new surfaces of integration are sections of spheres, but we assume that when the standoff range is large relative to the scene size, they may be well-approximated by planes.) Finally, we emphasize that we have defined the new projection function $p_{\theta,\psi}(u)$ in terms of the slant range u as opposed to a ground range $\bar{y}$, as was used to define $p_{\theta}(\bar{y})$.

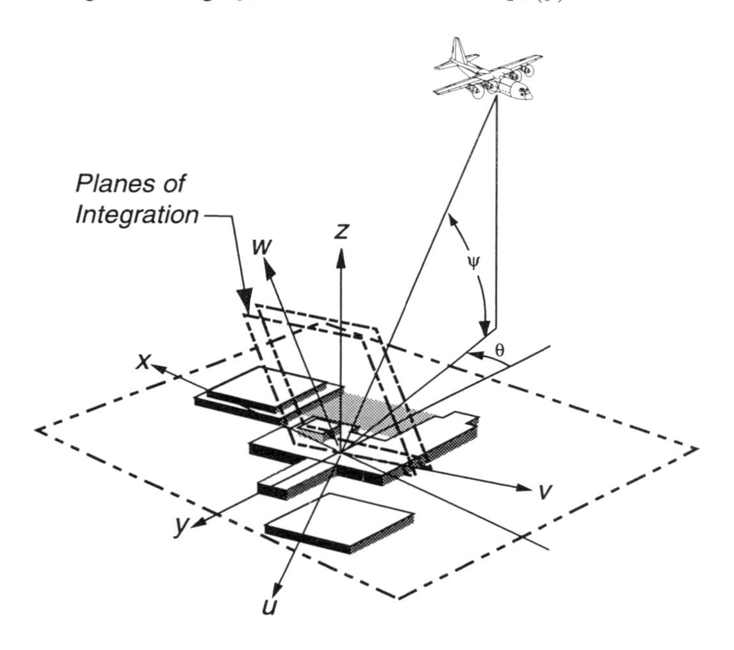

Figure 2.17 Three-dimensional SAR collection geometry. A projection function from an angular orientation of (θ, ψ) is obtained by two-dimensional integration of $g(x, y, z)$ across planes orthogonal to the u-vector, i.e., (v, w) planes.

A major result developed in Chapter 1 was that the deramped return signal (prior to range compression) from each transmitted chirp pulse represents a direct transduction of a certain portion of the Fourier transform of the scene reflectivity (Equation 1.38). That result now allows us to write an expression for the quadrature-demodulated version of the signal of Equation 2.42 as

$$\bar{r}_{\theta,\psi}(t) = \frac{A}{2} P_{\theta,\psi} \left[\frac{2}{c}(\omega_0 + 2\alpha(t - \tau_0)) \right] \tag{2.43}$$

where $P_{\theta,\psi}(U)$ is the Fourier transform of the projection function $p_{\theta,\psi}(u)$ and τ_0 is the delay term given by $\tau_0 = 2R/c$. According to the linear trace version of the projection-slice theorem, $\overline{r}_{\theta,\psi}(t)$ must therefore represent a trace (line) of $G(X, Y, Z)$ at angular orientation (θ, ψ). We showed in Chapter 1 that the limited duration of the transmitted pulse of Equation 2.41 implies that $P_{\theta,\psi}(U)$ will only be determined on a restricted interval of spatial frequencies given by

$$\frac{2}{c}(\omega_0 - \alpha\tau_c) \leq U \leq \frac{2}{c}(\omega_0 + \alpha\tau_c) . \tag{2.44}$$

As we have observed before, this *spatial* frequency range is proportional to the *temporal* frequency bandwidth (equal to $\alpha\tau_c/\pi$ Hertz) encompassed in the transmitted chirp.

The above analysis of an individual return pulse leads to the following description of a set of processed radar pulses collected over a range of angular orientations as the radar platform moves through the synthetic aperture. Each processed (demodulated) pulse produces values of the three-dimensional Fourier transform of $g(x, y, z)$ along a certain line segment, the direction of which is determined by the θ and ψ values associated with that pulse transmission point. The radial position and length of the segment are determined by the radar wavelength and bandwidth, respectively. As a result, a collection of pulses will sweep out a *ribbon* surface in this space, the precise shape of which is determined by the platform flight trajectory, i.e., the (θ, ψ) time history of the synthetic aperture. A ribbon of this kind, referred to as the *signal collection surface* [18], is shown in Figure 2.18. The spatial-frequency domain in which the collection surface exists is often referred to as the *phase-history domain*. This fundamental result that describes the collected SAR data as a surface in three-dimensional Fourier space was originally obtained by Walker [2] using a range-Doppler analysis. As demonstrated here, the derivation can be constructed using only the mathematics of three-dimensional tomography and some basic signal processing concepts. In particular, the notion of Doppler frequency is not used anywhere in our development.

The reader should carefully note four important points relevant to the synthesis of the collection surface depicted in Figure 2.18. First, the flight segment that produces the ribbon of phase-history data can be an arbitrary path in three-dimensional space. In particular, it is *not* assumed to be parallel to the x direction, nor is it assumed to be level, e.g., the aircraft could be climbing. Second, the coordinate system is chosen such that the projection of the center pulse of the aperture into the (X, Y) plane falls along the Y direction, which defines the direction of ground range. In Section 2.4.4, we more fully develop the ideas of ground-plane and slant-plane image reconstructions, and describe the associated geometries of each. Third, the look angles θ and ψ that correspond to each pulse transmission point along the flight path in the physical space are *the same angles* for which the processed return pulse data represent information in the spatial-frequency domain. This equivalence

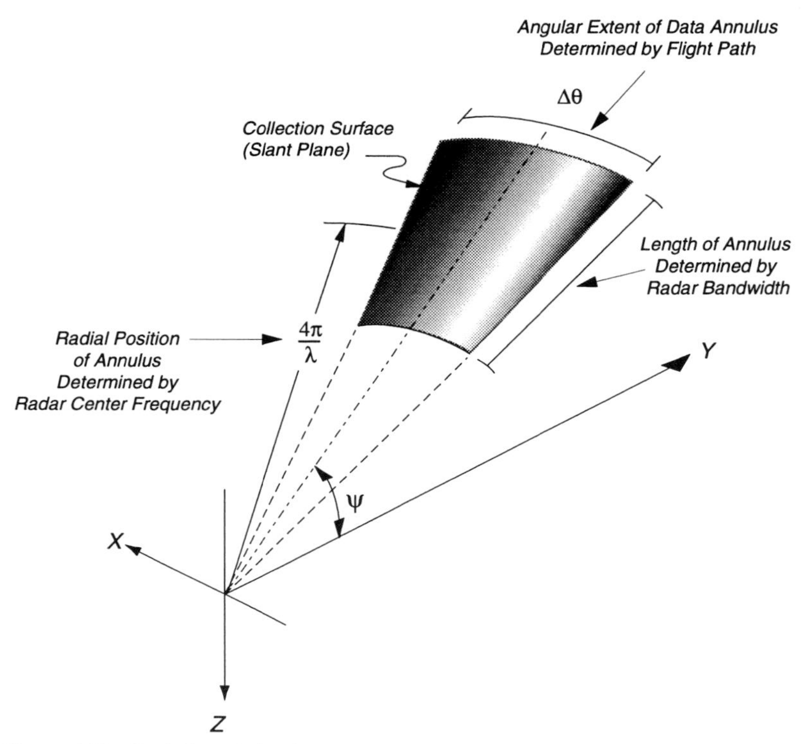

Figure 2.18 Depiction of the three-dimensional signal collection surface created in a spotlight-mode collection. Each deramped pulse (prior to range compression) yields one line segment of data in three-dimensional Fourier space. The angular orientation of the line in this phase-history domain is the same as the direction of the pulse transmission in the scene (image) space. This is a consequence of the linear trace version of the projection-slice theorem. Therefore, a collection of processed pulses sweeps out a ribbon surface, as shown.

of angular orientation of the pulse data in the two domains is a key concept in the spotlight-mode SAR tomographic paradigm. Fourth, the spatial extent of the ribbon surface is determined in the radial direction by the bandwidth of the transmitted chirp, while the angular extent is determined by the diversity of viewing angles inherent in the flight path, i.e., in the synthetic aperture. We should note at this juncture that although a direct analogy with x-ray CAT has allowed us to construct a mathematical model for the data collected by a spotlight-mode SAR, at least two additional important differences between SAR tomography and medical tomography have now emerged. First, in medical x-ray CAT the projection data are transduced directly in the image domain, i.e., the x-ray intensity ratio of Equation 2.18 represents a direct measurement of samples of a projection function of the x-ray attenuation profile $g(x, y)$. As we have demonstrated above, however, in a spotlight-mode SAR that employs chirp waveforms and the associated deramp processing, the processed return signal (prior to range compression) directly yields samples of the Fourier transform of a projection function of $g(x, y, z)$. Second, Figure 2.18 demonstrates that the Fourier data obtained by the spotlight-mode SAR are not centered at the origin of the transform space. Instead, they are *offset* in spatial frequency by an amount proportional to the radar center frequency. We discuss the ramifications of this fact later in this chapter.

Consider next the case of a straight-line flight path through the synthetic aperture, wherein the collection surface swept out in three-dimensional Fourier space is simply a plane, referred to as the *slant* plane. This models the situation for a typical airborne spotlight-mode SAR collection where the *out-of-plane* motion is insignificant. The slant plane is determined by the line of the flight path and the aim point (center) of the scene. It should be noted that the trajectory of a spaceborne SAR platform would not in general be described by a straight line, and as a result the collection surface would not be a slant plane. In addition, non-planar collection surfaces can also be produced unintentionally by aircraft SAR systems operating in conditions of high turbulence. The ramifications of this and other deviations from straight-line flight paths, both systematic and random, are addressed in Chapters 3 and 4.

The usual procedure for forming the final reconstructed two-dimensional SAR image of the scene in the case of planar collections is to compute the inverse Fourier transform of the data in the slant plane. An interpolation from the polar collection raster to a Cartesian grid is performed first to allow use of fast Fourier transform techniques. The following section outlines the major steps involved in this procedure, and shows the derivation of several expressions for key parameters used to describe the resulting image. These quantities include range and cross-range spatial resolutions, as well as sampling rates required along the synthetic aperture during the collection. A much more detailed description of the procedures used to design and implement a polar reformatting algorithm for image formation is presented in Chapter 3.

2.4.4 Formation of a Two-Dimensional Spotlight-Mode Image

In this section, we provide a general analysis of the procedure used to transform the phase-history data obtained on a slant plane into a two-dimensional SAR image of the scene. Chapter 3 discusses a number of practical issues concerning the actual implementation of an image-formation algorithm on a modern digital computer.

From the discussions of the preceding sections, it is clear that the phase-history domain description of a slant-plane collection is a set of samples lying on a polar raster imposed on an annulus in the slant plane, as shown in Figure 2.19. The most straightforward way to think of Fourier inversion of these data in creating a SAR image is to consider a set of coordinate axes (X', Y') in the slant plane. The center pulse of the aperture is used to define the Y' direction, with the X' axis orthogonal and lying in the plane. The Y' dimension then corresponds to *slant-range* spatial frequencies, so that a two-dimensional inverse Fourier transform of the data produces an image in a domain with x' and y' axes of cross-range and slant range. One alternative procedure would be to *project* the phase-history samples, as shown in Figure 2.19, onto the (X, Y) plane prior to performing the two-dimensional Fourier inversion. The effect of such a projection is to produce range spatial frequencies, Y, that correspond to *ground range* instead of slant range. Therefore, the SAR image formed by Fourier transformation has x and y axes of cross range and ground range. (Later, we will see why sometimes a ground-plane reconstruction can be advantageous when compared to the corresponding slant-plane image.)

Figure 2.20 shows the differences between these two basic forms of phase-history data prior to polar reformatting. An analysis of these diagrams allows us to compute a number of useful parameters regarding the corresponding image-domain reconstructions. We begin with an analysis of slant-plane data as depicted in the top diagram of Figure 2.20. The data in this case lie on a circular annulus where each radial segment has the same length, because the interval of spatial frequencies as given by Equation 2.44 is the same for every pulse. The entire annulus is offset from the origin by an amount equal to $2\omega_0/c = 4\pi/\lambda$. It is the extent of the data annulus in both dimensions that determines the spatial resolutions that can be achieved in the reconstructed image following Fourier inversion. Suppose that the polar-to-rectangular interpolation is performed so as to yield samples within a rectangular box as shown in Figure 2.20. The spatial bandwidth in the range dimension is

$$\Delta Y' = \frac{2}{c}(2\pi B_c) \; . \tag{2.45}$$

The nominal cross-range extent is determined by the radius $4\pi/\lambda$ and the angular extent $\Delta\theta$ of the annulus, so that we have

$$\Delta X' = 2\left(\frac{4\pi}{\lambda}\right)\sin(\Delta\theta/2) \; . \tag{2.46}$$

As we will see later in this section, the angular diversity $\Delta\theta$ is typically very small in spotlight-mode SAR collections. Under this small-angle assumption, the above expression for $\Delta X'$ can be approximated by

$$\Delta X' = \frac{4\pi}{\lambda}\Delta\theta \; . \tag{2.47}$$

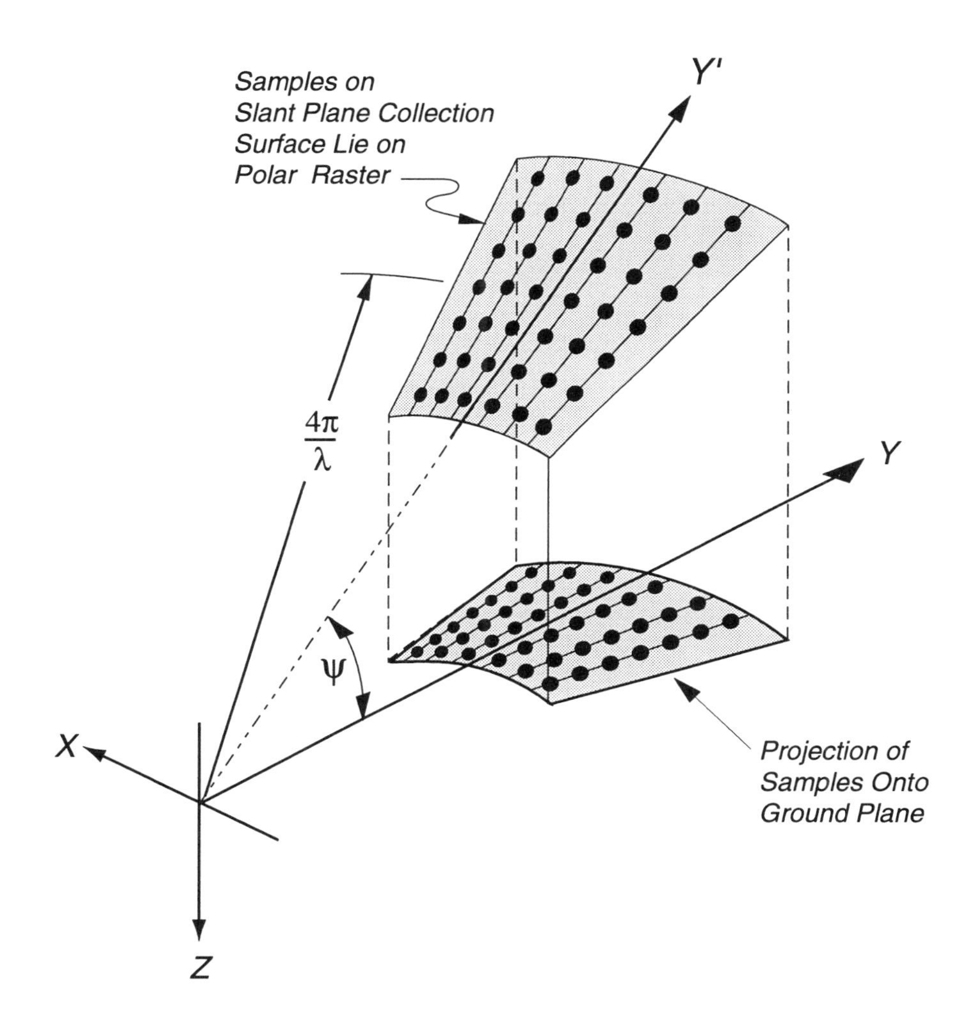

Figure 2.19 Projection of slant plane Fourier-domain samples onto ground plane. Two-dimensional Fourier inversion of these data following polar-to-rectangular interpolation leads to a ground-plane reconstructed image, as opposed to a slant-plane image.

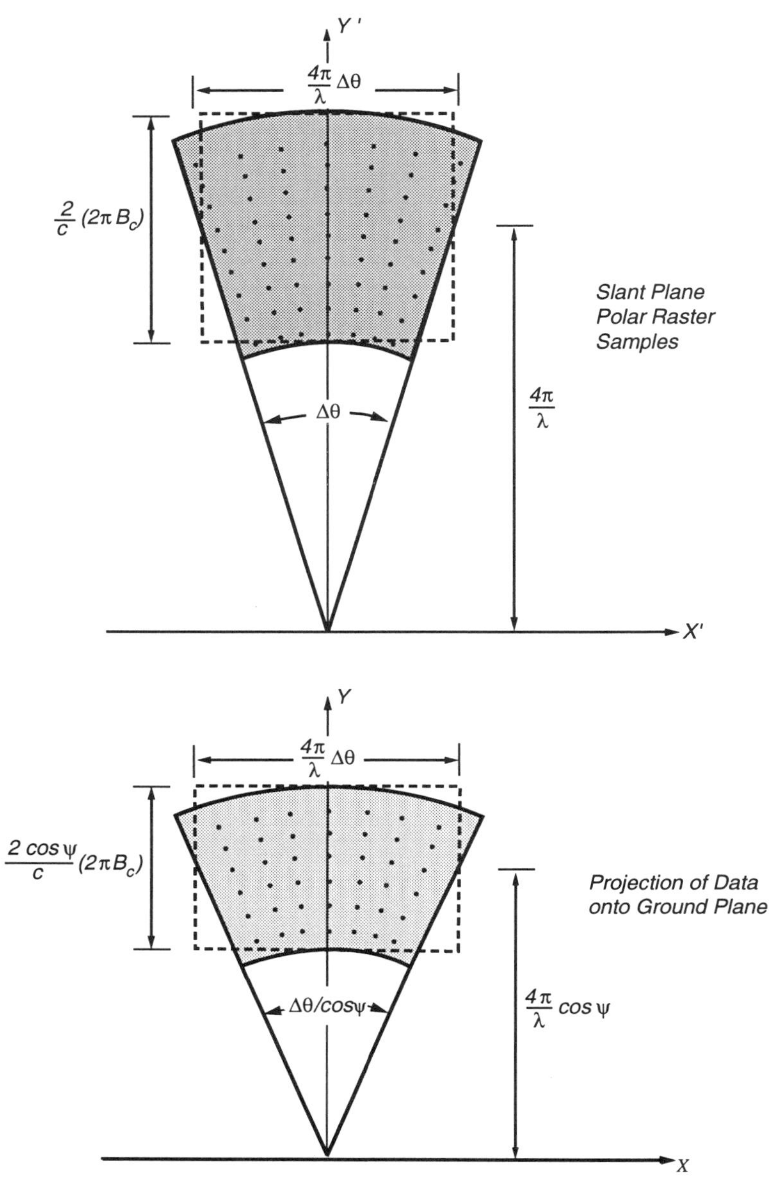

Figure 2.20 Fourier-domain polar raster samples in slant plane (top figure) and projected onto ground plane (bottom figure). The annulus of data for the ground-plane projection is contracted in the range spatial-frequency dimension by the factor of $\cos\psi$, as is the offset from the origin in this direction. Using a small angle approximation, it can be shown that the angle subtended is increased by the reciprocal of this factor.

The image-domain resolutions are easily obtained from the above expressions for spatial-frequency bandwidths. The resulting range resolution is

$$\rho_{y'} = \frac{2\pi}{\Delta Y'} = \frac{c}{2B_c} \tag{2.48}$$

while the cross-range resolution expression (for the small-angle approximation) is

$$\rho_{x'} = \frac{2\pi}{\Delta X'} = \frac{\lambda}{2\Delta\theta} \,. \tag{2.49}$$

At this point, we note that the above resolution expressions have lower bounds. Because the bandwidth of the radar can never exceed twice the center frequency, Equation 2.48 dictates that the range resolution can never be better than $\lambda/4$. Equation 2.46 in turn suggests that $\Delta X'$ cannot exceed $8\pi/\lambda$, so that the cross-range resolution is also bounded by $\lambda/4$. Of course, there are many practical reasons why real SAR systems typically do not achieve resolutions even close to these bounds.

Next, several calculations regarding the sampling rates of the phase-history domain data are in order. These are analogous to those made for the case of medical x-ray CAT in Section 2.3. The sampling rate along the synthetic aperture that is required to avoid spatial aliasing may be determined in the following way. We can calculate the minimum X' sampling interval in the phase-history domain required to reconstruct alias-free the scene patch of diameter $2L$. This is given by

$$\delta X' = \frac{2\pi}{2L} \,. \tag{2.50}$$

Therefore, the corresponding angular sampling interval (see top diagram of Figure 2.20) must be

$$\delta\theta = \frac{\delta X'}{(4\pi/\lambda)} = \frac{\lambda}{2(2L)} \,. \tag{2.51}$$

Because the angular orientations of a projection function in the physical space and its Fourier transform in the phase-history domain are the same, we can project this angular sampling interval out to the platform standoff range to compute the required along-track sampling interval (in meters) as

$$\delta A = R\,\delta\theta = \frac{\lambda R}{2(2L)} = \frac{D}{2} \tag{2.52}$$

where D is the physical antenna length that produces an angular beamwidth of $\beta = 2L/R$ and that therefore would illuminate the patch of diameter $2L$ from standoff range R. Equation 2.52 states that the minimum spacing of samples along the synthetic aperture to prevent aliasing in the reconstructed image is equal to one-half of the diameter of the physical antenna. It turns out that the same rate is required in a strip-map system, but in that case, the best cross-range resolution achievable is also given by $D/2$. With the spotlight mode, this resolution limit is lifted.

Next, we consider calculations of corresponding quantities for the image reconstructed from ground-projected phase-history samples. In general, the projection operation imposes a contraction of the Fourier-domain data along each line segment (i.e., radially) in the annulus, where the contraction factor is the cosine of the depression angle for the given pulse. As a result, the exact nature of the contraction is a function of the flight path, i.e., it depends on how the depression angle varies across the synthetic aperture. The simplest situation is when the flight path is level and the center pulse is launched perpendicular to the flight path, wherein the geometry is termed *broadside*. This collection modality is shown in the top diagram of Figure 2.21. The depression angle decreases for pulses further away from the aperture center.

For the broadside case, it is easily shown that the footprint of the ground-plane projection of phase-history samples is another annulus which is *uniformly* contracted in the Y dimension by an amount equal to $\cos\psi$, where ψ is the depression angle for the central pulse. (The spatially varying radial contraction turns out to be a constant contraction in the vertical dimension.) The bottom of Figure 2.20 shows the contraction of the data annulus for this mode. The upper and lower boundaries of the projected annulus are elliptical instead of circular, as is the case for the slant-plane data. Note that the Y offset as well as the Y extent of the data have contracted by $\cos\psi$. On the other hand, the angular diversity has increased by $1/\cos\psi$.

Using the dimensions shown in the bottom diagram of Figure 2.20, we can now calculate the expressions for range and cross-range resolution in the ground-plane reconstructed image as

$$\rho_y = \frac{2\pi}{\Delta Y} = \frac{c}{2B_c \cos\psi} \tag{2.53}$$

and

$$\rho_x = \frac{2\pi}{\Delta X} = \frac{\lambda}{2\Delta\theta} \,. \tag{2.54}$$

The ground-range resolution is scaled by $\cos\psi$, while the cross-range resolution remains unchanged from its slant-plane counterpart. As was discussed in Chapter 1, the difference in the size of ρ_y and $\rho_{y'}$ *does not* imply that the resolvability of two targets spaced some distance apart is improved in a slant-plane as opposed to ground-plane reconstruction. This is because the separation of the targets in slant range is also proportionally smaller. The only difference in this case between ground- and slant-plane reconstructions is a scaling of the range dimension.

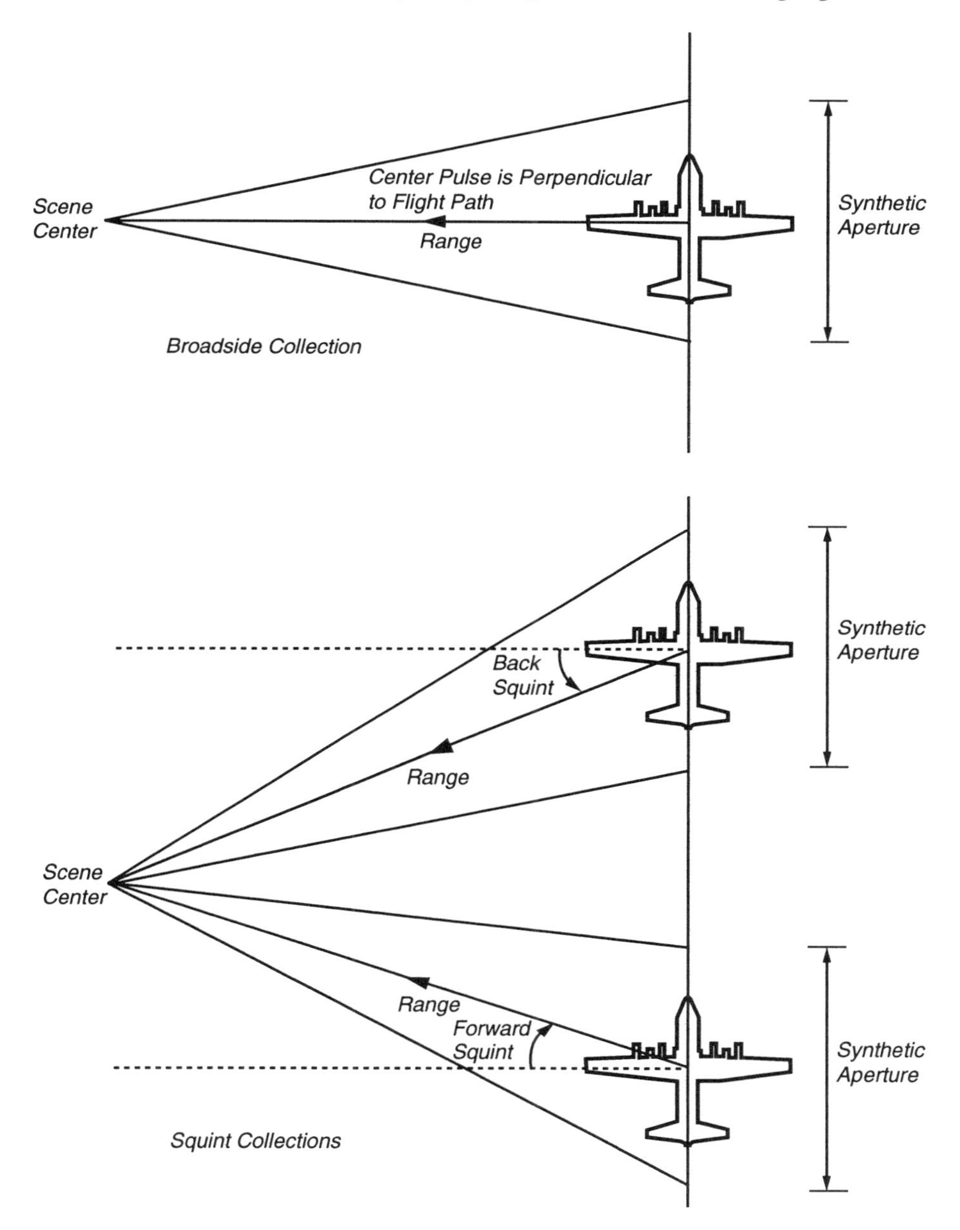

Figure 2.21 Collection geometries for broadside and squint imaging modalities in spotlight mode. For the broadside case, the center pulse in the synthetic aperture is perpendicular to the flight path. In the squint mode, it is not. In either case, the line connecting the center of the aperture and the scene patch center defines the range direction.

A more complicated situation arises, however, when at the aperture center the view from the antenna to the ground patch center is either forward or back. This collection modality is termed *squinted*, and is shown in the bottom diagram of Figure 2.21. Because the position and length of each data segment in Fourier space are both functions of the depression angle for that pulse, the precise shape of the upper and lower annulus boundaries depends on the flight-path geometry. As it turns out, there can be significant geometric differences between a slant-plane and ground-plane reconstructed image obtained from a squint-mode collection. In particular, while the slant-plane phase-history data always lie on an annulus (see top diagram of Figure 2.20), the projection of these data into the phase-history ground plane will in general be skewed instead of simply contracted in range. This subject is treated in detail in Chapter 3.

Some interesting issues concerning the reconstruction of SAR images are raised with a more careful examination of Equation 2.49. Consider a system with center frequency of 35 GHz that is capable of achieving a (slant) range resolution of 1 foot. This system requires a radar bandwidth of approximately 500 MHz. In order to achieve the same resolution in cross range, the angular diversity in the synthetic aperture required by Equation 2.49 is

$$\Delta\theta = \frac{\lambda}{2\rho_x} = \frac{0.0086m}{2 \cdot 0.3m} = 0.82^\circ \ . \tag{2.55}$$

A flight path spanning less than one degree of viewing angles can produce an azimuthal resolution of 1 foot. This calculation immediately raises the question of why a medical x-ray CAT scan must employ a full 180° of projection data, while a SAR can "get by" with a much narrower slice of data in the spatial-frequency domain. In fact, one might suspect from basic considerations of two-dimensional Fourier transforms that such a narrow slice of SAR phase-history data would *not* yield, upon inverse transformation, anything remotely resembling the true image. That is, even though it is the size of the slice of Fourier data that determines the resolution of the reconstructed image, the fact that such a restricted set of spatial frequencies is transduced in SAR would appear to be problematic. The reader is referred to an excellent paper by Munson and Sanz [19], who offer an explanation of this phenomenon. In brief summary, what their paper suggests is that a critical property of a SAR image lies in the fact that the SAR reflectivity function (image domain) typically possesses a phase function that is essentially uncorrelated. By using computer simulations, they demonstrate that when Fourier transformation occurs, this phase term acts to modulate the reflectivity magnitude information over the entire spatial-frequency plane, so that a small "chunk" of Fourier data taken anywhere can be inverted to give a reconstruction that in magnitude at least macroscopically resembles the original scene. The reconstruction is a *speckled* version of the scene reflectivity that does not match the true magnitude function point-for-point. These same properties also have relevance to the subject of holography, as is discussed in Chapter 3.

While we have seen that the mathematics of medical x-ray tomography provide a paradigm for the formulation of spotlight-mode SAR, the differences between the two types of imaging should be carefully recounted. They are that: 1) SAR utilizes *reflectivity* of microwaves, whereas x-ray CAT uses *transmissive* properties of x-rays; 2) use of the linear FM chirp and associated deramp processing in SAR provides Fourier-domain data directly at the output of the quadrature demodulator, whereas in x-ray CAT the image-domain projection data are transduced; 3) the Fourier data for SAR are *offset* in spatial frequency by $4\pi/\lambda$ (slant plane), whereas when the CAT projection data are Fourier transformed, they are effectively baseband, i.e., they are centered at the Fourier-domain origin; 4) the SAR Fourier data are only determined over an annulus that is angularly narrow, i.e., $\Delta\theta$ is typically only a few degrees, whereas in medical CAT views spanning a full 180° are employed, so that a circular region of the Fourier plane is filled in; 5) because the SAR is a coherent imaging system, a SAR image transduces the *complex* microwave reflectivity at every point, whereas a medical x-ray CAT scan computes the *real-valued* x-ray attenuation coefficient everywhere; 6) real-world spotlight-mode SAR image-formation systems employ polar reformatting for the reconstruction algorithm, while medical CAT scanners typically use the CBP algorithm; 7) a medical tomogram restricts the data collected to a thin slice, and thereby renders the reconstruction problem two-dimensional. A real section of earth surface, however, is typically not a flat slab. It may have significant height (terrain elevation) variations, so that a two-dimensional SAR reconstruction exhibits effects induced by the height profile of the reflectors. Thus, a real earth scene can only be properly treated by using a three-dimensional reflectivity model. This is the subject of Section 2.4.5.

One final note regarding summary point 6) in the paragraph above is in order. In spotlight-mode SAR image formation, the CBP algorithm generally takes more computing time than does polar reformatting. This is because in the reconstruction of an N x N phase-history array via polar reformatting, the number of multiplies required for each of the N^2 samples to perform the two-dimensional interpolation to a Cartesian grid is typically small (on the order of 16). This follows because the polar data in SAR are nearly on a Cartesian raster, due to the small diversity of angles typically involved, so that only a small number of neighboring samples need to be used in the interpolator.[10] The Fourier inversion time is small by comparison, so that the two-dimensional interpolation step dominates the total algorithm time. The result is that polar formatting for SAR has an operations count with an N^2

[10] In spotlight-mode SARs that use very low center frequencies, the angular diversity may be large enough that this approximation is not valid.

dependence. The operations count for CBP, on the other hand, increases as N^3. This is seen by considering that if N projections (pulses) are used, then N one-dimensional interpolations of the filtered projection functions are required for each of the N^2 reconstructed pixels. The dimensions of spotlight-mode SAR images are often large enough (N is typically on the order of thousands) to make CBP a much less attractive algorithm than polar reformatting. On the other hand, CBP is the algorithm employed in virtually all modern medical CAT scanners. The reader should consult references [14] and [15] for a discussion of this subject.

2.4.5 Projection Effects in the Reconstructed Two-Dimensional SAR Image: The Layover Concept

We have seen how tomographic principles can be used to derive a mathematical foundation that describes the collected phase-history data in spotlight-mode SAR and that allows certain key parameters concerning the reconstructed SAR image to be calculated. In this section we demonstrate another aspect of three-dimensional tomography that reveals the nature of the SAR image. This aspect considers the effects of representing a three-dimensional radar reflectivity function with a two-dimensional image.

Recall that the planar version of the projection-slice theorem (Equation 2.39) relates a planar projection of a three-dimensional structure to a planar slice of its Fourier transform. This implies that an inverse two-dimensional Fourier transform of the slant-plane phase-history data will produce an image that is a *projection of the three-dimensional target structure in a direction normal to the slant plane*. Each value in the image is the result of integrating the scene reflectivity function along a line perpendicular to the slant plane. Because we are assuming that the three-dimensional reflectivity function exists only on the surface of the scene, each line integral involved here will generally select a single point on that surface where it is intersected by the line of integration. When certain elevated targets (particularly man-made structures) are involved, however, the lines of integration may intersect the scene at multiple points. This will cause all such points to be superimposed on the same location in the image. This scenario is illustrated in Figure 2.22 where the elevated structure is superimposed on a certain region of the ground surface in the image. The elevated targets are said to be subject to *layover*.

The net result of the above analysis is that each target in the scene assumes a position in the SAR image that is determined by the target height and by the slant-plane orientation. Note from Figure 2.22 that the direction of the layover is prescribed as normal to the intersection of the slant plane and ground plane. This direction

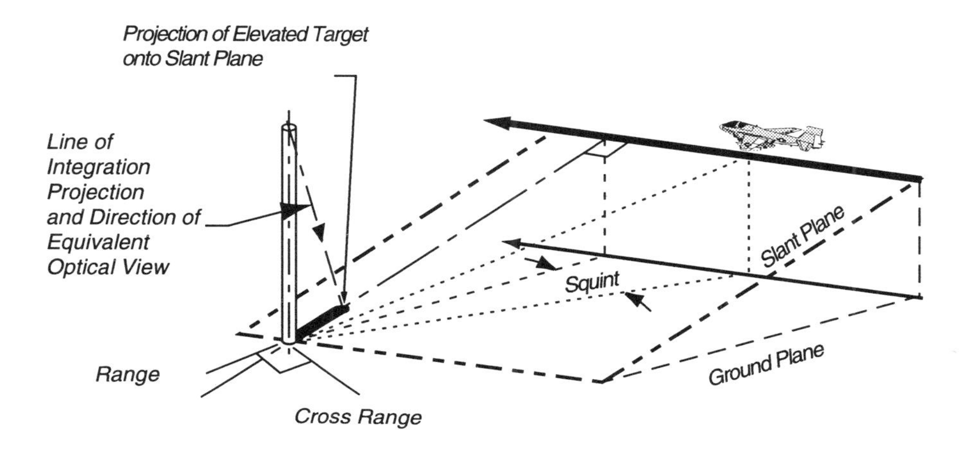

Figure 2.22 Projection of three-dimensional scene onto the SAR slant plane.

coincides with the range direction when the synthetic aperture is formed in level flight at broadside. In this situation, the effect is properly termed *range* layover. But for the most general imaging situation, such as the squinted geometry illustrated in Figure 2.22, it is clear that layover has components in both range and cross range.[11]

One interpretation of the layover effect is that the projection properties of the slant-plane SAR image are analogous to those of an optical image obtained by viewing the scene from a direction normal to the SAR slant plane, with the optical image plane (film plane) parallel to the slant plane. (The analogy is exact when the direction of optical illumination is coincident with the viewing direction for the SAR, and there is no superposition of targets in the SAR image.) This interpretation and our results regarding direction of layover are well known to SAR practitioners and can be established from the traditional view of a SAR as a range-Doppler imager. For example, an argument stated in that language could proceed as follows. Any two targets having the same range-Doppler coordinates are indistinguishable, by definition of what is transduced by the SAR, independent of where they actually reside in three-dimensional space. Therefore, the trajectory along which one would

[11] Throughout this book, we will use the term *layover* to denote the effect of height on the position of elevated targets in the SAR image, *whether or not superposition of multiple targets is involved.* Others choose to reserve use of the term *layover* strictly for those situations involving superposition, and use *foreshortening* to describe projection without superposition.

project elevated targets to determine their "laid-over" location in the two-dimensional SAR image would be determined by the intersection of the constant-range surfaces (spheres) and constant-Doppler surfaces (cones). One could then show that this amounts to a *projection line* which is normal to the slant plane. Once again, what the tomographic paradigm provides is not the uncovering of a new result, but an alternate means for obtaining and understanding an important established effect.

A computer simulation illustrates several aspects of the tomographic formulation derived and discussed above. Consider the SAR imaging scenario depicted in the top diagram of Figure 2.23. Here the target scene consists of ten isolated point targets: eight lie in a circle in a horizontal *ground plane*; one is located at the center of the circle; one lies at a point directly above the center target at an elevation equal to the radius of the circle. We assume a SAR collection geometry involving a straight line flight path and a *squinted* view of the scene that causes the slant plane to be oriented as shown in Figure 2.23. The resolution is assumed to be the same in range and cross range. Our simulation of this imaging process uses the three-dimensional tomographic paradigm as follows. First, we analytically compute the three-dimensional Fourier transform of each of the ten targets; second, we evaluate their sum on a Cartesian grid in the slant plane of the Fourier space. The inverse two-dimensional Fourier transform of the sampled data we obtain is identical (by the tomographic paradigm) to the slant-plane image created via the SAR image-formation process.

The image derived from the simulation is shown in the bottom of Figure 2.23, where the upper edge corresponds to the near-range direction. Notice that the circular array of targets traces out an *ellipse* in the image and that the elevated target appears at a location offset from the center target in both range and in cross range. The latter characteristic precisely constitutes the layover effect. The portrayal of a circle in the ground plane of the scene as an ellipse in the image represents a geometric distortion inherent in slant-plane SAR imagery. In particular, the scene appears foreshortened in the same direction as the layover. Both these effects result from three-dimensional target array projection onto the slant plane. These are exactly as predicted by the tomographic paradigm.

Figures 2.24, 2.25, and 2.26 demonstrate various aspects of the layover phenomenon when the images are reconstructed in the ground plane. The layover can still be viewed as projection in the direction normal to the slant plane. There will, however, be a minor modification: the plane *onto* which the projection is made is the ground plane instead of the slant plane. Figure 2.24 shows the case for a level flight, broadside collection, where the layover is purely in the range direction. Figure 2.25 shows how the same tall target (pole) will be laid over differently if the collection geometry involves a squint. A pair of real spotlight-mode SAR images is shown

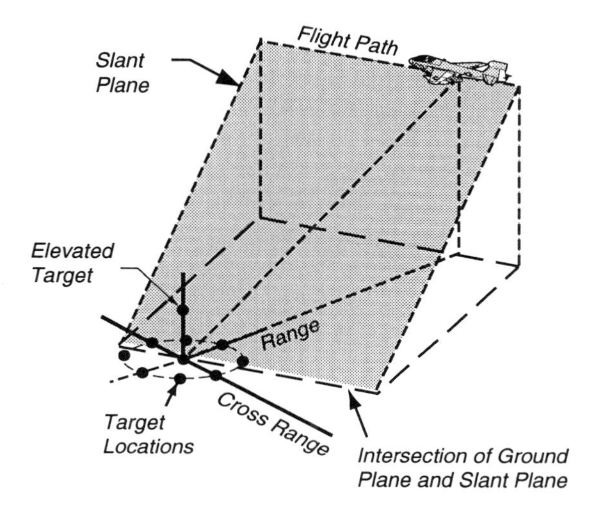

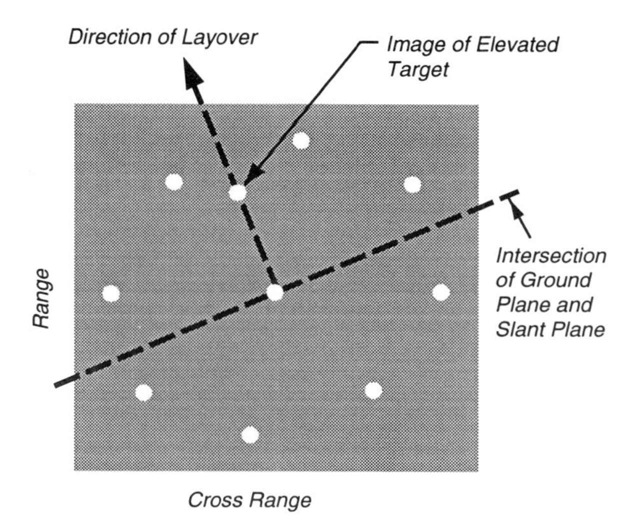

Figure 2.23 Simulated imaging geometry for squint collection. Target array and slant plane are shown above. Resulting simulated image demonstrating layover effects is shown below.

in Figure 2.26. In this figure, the two images were obtained with symmetric squint geometries: the first was obtained with a forward squint; the second was obtained using a backward squint angle of the same magnitude. In this particular scene, a fence line runs along a hill of significant height. The resulting layover looks quite different in the two scenes, because of the difference of the collection geometries. Note that in the left-hand image the fence appears to "bulge". The direction of the bulge is orthogonal to the flight line (consistent with our discussion of layover effects). The right-hand image, on the other hand, shows no fence bulge effect. In this case the layover direction coincides with the direction of the fence line: the fence lays over onto itself.

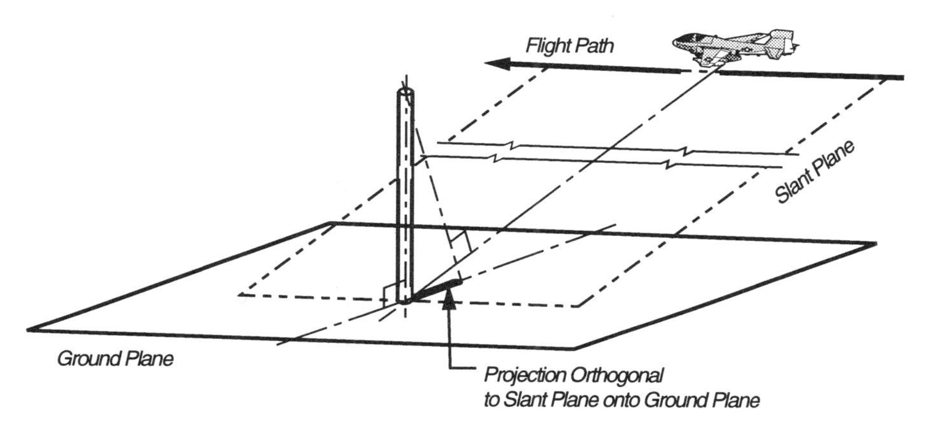

Figure 2.24 Layover in ground-plane imagery for case of broadside collection.

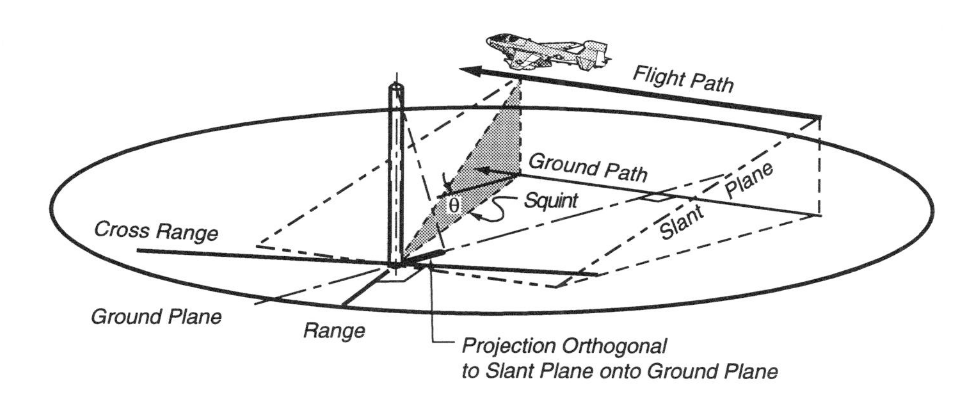

Figure 2.25 Layover in ground-plane imagery for case of squinted collection.

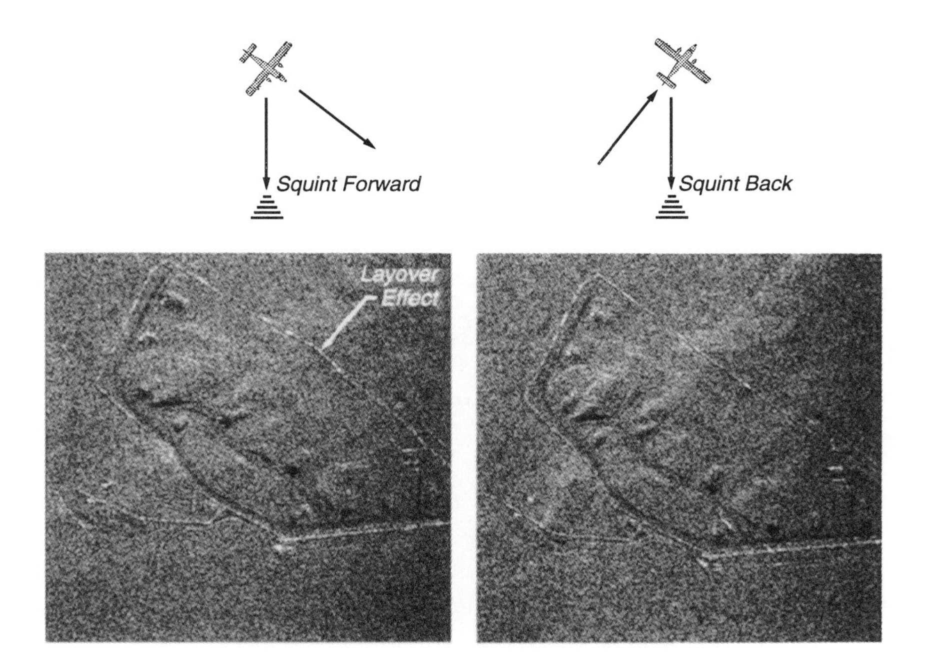

Figure 2.26 Layover effect in actual ground-plane SAR imagery. In this case, the two images were taken with symmetric squint angles, one forward and the other backward. A fence runs along a hill in the upper right portion of the scene. The image on the left shows that the direction of layover (orthogonal to the ground track line) is such that a "bulge" is injected into the fence on the highest part of the hill. For the image on the right, however, the layover direction (again, orthogonal to the ground track line) results in the fence laying over onto itself, leaving no bulge effect.

Summary of Critical Ideas from Section 2.4

- **A three-dimensional tomographic paradigm for spotlight-mode SAR can be constructed. A form of the projection-slice theorem called the linear trace version forms the foundation of the tomographic description of a phase-history data collection.**

- **The key result from the three-dimensional tomographic model is that a set of deramped pulses (prior to range compression) in a spotlight-mode collection forms a collection surface in the three-dimensional Fourier (spatial-frequency) domain.**

- **The shape of the collection surface is determined by the flight path of the SAR platform. This is because each pulse determines one line segment of the spatial-frequency data, with the angular orientation of the segment the same as the angular orientation of the platform relative to scene center when the pulse was transmitted.**

- **For the case of straight-line flight, the collection surface is a plane, referred to as the slant plane. This is the most common mode of collection.**

- **The reconstructed SAR image is formed from Fourier inversion of the data on the slant plane, or of the data projected to some other plane (e.g., the ground plane). In any case, the data are interpolated from the polar raster on which they are collected to a Cartesian raster, so that fast Fourier inversion algorithms can be used.**

- **The SAR images reconstructed from the slant plane and the ground plane exhibit different properties, including certain geometrical distortions.**

- **The resolution in both range and cross range of the SAR image is determined by the extent of the spatial-frequency data collected. The range extent is a function of the radar bandwidth, while the cross-range extent is a function of the angular diversity of the flight path, and the radar wavelength.**

- **The range resolution for a slant-plane reconstruction is**

$$\rho_{y'} = \frac{2\pi}{\Delta Y'} = \frac{c}{2B_c}.$$

Summary of Critical Ideas from Section 2.4 (cont'd)

- **The cross-range resolution (for small $\Delta\theta$) is given by**

$$\rho_{x'} = \rho_x = \frac{2\pi}{\Delta X'} = \frac{\lambda}{2\Delta\theta} \, .$$

- **The sampling rate required along the synthetic aperture so that spatial aliasing of the data does not occur is equal to one-half of the width of the physical antenna of the SAR.**

- **Two fundamental SAR imaging modalities are termed broadside and squint. In the broadside mode, the center pulse in the synthetic aperture is perpendicular to the flight path. In the squint mode, it is not. In either case, the line connecting the center of the aperture and the scene patch center defines the range direction.**

- **The layover effect for elevated targets in a spotlight-mode SAR image may be described in terms of tomographic projection. A variation of the projection-slice theorem called the planar slice version is used to derive this result. These projection effects in a SAR image are equivalent to those in an optical image taken from a perspective where the SAR slant plane is replaced with the optical film plane.**

2.5 THE COMPLEX RECONSTRUCTED SPOTLIGHT-MODE SAR IMAGE

In this section, we utilize the results of the previous section to derive a mathematical expression for the complex-valued reconstructed SAR image that is obtained in a spotlight-mode tomographic collection. This provides an interesting insight into what is actually transduced by a SAR image, and also establishes the foundation upon which the mathematics of *interferometric SAR* (IFSAR) will be built in Chapter 5.

We begin our development with a careful description of the reflectivity function, $g(x, y, z)$, which the SAR imaging system attempts to estimate. We suggested in Section 2.4.5 that we would confine our interest to those versions of $g(x, y, z)$ that involve only surface reflectivity, without allowing for penetration of the microwave energy below the surface. To this end, consider the model given by

$$g(x, y, z) = r(x, y) \cdot \delta(z - h(x, y)) \; . \tag{2.56}$$

Here, $r(x, y)$ represents the surface reflectivity density at ground coordinates (x, y), while $h(x, y)$ is the terrain elevation for the same point. The term $\delta(\cdot)$ is a Dirac delta function. Such a surface is shown in Figure 2.27.

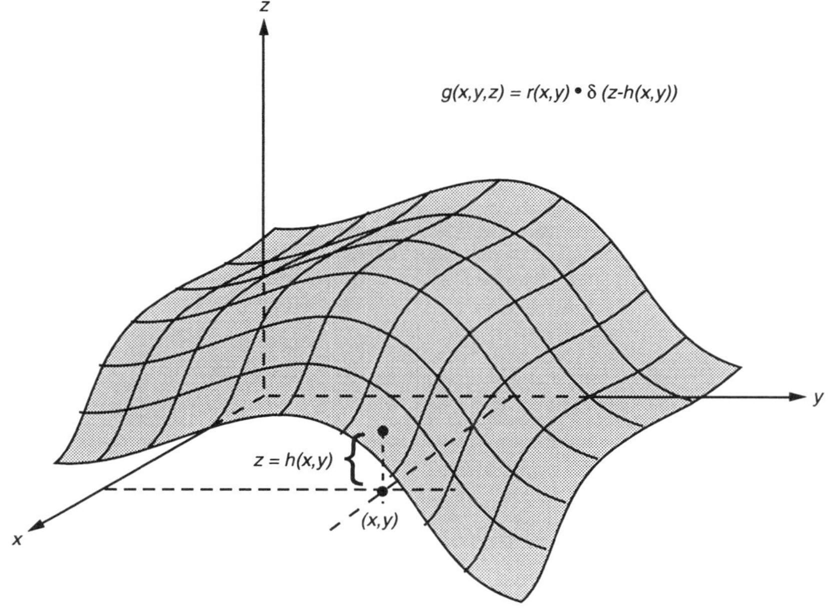

Figure 2.27 Three-dimensional model for reflectivity function.

An important expression will be that of the three-dimensional Fourier transform of $g(x, y, z)$. To this end, we calculate

$$G(X, Y, Z) = \mathcal{F}\{g(x, y, z)\} = \iiint g(x, y, z)\, e^{-j(xX+yY+zZ)}\, dx\, dy\, dz \quad (2.57)$$

where $\mathcal{F}\{\cdot\}$ denotes Fourier transformation. Substituting the expression for $g(x, y, z)$ yields

$$G(X, Y, Z) = \iiint r(x, y)\, \delta(z - h(x, y))\, e^{-j(xX+yY+zZ)}\, dx\, dy\, dz\ . \quad (2.58)$$

Performing the z integration first gives

$$G(X, Y, Z) = \iint r(x, y)\, e^{-jZh(x,y)}\, e^{-j(xX+yY)} dx dy\ . \quad (2.59)$$

Next, consider the situation of a planar collection surface, which occurs whenever the platform fight path is a straight line. In such a case, all the Fourier data lie in a plane given by

$$Z = \alpha X + \beta Y \quad (2.60)$$

with

$$\alpha = \tan \eta$$

and

$$\beta = \tan \psi\ .$$

Figure 2.28 depicts this slant plane, which passes through the origin of the three-dimensional Fourier space, as must any slant plane in a spotlight-mode collection.[12] Substituting the above expression for the slant plane into equation 2.59 yields

$$G_0(X, Y) = \iint r(x, y)\, e^{-j(\alpha X+\beta Y)h(x,y)}\, e^{-j(xX+yY)} dx dy\ . \quad (2.61)$$

Effectively, $G_0(X, Y)$ describes the Fourier transform values on the slant plane of interest in terms of the ground-plane spatial frequencies X and Y. In Section 2.4.4, we discussed this version of phase-history data as a projection of the slant-plane data onto the ground plane, as shown in Figure 2.19. Recall that the coordinate system shown there is always chosen so that the projection of the center pulse onto the (X, Y) plane lies along the Y axis, which defines the ground-range direction. The ground-plane phase-history samples lie on a finite polar raster where the Y

[12] The reader will note that we have changed our convention somewhat from the geometry described in Section 2.4.3. Here, we have defined the slant plane so that the range direction, y, points toward the SAR platform, instead of away from it (see Figure 2.28.) This choice of geometry turns out to be convenient in our development of an image-formation algorithm in Chapter 3.

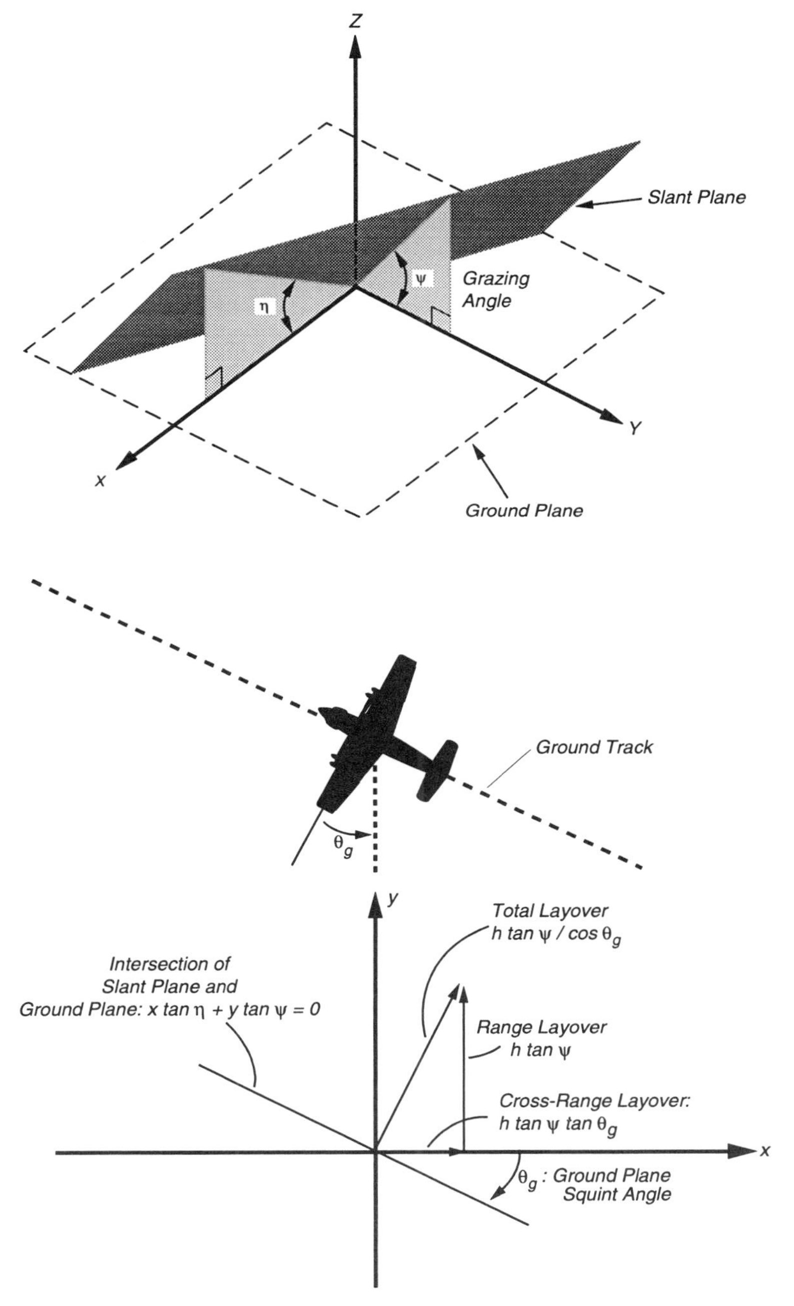

Figure 2.28 Geometries for slant plane and associated layover.

extent is defined by both the radar bandwidth and the nominal depression angle. The X extent is prescribed by the span of azimuthal viewing positions included in the platform flight path, as was illustrated in Figure 2.20. Finally, because the phase-history domain data are processed without the offset ($Y_0 = 4\pi \cos\psi/\lambda$) in the Y spatial-frequency dimension, they are related to $G_0(X, Y)$ by

$$G_1(X, Y) = G_0(X, Y + Y_0) \tag{2.62}$$

$$= \int\int r(x, y)\, e^{-j(\alpha X + \beta(Y+Y_0))h(x,y)}\, e^{-j(xX + y(Y+Y_0))}\, dx\, dy \tag{2.63}$$

$$= \int\int r_1(x, y)\, e^{-j(\alpha X + \beta Y)h(x,y)}\, e^{-j(xX + yY)}\, dx\, dy \tag{2.64}$$

where

$$r_1(x, y) = r(x, y)\, e^{-jyY_0}\, e^{-j\beta Y_0 h(x,y)}\, . \tag{2.65}$$

Note that $r_1(x, y)$ represents the original two-dimensional surface reflectivity function $r(x, y)$ with two phase functions imposed. One is a linear phase ramp in the y dimension, resulting from the spatial-frequency offset Y_0. The other phase term encodes the terrain height function $h(x, y)$ scaled by the quantity βY_0.

The X and Y spatial frequencies are limited in extent by the region of support in the $X - Y$ plane:

$$\begin{aligned} -\frac{\Delta X}{2} &\leq X \leq \frac{\Delta X}{2} \\ -\frac{\Delta Y}{2} &\leq Y \leq \frac{\Delta Y}{2} \end{aligned} \tag{2.66}$$

with (see bottom diagram of Figure 2.20)

$$\begin{aligned} \Delta X &= \frac{4\pi}{\lambda}\Delta\theta \\ \Delta Y &= \frac{2}{c}\cos\psi(2\pi B_c). \end{aligned} \tag{2.67}$$

By writing an analytical expression for the inverse Fourier transform of $G_1(X, Y)$, we obtain an equation for the reconstructed image $g_1(x, y)$ as follows:

$$g_1(x, y) = \mathcal{F}^{-1}\{G_1(X, Y)\} = \int\int G_1(X, Y)\, e^{j(xX + yY)}\, dX dY$$

$$= \int_A\int \left\{ \int\int r_1(\tilde{x}, \tilde{y})\, e^{-j(\alpha X + \beta Y)h(\tilde{x},\tilde{y})}\, e^{-j(\tilde{x}X + \tilde{y}Y)} d\tilde{x} d\tilde{y} \right\} e^{j(xX + yY)}\, dX\, dY\, . \tag{2.68}$$

In the above equation, the integration over region A implies the limits on X and Y as prescribed by Equation 2.67. The implied limits on the inner integral are over the patch of radius L. Changing the order of integration then yields

$$g_1(x,y)=\iint\left[\iint_A e^{j\{X[-\alpha h(\tilde{x},\tilde{y})-\tilde{x}+x]+Y[-\beta h(\tilde{x},\tilde{y})-\tilde{y}+y]\}}dX\,dY\right]r_1(\tilde{x},\tilde{y})d\tilde{x}d\tilde{y}\;. \tag{2.69}$$

Because the result of the integration on X and Y is a two-dimensional *sinc* function $s_A(x,y)$ Equation 2.69 becomes

$$g_1(x,y) = \iint s_A(-\alpha h(\tilde{x},\tilde{y}) - \tilde{x} + x, -\beta h(\tilde{x},\tilde{y}) - \tilde{y} + y)\, r_1(\tilde{x},\tilde{y})\, d\tilde{x}\, d\tilde{y} \tag{2.70}$$

with the *sinc* function given by

$$s_A(x,y) = \Delta X \Delta Y \operatorname{sinc}\left[\frac{x\Delta X}{2\pi}\right] \operatorname{sinc}\left[\frac{y\Delta Y}{2\pi}\right]\,. \tag{2.71}$$

The width of the mainlobe (the distance between first zero-crossings) of the *sinc* function is $4\pi/\Delta X$ by $4\pi/\Delta Y$. If we now assume that the terrain elevation profile $h(x,y)$ varies slowly enough to be considered constant over this mainlobe width, then Equation 2.70 may be interpreted as a convolution:

$$\begin{aligned} g_1(x+\Delta x, y+\Delta y) &= s_A(x,y) \otimes r_1(x,y) \\ &= s_A(x,y) \otimes \left[r(x,y)e^{-j\beta Y_0 h(x,y)}e^{-jyY_0}\right] \end{aligned} \tag{2.72}$$

with

$$\begin{aligned} \Delta y &= \beta\, h(x,y) = (\tan\psi)\, h(x,y) \\ \Delta x &= \alpha\, h(x,y) = (\tan\eta)\, h(x,y)\,. \end{aligned} \tag{2.73}$$

At this point, two important aspects of Equation 2.72 should be examined, as it is our defining image equation for spotlight-mode SAR. First, it is instructive to review Equations 1.16 through 1.18 and Equation 1.21 of Chapter 1. There, we first analyzed the one-dimensional reconstruction of the reflectivity function from CW burst return data. That analysis shows that the processed return can be interpreted as a convolution of a kernel function (envelope of the burst) with the reflectivity function multiplied by a linear phase term involving the carrier, $-\omega_0 t$. This, in turn, lead to an equivalent description of the signal in Fourier space as offset narrowband data (Equation 1.21). It should not be surprising, therefore, that when we Fourier invert the offset data $G_1(X,Y)$ above, the result is the same. That is, for the case of a flat scene, i.e., $h(x,y) = 0$ everywhere, Equation 2.72 is of the same form as Equation 1.18, with the spatial carrier frequency, Y_0, now in the linear phase term.

The convolving kernel is the *sinc* function corresponding to the aperture region A. The reconstructed image in this case is therefore simply a narrowband version of the surface reflectivity function $r(x, y)$. The *height-dependent phase term*, $\beta Y_0 h(x, y)$, prescribes how the image is altered when the terrain is not flat. This phase term turns out to be very important to our formulation of interferometric SAR, presented in Chapter 5.

Second, note that Equation 2.72 states that a reflectivity value for any position in the physical scene will be translated in the reconstructed SAR image to a new position, with the translations in range and cross range given by Equation 2.73. These shifts in range and cross-range *quantify the layover effect* that was qualitatively explained in Section 2.4.5. (They will be important in Chapter 5 to our development of image registration for SAR interferometry and for SAR stereoscopy.) The bottom diagram of Figure 2.28 shows these layover components and their relationship to the imaging geometry parameters.

Summary of Critical Ideas from Section 2.5

- **An expression for the complex reconstructed spotlight-mode SAR image may be derived as**

$$g(x + \Delta x, y + \Delta y) = s_A(x, y) \otimes \left[r(x, y) e^{-j\beta Y_0 h(x,y)} e^{-jyY_0} \right]$$

where $r(x, y)$ is the terrain reflectivity, $h(x, y)$ is the terrain height profile, and $s_A(x, y)$ is the *sinc* function determined by the aperture support region A. This equation shows how terrain height is encoded as phase in the reconstructed image, a concept vital to SAR interferometry (see Chapter 5).

- **The range and cross-range layover terms (vital to the development in Chapter 5 of SAR stereoscopy) are**

$$\begin{aligned} \Delta y &= \beta\, h(x, y) = (\tan \psi)\, h(x, y) \\ \Delta x &= \alpha\, h(x, y) = (\tan \eta)\, h(x, y)\ . \end{aligned}$$

2.6 LIMITATIONS OF THE TOMOGRAPHIC PARADIGM

The tomographic paradigm for SAR imaging described above allowed us to present the concepts of synthetic aperture radar in a clear and concise way within a signal processing framework. This paradigm also serves as the basis for developing other SAR concepts in subsequent chapters. However, it is important to recognize that there are certain limitations to the scope of the tomographic description of the SAR imaging process, or indeed any description that results in portraying the collected SAR data as samples of the three-dimensional Fourier transform of the scene function, e.g., Walker's original description of spotlight-mode SAR. In Appendix B, we derive an exact expression for the radar return signal as dictated by the spatial geometry of the collection and compare this result with the signal obtained by sampling the Fourier domain as per the tomographic model. This derivation involves a single point target in the scene, recognizing that any scene may be considered to be the superposition of a large number of such point targets. We concentrate on the phase of the return signal because it is the phase that is of utmost importance in the coherent averaging involved in forming an image from the data. The resulting point target *phase history* is shown to contain two second order components not predicted by the tomographic theory, the presence of which will place certain restrictions on the size and resolution of the image that may be formed by Fourier processing.

These two second order components in the phase history of a point target arise when two fundamental assumptions in the tomographic derivation break down. The first assumption is that the surfaces of integration in the formation of a projection function are planes, as opposed to spherical surfaces (see Figure 2.17). This approximation requires that the distance from the radar platform to the scene center be large relative to the diameter of the scene patch, so that the curvature of the wavefronts is negligible. We will see that when the patch size becomes large enough, a quadratic phase term appears. The second assumption is that the residual phase term in the quadrature demodulation process, as given by Equation 1.37, is essentially zero. This skew term, in fact, is not negligible when the patch size is large enough. Again, a resulting quadratic phase term appears in the phase history.

The effect of a quadratic phase term in the phase history is to convolve the image with a kernel consisting of the Fourier transform of a complex exponential having that quadratic phase. The image will be blurred (in the cross-range dimension in this case) by an amount commensurate with the width the convolving kernel, which in turn depends on the peak amplitude of the quadratic. Figure 2.29 shows the cross-range blurring effect for quadratic phase errors of varying amplitude (peak phase). The plot corresponding to a zero-amplitude phase error is, of course, the ideal point-target response function, or impulse response function (IPR), for the given aperture, or size of phase-history patch processed. (This is in the absence of aperture weighting, which we will discuss at length in Chapter 3.) The remaining plots show clearly how a quadratic phase error of sufficient amplitude can degrade image quality. As a general rule, however, we will consider the defocus effect corresponding to a quadratic phase amplitude of $\pi/4$ radians or less to be negligible.

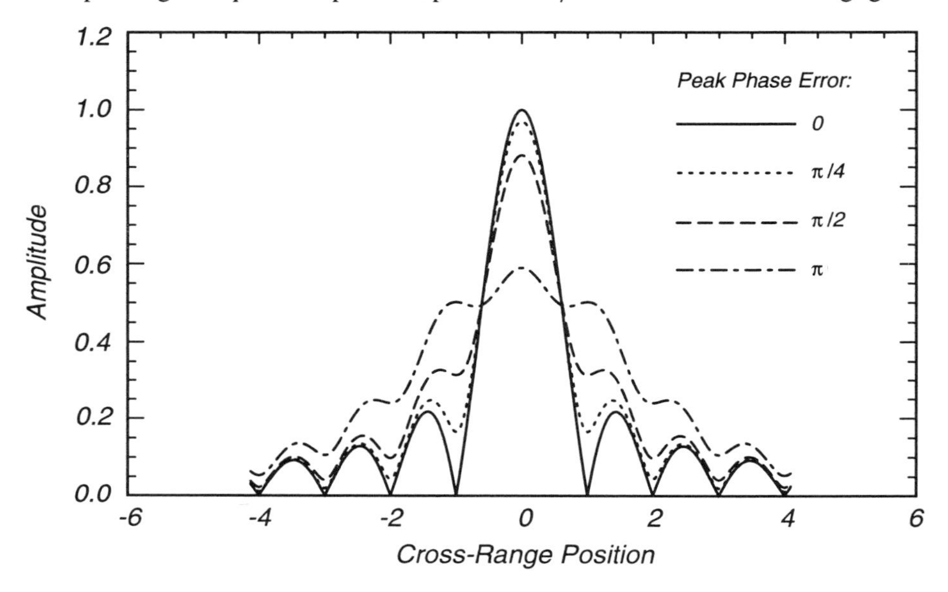

Figure 2.29 Quadratic phase-error blur functions for various levels of peak phase.

Below, we present quantitative relationships that describe the effect of the quadratic phase term generated by the breakdown of the two assumptions in the tomographic paradigm. These relationships take the form of a maximum image patch size that is allowable in order for the effects of quadratic defocus in the reconstructed image to be negligible. The derivation of these equations is performed by calculating the exact expression for the phase of the return from an ideal point target, including the effects of range curvature and the residual deramp skewing. The details of the derivations of these expressions are given in Appendix B.

For the range curvature effect, the maximum image-domain patch radius is given by

$$L_{curv} = \rho_x \sqrt{\frac{2r_0}{\lambda}} \tag{2.74}$$

where ρ_x is the cross-range resolution, r_0 is the operational slant range, and λ is the nominal radar wavelength. For a SAR system with $\lambda = 3$ cm, $r_0 = 10000$ m, and $\rho_x = 1$ m, we calculate a maximum patch radius of 816 m.

For the deramp residual (skew) term, the maximum patch size is given by

$$L_{skew} = \rho_x \frac{\omega_0}{2\sqrt{\pi\alpha}} = \rho_x \frac{f_0}{\sqrt{\dot{f}}} \,. \tag{2.75}$$

Here $f_0 = \omega_0/2\pi$ is RF center frequency in Hertz and $\dot{f} = \alpha/\pi$ is chirp rate expressed in Hertz per second. Note that this restriction does not depend on the standoff range. For the same operating parameters listed above and assuming $\dot{f} = 7 \times 10^{12}$ Hz/sec, the calculated maximum patch radius is 3780 m. In this case, the restriction due to wavefront curvature is clearly more severe than that due to the deramp residual term.

Summary of Critical Ideas from Section 2.6

- **Two of the assumptions that are critical to the development of the spotlight-mode SAR tomographic model are: the wavefronts are planar; the residual phase term (skew) from the deramp process is negligible.**

- **The first assumption requires that the distance from the radar platform to the scene center is large relative to the diameter of the scene patch, so that the curvature of the wavefronts is negligible. When the patch size becomes large enough, a quadratic phase term appears. An expression for the size of the patch radius beyond which the quadratic phase term results in unacceptable degradations in the reconstructed image quality is:**

$$L_{curv} = \rho_x \sqrt{2r_0/\lambda} \,.$$

- **The skew term of the second assumption is not negligible when the patch size is large enough. Again, a resulting quadratic phase term appears in the phase history. An expression for the size of the patch radius beyond which this quadratic phase term results in unacceptable degradations in the reconstructed image quality is:**

$$L_{skew} = \rho_x \omega_0/(2\sqrt{\pi\alpha}) = \rho_x f_0/\sqrt{\dot{f}} \,.$$

2.7 TOMOGRAPHIC DEVELOPMENT OF BISTATIC SAR

In this section we demonstrate that the tomographic paradigm can be employed to describe a variation on the collection modality of spotlight-mode SAR. This collection scenario is known as *bistatic* SAR. It involves a radar transmitter and receiver that are not collocated, as they are in the traditional *monostatic* mode that we have described in the previous sections of this chapter. Figure 2.30 depicts the bistatic collection geometry, wherein two aircraft flying in possibly different directions act as transmitter and receiver. (Our development of bistatic SAR parallels that of Arikan [20].)

In this two-dimensional simplification, the contours of constant-range are curves where the sum of the distances from transmitter to target and target to receiver are constant. These contours turn out to be ellipses, with the foci located at the positions of the transmitter and receiver, as shown in Figure 2.30. We can argue, as we did in the monostatic case, that for standoff distances that are large compared to the patch radius, the elliptical contours may be treated as segments of straight lines across the scene. As a result, the return waveform transduces integrated reflectivities along these straight lines, as a function of the incremental transmitter-target-receiver distance, ρ. It can easily be demonstrated that this is equal to a tomographic projection function obtained along a direction that bisects the so-called *bistatic angle*, i.e., the angle formed between the transmitter-target and the receiver-target lines. The appropriate range variable, u, along this direction is related to ρ as

$$\rho = 2u\cos(\beta/2) \ . \tag{2.76}$$

Figure 2.31 depicts this geometrical relationship. We know from the tomographic development of the monostatic case that each quadrature-demodulated received pulse will transduce certain values of the Fourier transform $G(X, Y)$ of the scene reflectivity. Because of the relationship between ρ and u (Equation 2.76), however, the spatial frequencies transduced will be scaled by the factor $\cos\beta/2$. The demodulated return will be given by

$$\tilde{r}_c(t) = A_2 \int_{-u_1}^{u_1} g(u) \, \exp\left[j \left[-\frac{2u\cos(\beta/2)}{c}(\omega_0 + 2\alpha(t - \tau_0)) \right] \right] du \ . \tag{2.77}$$

When the transmitter and receiver both fly in straight-line segments (as shown in Figure 2.32), a set of projection functions that span a range of angles determined by the bisectors of all the bistatic angles involved will be determined. Each segment in Fourier space will be scaled by the cosine of half of the bistatic angle as prescribed by Equation 2.77. Therefore, the resulting annulus of data will be skewed (see Figure 2.32).

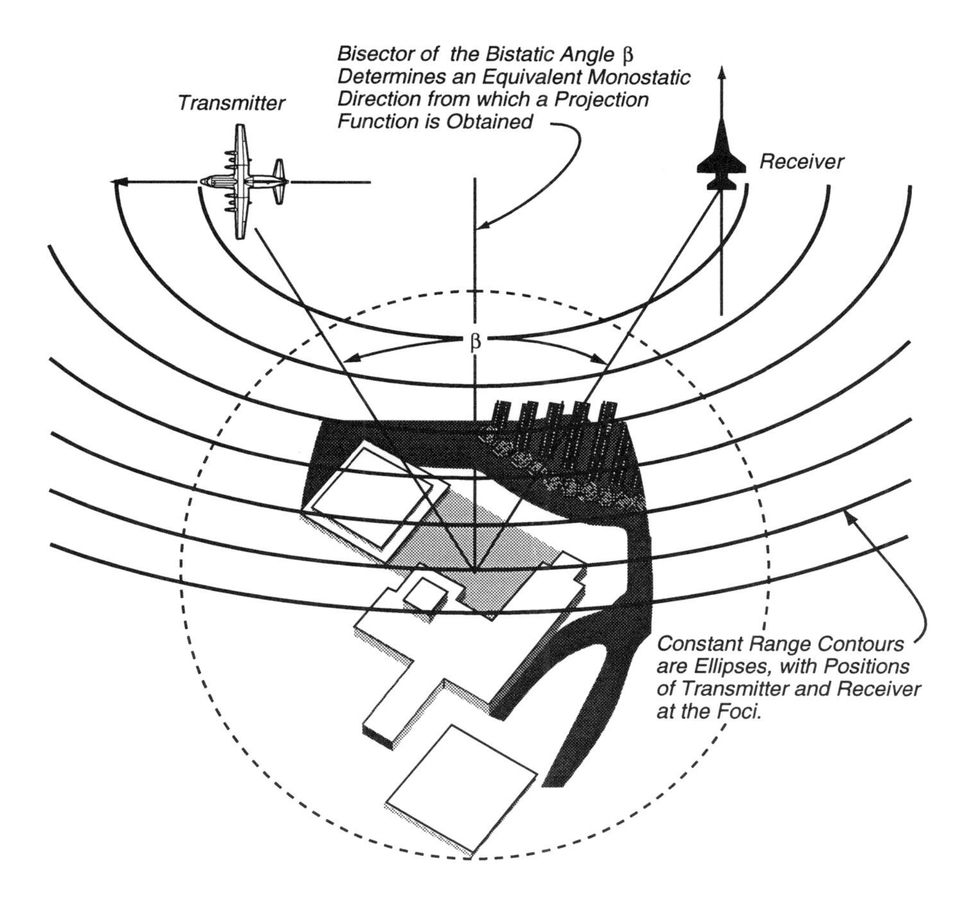

Figure 2.30 Bistatic modality for SAR imaging. Constant range contours are ellipses, with the transmitter and receiver positions as the foci. For large standoff ranges compared to the patch size, these can be treated as straight lines.

The subject of bistatic SAR is one largely of research interest, because there is a major practical problem encountered in implementing such a system. Specifically, knowledge of the correct demodulation time for each received pulse depends on precise knowledge of both transmitter and receiver positions. Stated another way, the transmitter and receiver must be synchronized in some manner. In general, this requires either a side communication channel between the two that will provide this critical timing information, or the use of highly stable clocks (e.g., atomic clocks) on both platforms.

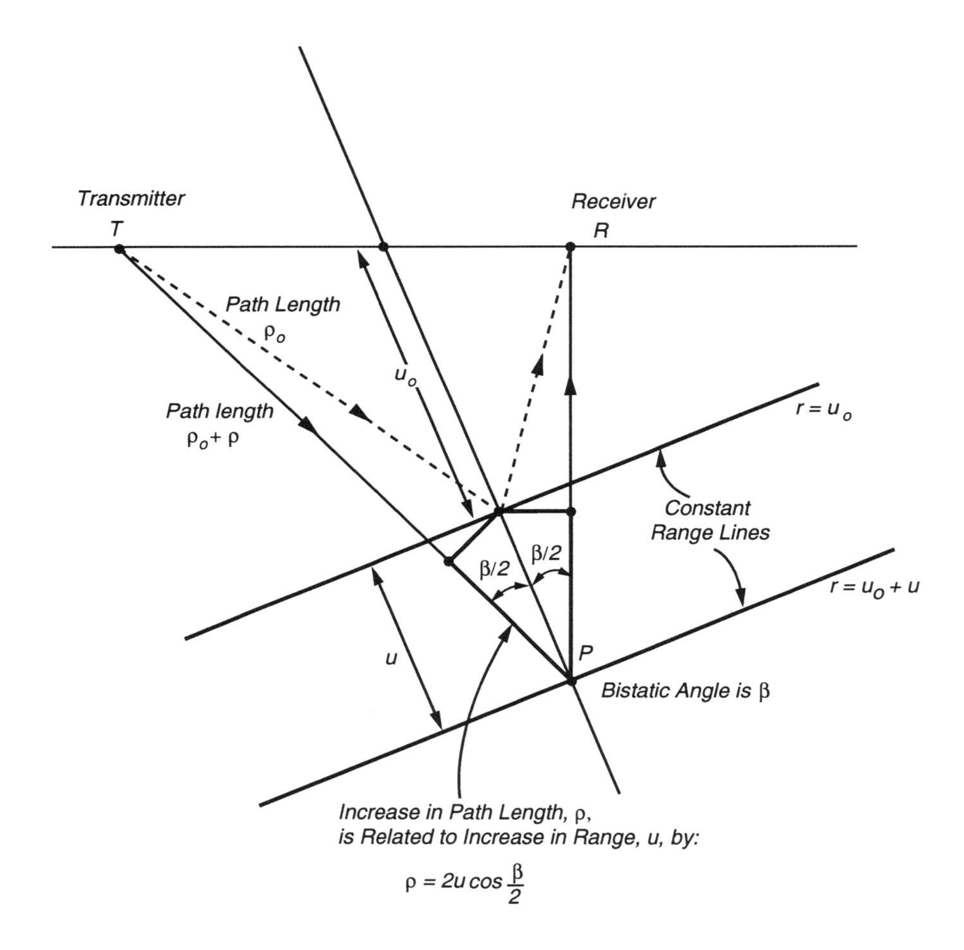

Figure 2.31 Relationship between equivalent monostatic range, u, and transmitter-target-receiver path length, ρ, in bistatic SAR. This factor affects scaling of Fourier-domain data.

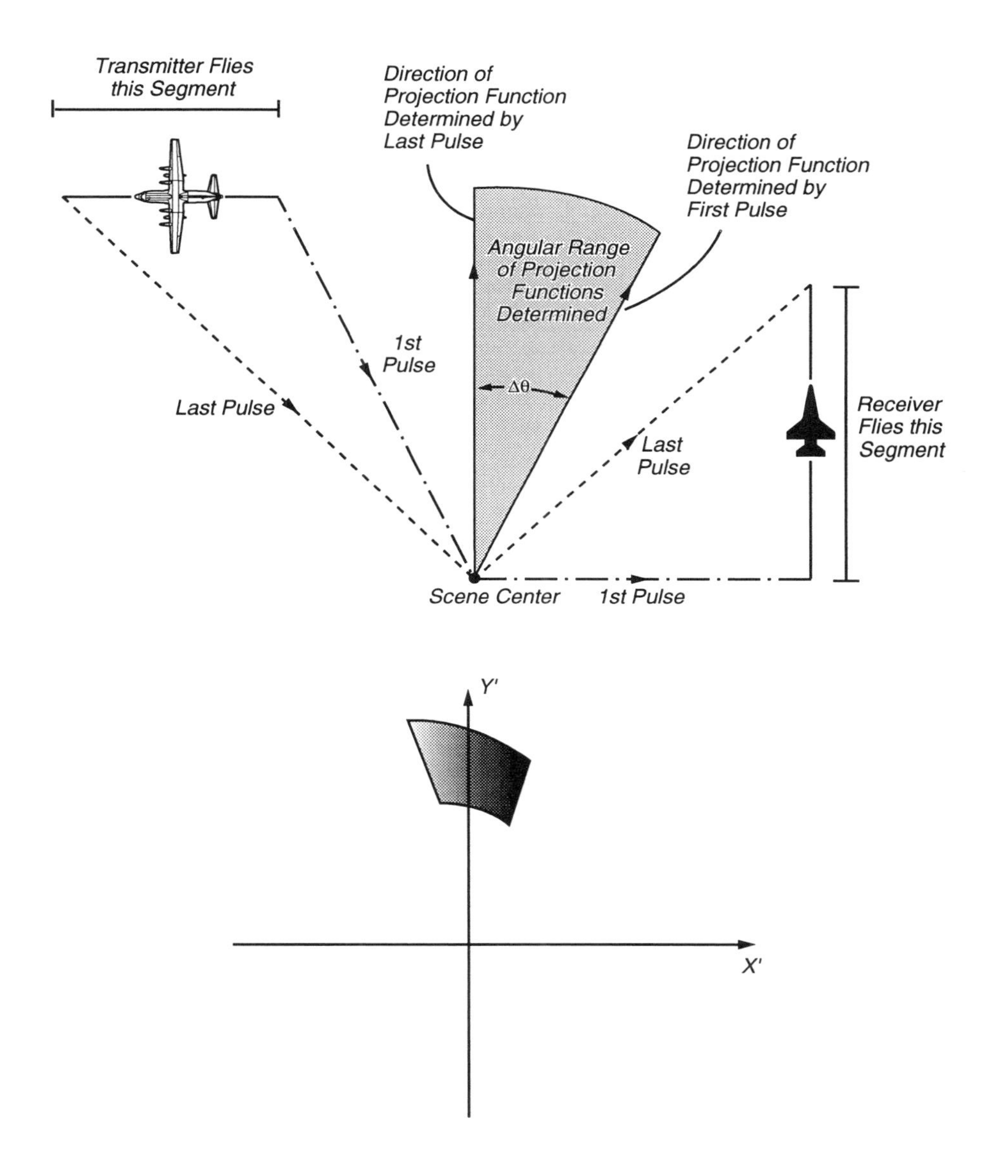

Figure 2.32 Collection of bistatic SAR data. Directions of equivalent monostatic projection functions determined in bistatic collection (top diagram). Bottom diagram depicts resulting skewed annulus of phase-history data. The length and position of each segment of Fourier data are scaled by the factor $\cos \beta/2$.

REFERENCES

[1] D. C. Munson, J. D. O'Brien, and W. K. Jenkins, "A Tomographic Formulation of Spotlight-Mode Synthetic Aperture Radar," *Proceedings of the IEEE*, Vol. 71, No. 8, pp. 917-925, August 1983.

[2] J. Walker, "Range-Doppler Imaging of Rotating Objects," *IEEE Transactions on Aerospace and Electronic Systems*, Vol. AES-16, No. 1, pp. 23-51, January 1980.

[3] C. V. Jakowatz, Jr. and P. A. Thompson, "A New Look at Spotlight-Mode Synthetic Aperture Radar as Tomography: Imaging Three-Dimensional Targets," *IEEE Transactions on Image Processing*, May 1995.

[4] C. A. Wiley, "Synthetic Aperture Radars - A Paradigm for Technology Evolution," *IEEE Transactions on Aerospace and Electronic Systems*, AES-21, pp. 440-443, 1985.

[5] C. A. Wiley, "Pulsed Doppler Radar Methods and Apparatus," United States Patent, No. 3,196,436, 1965 (originally filed, 1954).

[6] L. J. Cutrona, W. E. Vivian, E. N. Leith, and, G. O. Hall, "A High Resolution Radar Combat Surveillance System," *IRE Trans. Military Elect.*, MIL-5, pp. 127-131, 1961.

[7] J. C. Curlander and R. N. McDonough, *Synthetic Aperture Radar: Systems and Signal Processing*, John Wiley and Sons, Inc., New York, 1991.

[8] J. A. Parker, *Image Reconstruction in Radiology*, CRC Press, 1990.

[9] G. N. Hounsfield, "A Method and Apparatus for Examination of a Body by Radiation Such as X or Gamma Radiation," Patent Specification 1283915, The Patent Office, London, England, 1972.

[10] G. N. Hounsfield, "Computerized Transverse Axial Scanning Tomography: Part I, Description of the System," *British Journal of Radiology*, 46, pp. 1016-1022, 1973.

[11] W. H. Oldendorf, "Isolated Flying-Spot Detection of Radio-Density Discontinuities; Displaying the Internal Structural Pattern of a Complex Object, *IRE Transactions on Biomed. Electronics*, BME-8, pp. 68-72, 1961.

[12] A. M. Cormack, "Representation of a Function by its Line Integrals, with Some Radiological Applications," *Journal of Applied Physics*, 34, pp. 2722-2727, 1963.

[13] R. M. Mersereau, "Recovering Multidimensional Signals from Their Projections," *Computer Graphics and Image Processing*, Vol. 1, pp 179-195, October 1973.

[14] G. T. Herman, *Image Reconstruction from Projections: The Fundamentals of Computerized Tomography*, Academic Press, 1980.

[15] A. C. Kak and M. Slaney, *Principles of Computerized Tomographic Imaging*, IEEE Press, 1988.

[16] J. Radon, "Uber die Bestimmung von Funktionen durch Ihre Integralwerte Langs Gewisser Mannigfaltigkeiten," *Ber. Verb. Saechs., Akad. Wiss., Leipzig, Math. Phys. Kl.*, 69, pp. 262-277, 1917.

[17] F. Natterer, *The Mathematics of Computerized Tomography*, John Wiley & Sons, 1986.

[18] D. A. Ausherman, A. Kozma, J. L. Walker, H. M. Jones, and E. C. Poggio, "Developments in Radar Imaging," *IEEE Transactions on Aerospace and Electronic Systems*, Vol. AES-20, No. 4, pp 363-400, July 1984.

[19] D. C. Munson and J. L. C. Sanz, "Image Reconstruction from Frequency-Offset Fourier Data," *Proceedings of the IEEE*, Vol. 72, pp. 661-669, June 1984.

[20] O. Arikan, "Investigation of Topics in Radar Signal Processing", Ph.D. Thesis, University of Illinois, 1990.

3

ASPECTS OF SPOTLIGHT-MODE IMAGE FORMATION

3.1 INTRODUCTION

We have said that forming an image from the collected SAR data requires nothing more than computing a two-dimensional Fourier transform. While this simple idea is in fact the essence of image formation, its implementation leads to a number of important issues that need to be addressed. In this chapter, we discuss several practical aspects of the process of transforming SAR phase-history data into imagery.

Section 3.2 begins the chapter with a discussion of two aspects of image formation via Fourier inversion: image resolution and area coverage. Section 3.3 describes holographic properties of spotlight SAR data and discusses the similarity of SAR images to coherent optical holograms. It is then demonstrated how these holographic properties can be exploited to change the displayed image resolution or area coverage, and how to reduce coherent speckle by multi-look processing. Section 3.4 explains and analyzes the image properties resulting from image formation with non-polar-formatted data. This concept is important because it indicates to what extent a simple 2-D fast Fourier transform (FFT) of the non-polar-formatted data produces an acceptable image, and how drastically overall quality is degraded as system resolution increases.

Section 3.5 delves into various aspects of the polar resampling problem that include why it is required and how it is accomplished with an eye towards practical implementation. The resampling (interpolation) problem leads to a discussion of the relevant theory of resampling and interpolation and justifies the notion that resampling can be viewed as a linear digital filtering operation. This digital filtering process is presented with emphasis on practicality of implementation. A simple pictorial explanation of the mechanics of interpolation relates the concepts to the mathematics.

Image formation is ultimately accomplished by use of Fourier transforms. Section 3.6 discusses the properties of discrete versus continuous Fourier transforms, aperture weighting for sidelobe control, and image oversampling issues. With the notion of image formation by Fourier transformation firmly established, explanations of the relationships between SAR data acquisition, subsequent image formation, and resulting image properties are given. Section 3.7 discusses these issues in the context of slant-plane image formation.

When spotlight-mode data are collected in straight and level flight, the subsequent image-formation processing is straightforward in concept and relatively simple to implement. However, the real world does not always present such ideal conditions. Data are often collected in a non-planar manner as a result of air turbulence, aircraft maneuvers, or curved trajectories associated with spaceborne platforms. Section 3.8 discusses the important topic of out-of-plane correction, a method by which such data can be compensated to maintain focus in the reconstructed image. Out-of-plane corrections then afford an explanation of ground-plane image formation as a natural and easily obtainable byproduct. Section 3.9 discusses these issues and the resulting image and phase-history properties.

In various places throughout this chapter, some of the actual implementation details have been purposely omitted. Also, some image properties related to the choice of certain processing parameters are not described. We chose instead to focus the discussions on the more fundamental issues and concepts. However, in Section 3.10 we do summarize the extremely important subjects of spatial frequencies, image scale factors, image dimensions, and resolution as they are affected by certain details of implementation, as well as by actual radar, geometry, and sampling parameters.

Finally, Section 3.11 puts much of what we have discussed about image formation in perspective by describing a typical, image-formation procedure. That is, we describe the steps that a radar engineer might follow to form a high-resolution spotlight SAR image from the received data. A typical image-formation session outlines initial requirements and assumptions, and key issues are addressed.

3.2 IMAGE RESOLUTION AND AREA COVERAGE

Among the more fundamental and important aspects of image formation are the resolution and the area of terrain covered by the image that can be formed from a particular SAR collection. In general, it is desirable to produce an image that resolves as much detail as possible over as large an area as possible. Consider first the aspect of image resolution. The fact that the data transduced by the SAR span a limited region of the spatial-frequency domain of the scene implies that the image formed by Fourier transforming these data will be limited in the fine detail it conveys about the actual scene. This constitutes a limit in what is referred to as the *resolution* of the image. The relationship between the size (extent) of the spatial frequency region transduced and the resolution of the resulting image can be derived from the properties of Fourier transform. Although we introduced some of these concepts in Section 2.4.4, we revisit the subject in somewhat greater detail here.

Let $f(x', y')$ denote a two-dimensional scene function that is to be reconstructed in an image, and let $F(X', Y')$ represent the two-dimensional Fourier transform of $f(x', y')$. We will assume that the scene is unrestricted in its spatial-frequency content, which implies that $F(X', Y')$ will be of infinite extent in the X', Y' spatial frequency domain. Suppose that the region of spatial-frequency data available to be processed, referred to as the *region of support*, or *aperture* is confined to a rectangle of dimensions $\Delta X', \Delta Y'$ in the slant plane (see top diagram of Figure 2.20). The observed data can then be expressed by the function

$$G(X', Y') = \mathcal{R}_A(X', Y') F(X', Y') \tag{3.1}$$

where

$$\mathcal{R}_A(X', Y') = \begin{cases} 1, & |X'| \leq \Delta X'/2 \quad \& \quad |Y'| \leq \Delta Y'/2 \\ 0, & \text{otherwise} \end{cases} \tag{3.2}$$

defines a rectangular window as shown in Figure 3.1.

The image produced by computing a two-dimensional (inverse) Fourier transform[1] of the observed data can be written as

$$\begin{aligned} g(x', y') &= \int_{-\infty}^{\infty} \int_{-\infty}^{\infty} G(X', Y') e^{j(x'X' + y'Y')} dX' \, dY' \\ &= \int_{-\infty}^{\infty} \int_{-\infty}^{\infty} \mathcal{R}_A(X', Y') F(X', Y') e^{j(x'X' + y'Y')} dX' \, dY' \, . \end{aligned} \tag{3.3}$$

[1] We make reference throughout this chapter to the Fourier transform without always distinguishing between forward and inverse transforms.

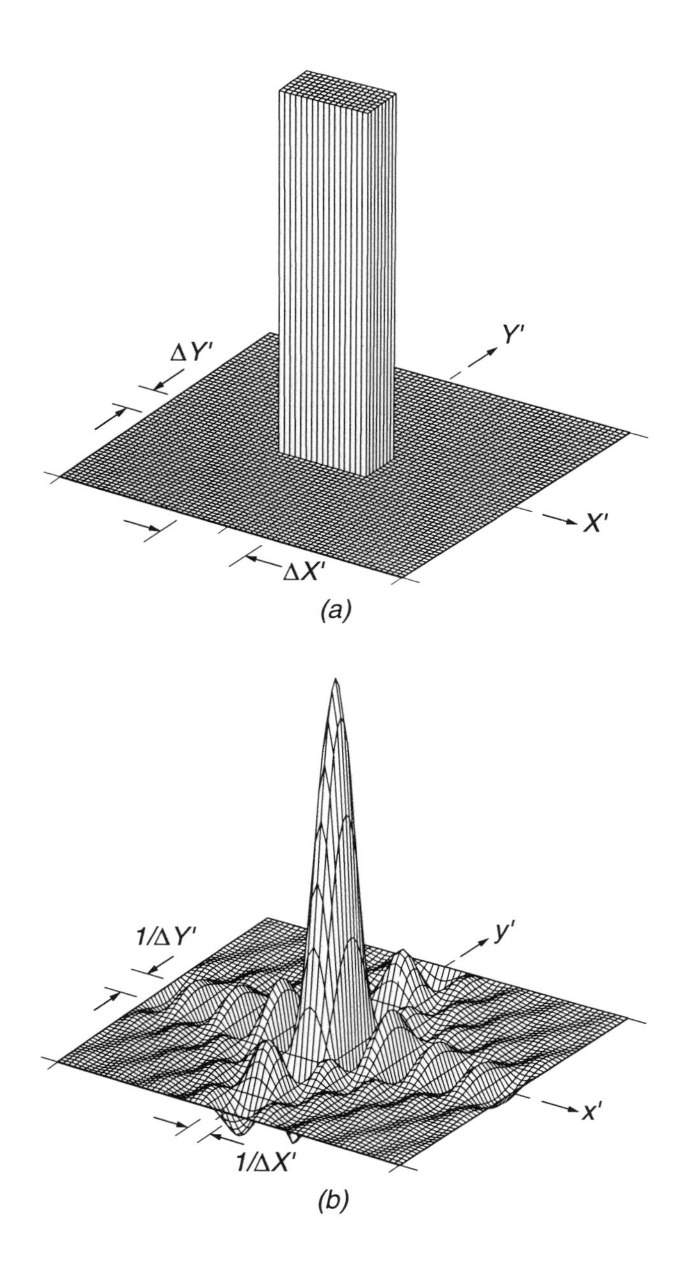

Figure 3.1 A rectangular window and its associated frequency response. (a) The rectangular window defining the phase-history region of support. (b) The two-dimensional *sinc* function obtained by Fourier transforming the rectangular window.

A well-known property of Fourier transforms is that the transform of a product is equal to the convolution of the transforms of the individual terms in the product. Thus, we can write

$$g(x', y') = s_A(x', y') \otimes f(x', y') \tag{3.4}$$

where $\otimes$ denotes two-dimensional convolution and $s_A(x', y')$ is the two-dimensional Fourier transform of $\mathcal{R}_A(X', Y')$. With $\mathcal{R}_A(X', Y')$ as defined by Equation 3.2, it is easy to show that its Fourier transform is of the form

$$s_A(x', y') = \frac{\sin(\Delta X' x'/2)}{x'/2} \, \frac{\sin(\Delta Y' y'/2)}{y'/2} \, . \tag{3.5}$$

Using the familiar *sinc* function, defined by $\operatorname{sinc} \xi = [\sin(\pi\xi)]/\pi\xi$, we have that

$$s_A(x', y') = \Delta X' \Delta Y' \, \operatorname{sinc}(\Delta X' x'/2\pi) \, \operatorname{sinc}(\Delta Y' y'/2\pi) \, . \tag{3.6}$$

This product of *sinc* functions in the x' and y' coordinates constitute what we will refer to as a *two-dimensional* sinc *function*, a plot of which is shown in Figure 3.1(b). The dimensions of the mainlobe of the 2-D *sinc* function are nominally expressed by the peak-to-first-null distance in the x' and y' directions, respectively. By setting the two *sinc* function arguments in Equation 3.6 to unity, we can see that these dimensions are reciprocally related to the dimensions of the phase-history region of support as follows:

$$\begin{aligned} \rho_{x'} &= 2\pi/\Delta X' \\ \rho_{y'} &= 2\pi/\Delta Y' \, . \end{aligned} \tag{3.7}$$

Equation 3.4 above states that the image produced is the result of convolving the true scene function with a particular 2-D *sinc* function. Thus, the image is a filtered, or smoothed, rendition of the actual scene. As such, it reveals something less than all the detail present in the scene. Specifically, a point target (represented by an impulse function) in the scene will appear in the image as a 2-D *sinc* function with mainlobe dimensions $\rho_{x'} \times \rho_{y'}$. We will say, therefore, that the resolution of the image is specified by $\rho_{x'}$ and $\rho_{y'}$ in the two orthogonal dimensions.

In the context of SAR imaging, we will identify x' and y' with the azimuth (cross-range) and range dimensions, respectively. The region of support of the transduced data in the corresponding spatial-frequency domain has dimensions (see Equations 2.45, 2.47 and Figure 2.20)

$$\begin{aligned} \Delta X' &= 4\pi\Delta\theta/\lambda \\ \Delta Y' &= 4\pi B/c \end{aligned} \tag{3.8}$$

where B is the radar (chirp) bandwidth, $\Delta\theta$ is the angular diversity of the synthetic aperture, and λ is radar wavelength. Substitution of these spatial bandwidths into

Equations 3.7 yields the nominal image resolution expressions

$$\begin{aligned} \rho_{x'} &= \lambda/2\Delta\theta \\ \rho_{y'} &= c/2B \end{aligned} \tag{3.9}$$

for the azimuth (cross-range) and range dimensions, respectively.

The fundamental relationships expressed by Equations 3.7 and 3.9 will be used throughout the chapter to describe the *nominal* resolution of the image that can be formed from a particular collection. We will see, however, that in practice we will generally produce images with somewhat reduced resolution in order to improve image quality in other respects.

Let us now turn to the matter of the area of coverage or size of the image. The limitation on the size of the image formed is a consequence of the fact that the signal transduced by the SAR is not in fact the continuous one represented in Equation 3.1, but instead is an array of discrete samples. The discrete sampling in cross range occurs because the radar transmits discrete pulses; while in range, it is the explicit result of digitizing the video signal. If we think of the spatial-frequency data transduced by the SAR in terms of a time series in each dimension (sometimes referred to as *fast time* for the range dimension and *slow time* for the cross-range dimension), then the image produced by Fourier transforming corresponds to a (two-dimensional) spectral *analysis* of the data. This view of the imaging process is a departure from our earlier stated view that the image is a *synthesis* of the scene from its spatial-frequency data. The analysis perspective has the advantage of allowing us to use the language and results of time series analysis theory to describe certain characteristics of the image.

A fundamental result from sampling theory (see Section 3.5.1) states that a continuous signal can be represented by and reconstructed from a sequence of discrete samples if (1) the signal is bandlimited and (2) the sample rate exceeds the signal bandwidth. If the signal to be represented is not bandlimited, it must be made so by pre-filtering before it is sampled. Then, only the portion of the spectrum retained after filtering can be reconstructed from the sampled data. In the SAR context, this implies that the maximum extent of the scene (analogous to the signal spectrum) that can be reconstructed in the image is equal to the rate at which the radar data are sampled in both dimensions. (It also implies that certain conditions relating to sample rates and signal bandwidth must have been satisfied in the collection process in order for the data to be valid.) Suppose now that the region of support shown in Figure 3.1(a) is spanned by a discrete array consisting of N_x samples in the cross-range dimension and N_y samples in the range dimension. Then, the maximum image dimensions, proportional to the density of phase-history sampling in each direction, are

$$\begin{aligned} D_x &= 2\pi N_x / \Delta X' = \lambda N_x / 2\Delta\theta \\ D_y &= 2\pi N_y / \Delta Y' = c N_y / 2B \ . \end{aligned} \tag{3.10}$$

These expressions actually give the upper bounds on image dimensions. The practical limits will be somewhat smaller because the filtering that precedes sampling necessarily attenuates those data near the cutoff frequency, which correspond to the edges of the image (see Section 3.5.2). Section 3.10 will deal with some of these issues in more detail.

Summary of Critical Ideas from Section 3.2

- **The relationship between the size, or extent, of the spatial frequency region transduced and the resolution of the resulting image can be derived from the properties of Fourier transform.**
- **The image resolution, or the nominal peak-to-null width of the impulse response, is inversely proportional to the dimensions of the phase history region of support.**
- **Because a finite piece of the phase history is transduced, the resulting image is a filtered, or smoothed, rendition of the actual scene. As such, it will reveal something less than all the actual detail present in the scene.**
- **The limitation on the size of the image formed is a consequence of the inherent sampled nature of the phase history. The area covered by an image is proportional to the density of phase-history sampling.**

3.3 HOLOGRAPHIC PROPERTIES AND MULTI-LOOK CONCEPTS

In coherent optics, holography refers to a method of reconstructing an image of an object from a record of its interference or diffraction pattern [1, 2]. This record is called a hologram and contains all the information needed to allow reconstruction of the original illuminated object wavefronts into a three-dimensional image in some volume of space.

As we have seen, spotlight SAR records a slice of the three-dimensional Fourier transform of the object's complex reflectivity. This slice, called the phase history, contains all the information necessary to reconstruct an image (complex reflectivity) of the illuminated terrain as seen by the SAR.

It is appropriate to make a connection between optical holography and synthetic aperture radar. The far-field diffraction pattern of coherent optics loosely corresponds to the complex reflectivity (image) in SAR. Furthermore, the optical diffraction pattern is the Fourier transform of the complex-amplitude aperture distribution. In spotlight SAR, the complex image is related to the complex phase history through a Fourier transform. The larger the aperture or hologram (corresponding to the size of the phase history), the finer the resolution of the corresponding reconstructed imagery. Additionally, the higher the spatial-frequency content of the hologram (phase history), the further the spatial extent or field-of-view (area coverage) of the diffraction pattern (SAR image). We have already alluded to some of these resolution versus area coverage issues in Section 3.2. Further discussion will be forthcoming in Section 3.5.

The spotlight SAR phase history can be considered a hologram in that any piece of the phase history is sufficient to reconstruct an image of the entire scene.[2] The resolution will be dependent on the size of the piece used. The impulse-response function (IPR) depends on the support of the aperture piece selected. The IPR becomes a 2-D *sinc* function for a rectangular support, a 2-D Airy pattern for circular apertures [3], and more complicated patterns for more complicated aperture supports. Two-dimensional weighting of these aperture distributions alters the relationship between sidelobe structure and mainlobe width.

[2] Specifically, it is sufficient to reconstruct an orthogonal projection of the 3-D complex scene reflectivity onto a plane containing the scene center and the flight path of the SAR. Unlike optical holography, no 3-D imaging is possible with single slices of Fourier-domain SAR data. However, 3-D information becomes available with two or more slices of Fourier space. This idea forms the basis for interferometric SAR (see Chapter 5).

In summary, the Fourier transform relationship between phase history and SAR image corresponds to the transform relationship between aperture distribution and diffraction pattern and forms a connecting link between holography and spotlight SAR. Some of these holographic properties are illustrated in Figure 3.2.

Close scrutiny of any SAR image will show that the brightness distribution is not smooth and continuous but is instead composed of a complicated granular pattern of bright and dark spots called speckle. Speckle patterns occur in any form of coherent imaging (e.g., optical holography, SAR, etc.) where objects illuminated by coherent radiation have surface features that are rough on the scale of the illuminating wavelength. Speckle results from random phasor sums from many scattering centers within a given resolution cell and bears little resemblance to any large-scale scattering properties of the illuminated terrain. Figure 3.3 shows a typical speckle amplitude[3] pattern that results from the coherent imaging of a uniform rough surface.

Speckle consists of a superposition of randomly phased, diffraction-limited impulse-response functions, therefore, the nominal speckle size indicates the achievable diffraction-limited resolution of a particular image [4]. It is interesting that this phenomenon exists whether or not the coherent image is focused.[4] Focusing only affects the phase of coherent scatterers over the aperture and cannot affect the statistical properties of waves scattered by a rough surface. Therefore, close scrutiny of speckle size in a SAR image will indicate the achievable resolution. Qualitative comparison of the smallest speckle with the sharpness of a point reflector will indicate whether diffraction-limited[5] resolution has been achieved.

Figure 3.4 shows a comparison of two speckle patterns (displayed using log-scaled amplitude) for which the diffraction-limited image resolutions differ by a factor of two. A large amplitude diffraction-limited point response (2-D sinc function) is embedded in each image for comparison. Note that the mainlobe width of the impulse response is approximately equal to the nominal speckle size.

To verify the conjecture that speckle is unaffected by focus, notice that the nominal speckle size has not changed in Figure 3.5 (compared to Figure 3.4a), even though

---

[3] Throughout this chapter, we will use the terms amplitude, intensity, and brightness. As applied to complex functions, amplitude refers to the magnitude, whereas intensity refers to squared magnitude. Brightness typically refers to an appropriate means of scaling the amplitude or intensity for display purposes.

[4] A SAR image can appear defocused if deterministic phase errors in the phase-history data are not properly compensated during image formation. (Refer to Chapter 4 for a thorough discussion of phase errors and correction.)

[5] Diffraction-limited resolution refers to the finest resolution achievable from a finite data aperture free of any phase errors or aberrations.

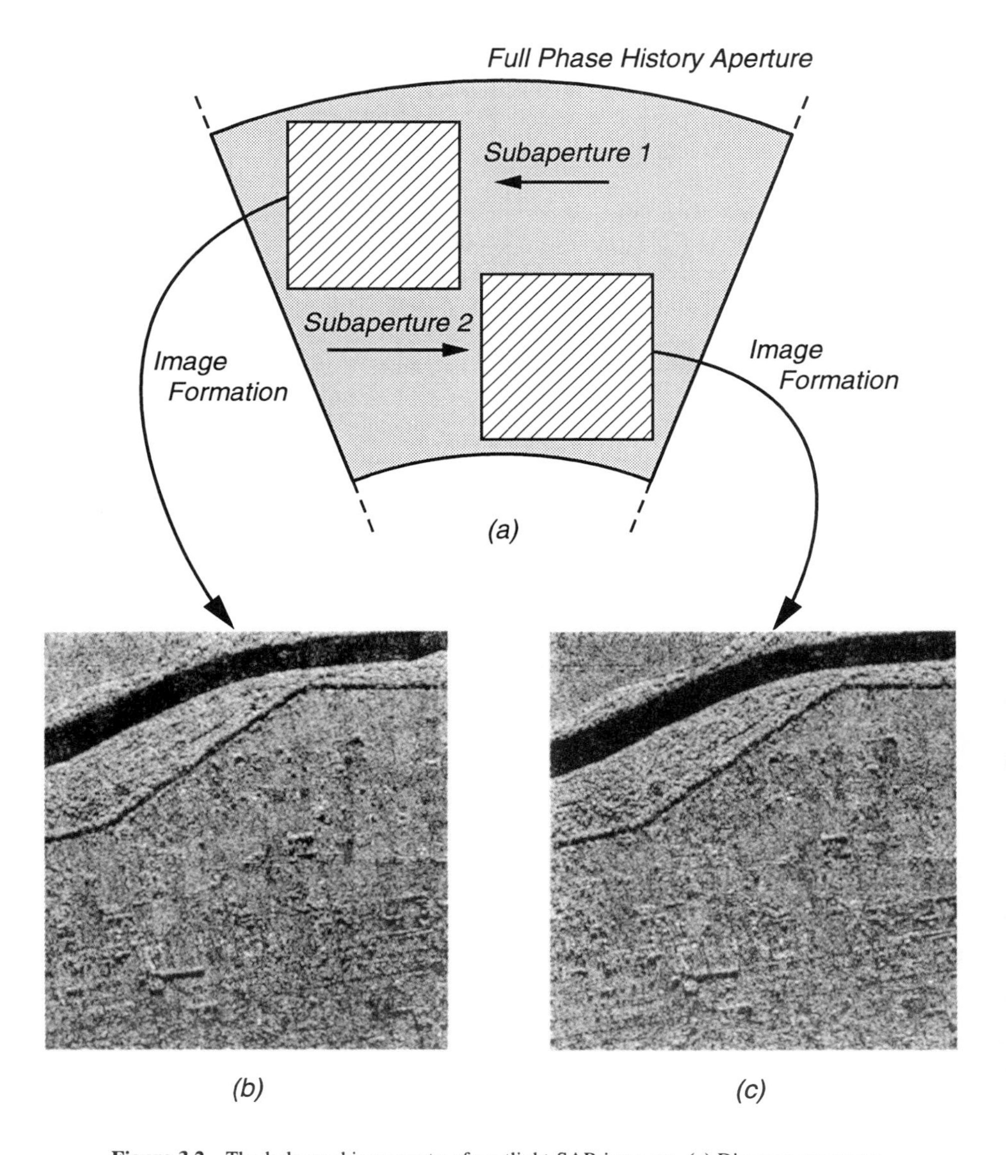

Figure 3.2 The holographic property of spotlight SAR imagery. (a) Diagram representing the pieces of Fourier space selected to form images. (b) Lower-resolution full-area-coverage image formed from the first selected subaperture of Fourier space. (c) Image formed from the second selected subaperture. Note that the images in (b) and (c) are virtually identical except for the speckle. Fourier space acts like a hologram in that any selected subaperture is sufficient to form a full image whose resolution is dependent on the dimensions of the subaperture selected. Independent subapertures produce images with uncorrelated speckle, a property that can be exploited with multi-look concepts.

Figure 3.3 A typical speckle amplitude pattern resulting from coherent imaging of a uniform rough surface.

the point response is defocused by application of a 1-D cross-range quadratic (25 radians peak) phase error in the aperture domain.

The probability density function of the speckle intensity I, observed in the image, obeys negative exponential statistics [5]:

$$p_I(I) = \begin{cases} \frac{1}{\bar{I}} \exp \frac{-I}{\bar{I}} & I \geq 0 \\ 0 & \text{otherwise} \end{cases} \tag{3.11}$$

where $\bar{I}$ is the mean intensity. It can be shown that the standard deviation of the speckle intensity σ_I is equal to the mean $\bar{I}$. Therefore, fluctuations of intensity about the mean are quite large which can make visual interpretation difficult. In most applications, speckle is an undesirable phenomenon for it masks subtle image brightness transitions, destroys apparent brightness connectivity within common scattering features, alters visual resolvability of non-point-like features, and reduces image interpretability in general. (Notice how the speckle begins to interfere with the lower level sidelobes as seen in Figure 3.4b.)

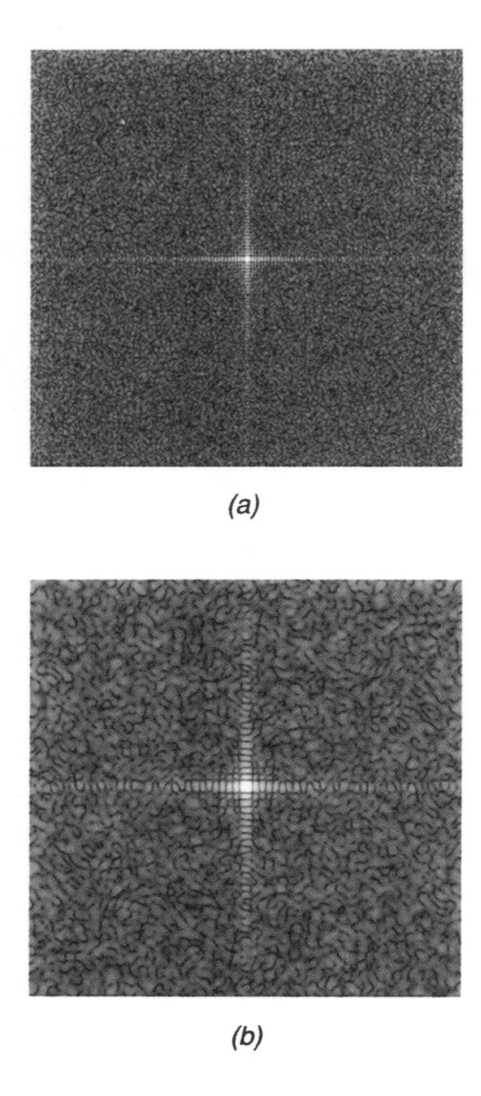

Figure 3.4 Comparison of speckle patterns with resolutions differing by a factor of two. (a) Nominal resolution. (b) Low resolution. Speckle consists of a superposition of randomly phased diffraction-limited impulse-response functions, and the nominal speckle size gives an indication of the achievable diffraction-limited resolution of a particular image. The diffraction-limited impulse-response function is a 2-D *sinc* function for these images.

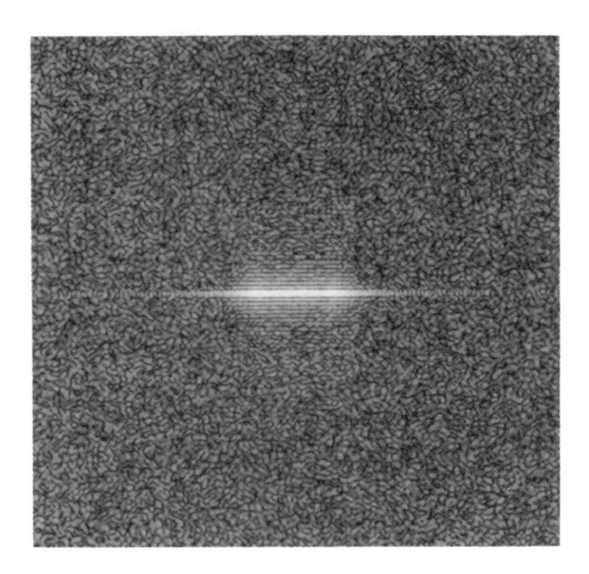

Figure 3.5 A quadratic phase error (25 radians peak) defocuses the coherent point response, but not the speckle size. The nominal speckle size indicates the diffraction-limited resolution potentially achievable *if* the focusing error could be removed.

We discussed some of the holographic properties of the SAR phase history in Section 3.3 and learned that independent portions of the phase history can be used to form independent images of the entire scene. We will now take advantage of this property and illustrate how this simple procedure allows mitigation of speckle, albeit with some tradeoffs.

Further analysis of speckle statistics indicates that the mean intensity distribution $\bar{I}(x, y)$ in a coherent image is identical to the intensity that is observed if the object were illuminated with spatially *incoherent* waves with the same statistical properties. Incoherent illumination can be thought of as a rapidly varying random sequence of spatially coherent wavefronts with highly complex phase structure. The structure of the wavefronts at any fixed spatial location are independent from one time to the next. Thus, (assuming equal bandwidths are involved) the time-integrated intensity is identical to the ensemble average intensity.

Typically, we do not have at our disposal an ensemble of independent SAR images of common terrain over which to average the intensities. However, we do have the option of partitioning the phase history into non-overlapping sections (with common aperture dimensions) and non-coherently[6] summing the image intensities formed from each phase-history piece. Therefore, rather than taking "multiple looks" at the terrain, we obtain a similar effect by taking multiple looks from within the original phase-history domain (see Figure 3.2). Of course, the non-coherently integrated image formed in this fashion suffers a loss of resolution, because each image in the sum comes from a physically smaller portion of the phase history. However, the reduction in speckle fluctuations can more than make up for that loss in terms of increased image interpretability.

In practice, it is quite reasonable to expect acquisition of spotlight data over a significantly larger angular aperture (cross-range dimension) than what is required to match the aperture range dimension for subsequent image formation with equal range and cross-range resolutions. Therefore, many spotlight SAR acquisitions have some potential for multi-look processing. If it is possible to sacrifice considerable resolution for some applications, a significant number of looks can be processed from one acquisition.

We illustrate some of these concepts with the following simulations shown in Figure 3.6. Figure 3.6(a) shows the result of two-look processing whereby two equal resolution images of a point target on a uniform clutter background were non-coherently summed in amplitude (magnitude). A comparison of Figure 3.6(a) with the single-look image of Figure 3.4(a), shows a slight reduction of speckle as expected. Images formed with four-look, eight-look, and sixteen-look processing are shown in Figures 3.6(b) through 3.6(d), respectively. A progressively decreasing speckle appearance is evident along with progressively increasing discrimination of the low-level sidelobe structure of the point target in the multi-look images. Figure 3.7 illustrates the same concept with actual SAR imagery. Note the rather dramatic improvement of the nine-look image compared to the single-look image. It should be emphasized that multi-look processing implies that extra Fourier data are available for the multiple looks (as was the case for our examples). In practice, it is sometimes advantageous to do multi-look processing even though extra Fourier data are not available. In this case, partitioning the available aperture into independent pieces sacrifices resolution; nevertheless, multi-look processing can produce smoother and more interpretable images [6][7].

[6] Non-coherent integration refers to summing the image intensities (or amplitudes) rather than summing the complex reflectivities.

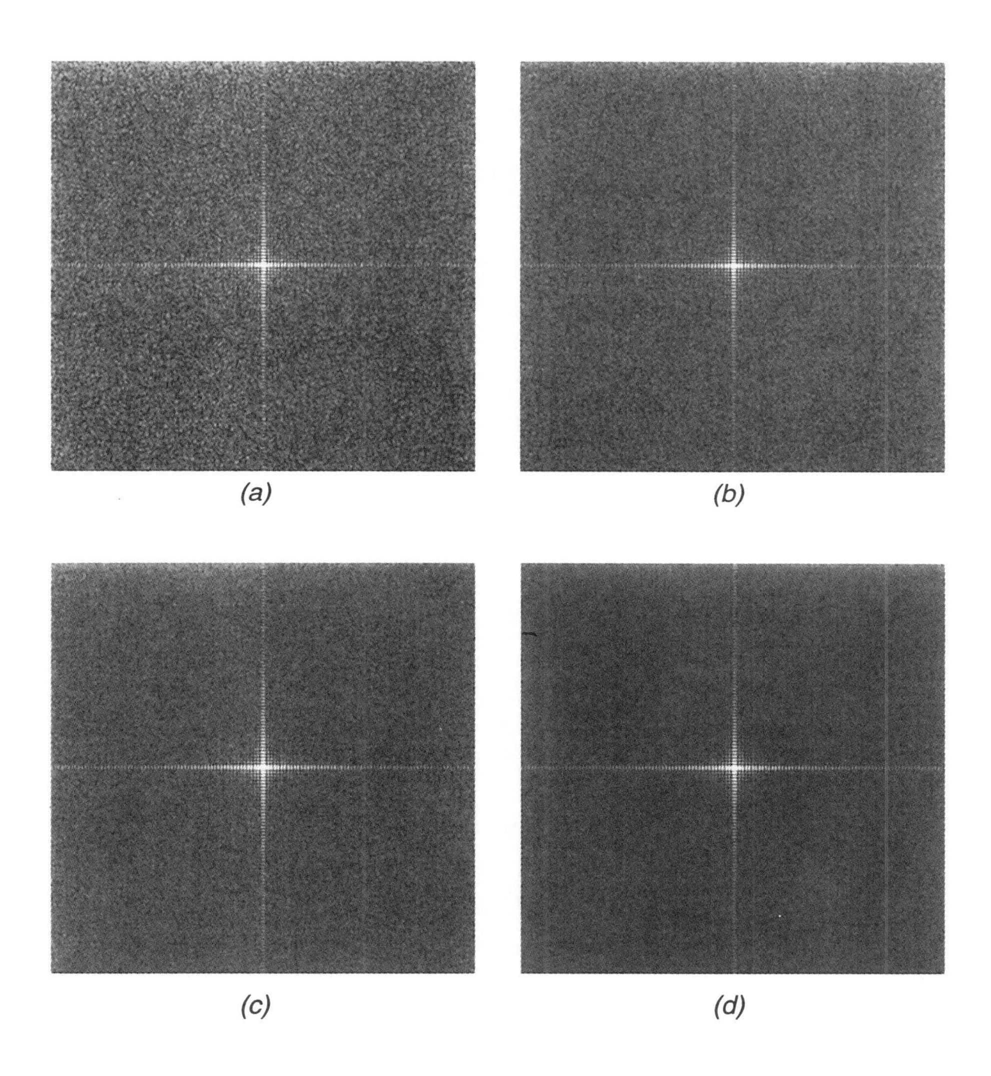

Figure 3.6 Multi-look results from synthetic imagery. (a) A two-look speckle pattern and single-point response. This image was formed by non-coherently summing two equal resolution images formed from two non-overlapping (i.e., independent) regions of the phase history. Each phase-history piece used in this construction covered one-quarter of the total phase history in each dimension; thus, a maximum of sixteen independent pieces were available. (b) Four-look speckle pattern and single-point response. (c) Eight-look speckle pattern and single-point response. (d) Sixteen-look speckle pattern and single-point response. The entire phase history was used for the sixteen-look image, but the resulting resolution is one-fourth (in each dimension) of that ultimately achievable with single-look processing.

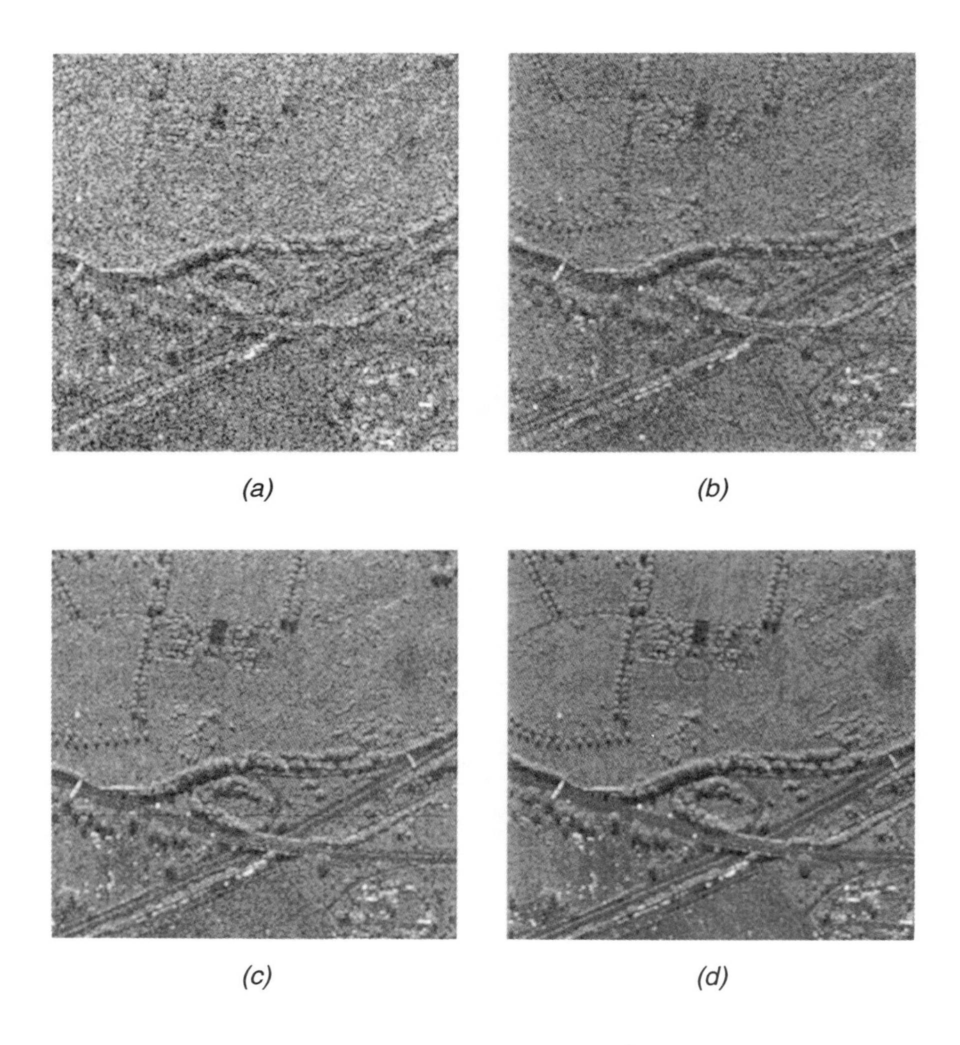

Figure 3.7 Multi-look results from actual SAR imagery. (a) Single-look image. (b) Two-look image. (c) Three-look image. (d) Nine-look image. Note the dramatic reduction of speckle and the greatly improved interpretability of this multi-look image compared to a single look.

Summary of Critical Ideas from Section 3.3

- **The spotlight SAR phase-history slice can be considered a hologram because any piece of the phase history is sufficient to reconstruct an image of the entire scene, with resolution proportional to the size of the piece used.**

- **Coherent SAR imagery suffers from speckle phenomena. Uniform reflecting surfaces that are rough on the scale of a wavelength produce a highly granular structure in the subsequent image.**

- **The speckle intensity variations are equal to the mean which can make visual interpretation difficult.**

- **Speckle statistical properties are unaffected by phase errors. The nominal speckle size is indicative of the ultimately achievable diffraction-limited resolution of the system without regard to the actual achieved resolution (possibly defocused or otherwise degraded because of phase errors).**

- **The holographic property can be exploited by multi-look processing for speckle reduction. Independent images formed from independent pieces of the phase history can be non-coherently summed to reduce the speckle variance.**

3.4 PROPERTIES OF NON-POLAR-FORMATTED DATA

We have established that the SAR phase-history data consist of a two-dimensional slice of the Fourier transform of the scene function. It follows that we can form an image of the scene simply by computing an inverse Fourier transform of the data. The difficulty in doing this is that the data have been collected on a polar grid, and we do not have an efficient algorithm for computing Fourier transforms of polar-sampled data directly. The ultimate solution will be to resample the data onto a rectangular grid and then to apply the Fast Fourier Transform (FFT) algorithm. We will address polar-to-rectangular resampling in Section 3.5, but first let us consider a simpler alternative. Suppose we simply pretend that the data, as they are collected, are already on a Cartesian grid and proceed to compute the FFT directly. This method will greatly simplify the image-formation process by eliminating the polar-to-rectangular resampling step. In this section we will establish the limits within which an acceptable image can be produced by this shortcut method. We will then show the consequences of using this method beyond the established limits.

Consider an array of SAR phase-history data sampled on a polar grid in the slant plane, such as that seen in Figure 3.8. Here, the samples are assumed to lie at positions spaced uniformly in the polar k, θ coordinates.[7] The polar coordinates are related to the Cartesian X', Y' coordinates by the standard trigonometric formulas

$$\begin{aligned} X' &= k \sin\theta \\ Y' &= k \cos\theta \, . \end{aligned} \tag{3.12}$$

However, when θ is small and k deviates only slightly from its central value k_0 this coordinate transform can be approximated with the linear relationships

$$\begin{aligned} X' &\cong k_0\theta \\ Y' &\cong k \, . \end{aligned} \tag{3.13}$$

This suggests that it could be acceptable to process the polar sampled data as if they were actually arrayed on a rectangular grid. To depict this situation, let us define a set of pseudo-rectangular coordinates (including an offset by $Y' = k_0$ to the center of the polar annulus) on the polar grid by

$$\begin{aligned} \mathcal{X} &= k_0\theta \\ \mathcal{Y} &= k - k_0 \, . \end{aligned} \tag{3.14}$$

[7] In practice (for a straight-line flight path and constant PRF) the polar data are typically uniform in $\tan\theta$, instead of θ; however, this is a subtle distinction that does not alter the results derived in this section.

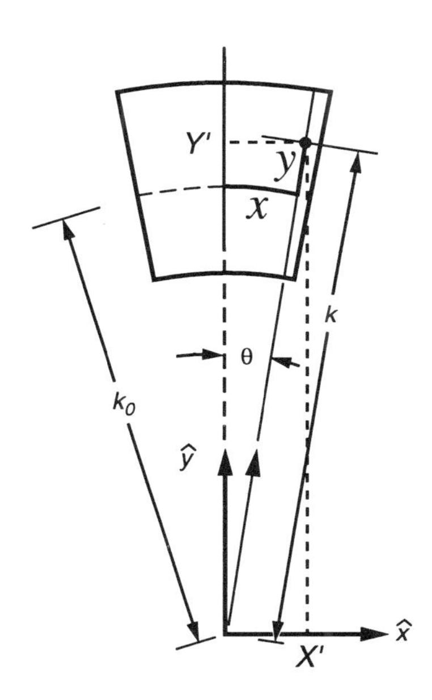

Figure 3.8 The SAR polar sampling grid. Here, $\hat{x}$ and $\hat{y}$ refer to the slant-plane unit vectors. The nominal polar radius, $k_0 = 4\pi/\lambda$, and an arbitrary polar value, k, are measured in units of radian spatial frequencies derived from actual SAR parameters. Section 3.10 addresses these issues in more detail.

The relationship of these coordinates to the actual rectangular coordinates is given by

$$\begin{aligned} X' &= (\mathcal{Y} + k_0)\sin(\mathcal{X}/k_0) \\ Y' &= (\mathcal{Y} + k_0)\cos(\mathcal{X}/k_0) \end{aligned} \tag{3.15}$$

or, in terms of series expansions,

$$\begin{aligned} X' &= \mathcal{X} + \mathcal{X}\mathcal{Y}/k_0 + ... \\ Y' &= k_0 + \mathcal{Y} - \mathcal{X}^2/2k_0 + ... \end{aligned} \tag{3.16}$$

where the omitted terms are third order and above.

Consider now the phase history from a point-target scatterer located at spatial coordinates (x_0', y_0') in the scene. The phase function produced by such a target (omitting particular high-order effects) is shown in Appendix B (Equation B.19) to be a linear function of the Cartesian spatial-frequency coordinates X' and Y' of the form

$$\phi_1 = x_0'X' + y_0'Y' \; . \tag{3.17}$$

The Fourier transform of $e^{j\phi_1}$ with respect to X' and Y' will reconstruct an image of the target at (x_0', y_0'). But in terms of the polar grid coordinates $\mathcal{X}$ and $\mathcal{Y}$, this point-target phase function (using Equation 3.16) is

$$\phi_1 = y_0' k_0 + x_0'\mathcal{X} + y_0'\mathcal{Y} + x_0'\mathcal{X}\mathcal{Y}/k_0 - y_0'\mathcal{X}^2/2k_0 + \ldots \tag{3.18}$$

and the complex video signal to be transformed is

$$S(\mathcal{X},\mathcal{Y}) = e^{j\phi_1} = e^{jy_0'k_0} \times e^{j(x_0'\mathcal{X}+y_0'\mathcal{Y})} \times e^{jx_0'\mathcal{X}\mathcal{Y}/k_0} \times e^{-jy_0'\mathcal{X}^2/2k_0} \times \ldots. \tag{3.19}$$

The result of Fourier transforming $S(\mathcal{X},\mathcal{Y})$ with respect to the polar coordinates $\mathcal{X}$ and $\mathcal{Y}$, instead of X' and Y', will be an image that is defocused as a consequence of nonlinear phase terms. (The constant phase term in Equation 3.18 is an inconsequential result of baseband translation common to all image formation methods.) The effect of the quadratic phase-error terms will be to convolve the ideal image with the Fourier transforms of the complex exponential terms $\exp[jx_0'\mathcal{X}\mathcal{Y}/k_0]$ and $\exp[-jy_0'\mathcal{X}^2/2k_0]$, which will cause the image to be defocused. When x_0' and y_0' are sufficiently small (the target is close to the center of the scene) and/or when k_0 is very large (radar wavelength is small) the defocusing effect tends to be insignificant and a well-focused image is formed. Conversely, the image will become seriously defocused when these conditions are not met.

Let us examine in detail the effects of the two quadratic phase-error terms in Equation 3.18. The $\mathcal{X}^2$ term represents a parabolic phase error of the form shown in Figure 3.9. A phase error of this form causes a cross-range defocus, or blur, in the image of the form shown in Figure 3.10. The extent of the blur is proportional to the location (y_0') of the target in range relative to the center of the scene. The $\mathcal{X}\mathcal{Y}$ term in Equation 3.18 represents a hyperbolic phase error of the form shown in Figure 3.11. This phase function results in a two-dimensional blur (sometimes described as astigmatism) as shown in Figure 3.12. The amount of blur in this case is proportional to the cross-range location (x_0') of the target. Appendix E provides additional information concerning the effects of these phase errors on the phase-history signals and the resulting point responses.

Consider the effect of the $\mathcal{X}^2$ term above. A practical guideline is that a phase error of this type can be ignored if its peak amplitude does not exceed $\pi/4$ radians across the aperture. This leads to the requirement that

$$\left| \frac{y_0'(\Delta X'/2)^2}{2k_0} \right| \le \frac{\pi}{4} \tag{3.20}$$

where $\Delta X'$ represents the width of the aperture in the cross-range dimension. If we now substitute Equation 3.7 (relating aperture extent to cross-range resolution $\rho_{x'}$)

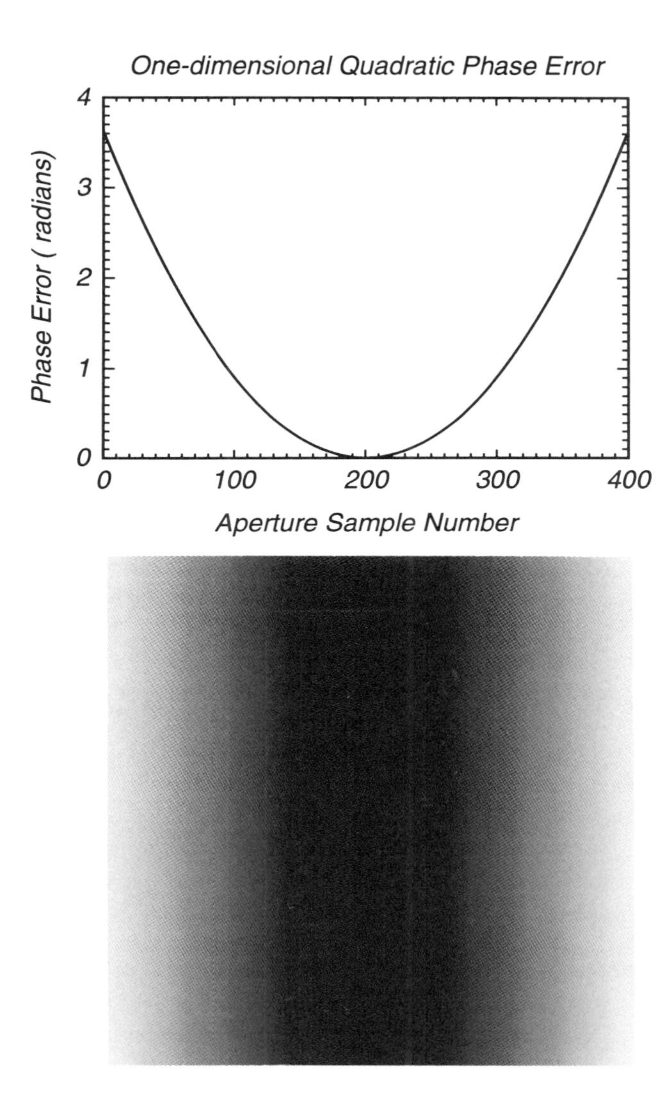

Figure 3.9 A parabolic $\mathcal{X}^2$-type phase error. The horizontal (cross-range) variation is shown in the 1-D plot at the top; the corresponding two-dimensional brightness function is shown at the bottom. This one-dimensional phase error results in a one-dimensional defocusing of the subsequent point target response. (See Figure 3.10.)

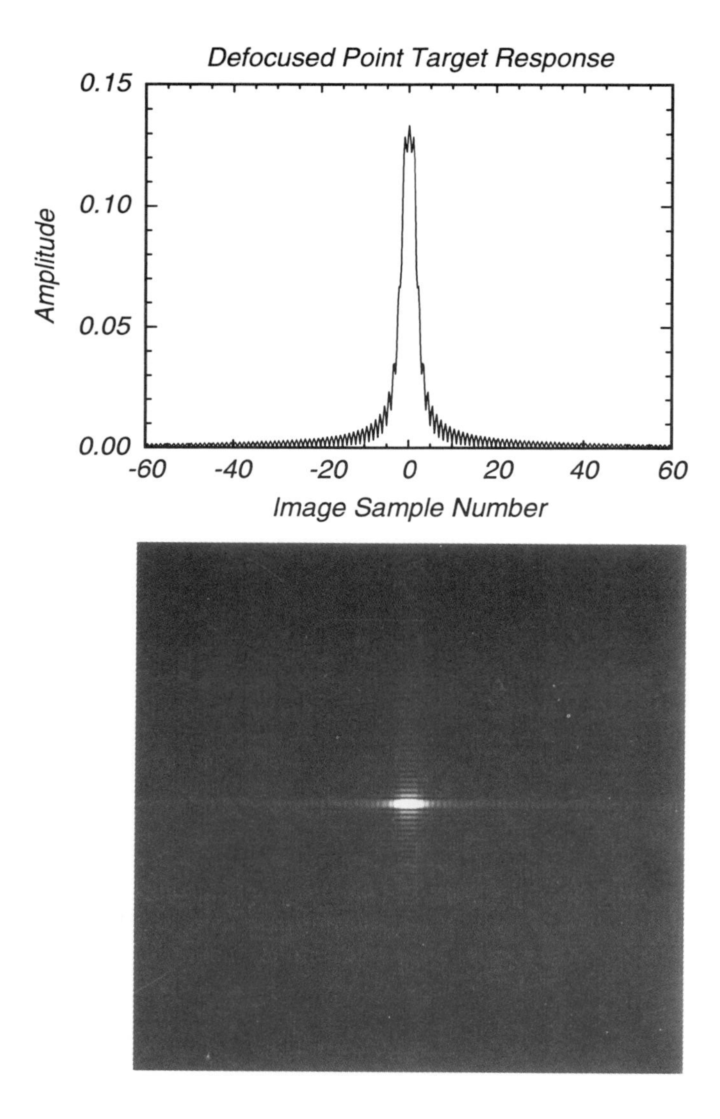

Figure 3.10 The one-dimensional defocusing of a point target by an $\mathcal{X}^2$-type phase error. The 1-D cross-range response is shown in the plot at the top; the 2-D image response is shown at the bottom.

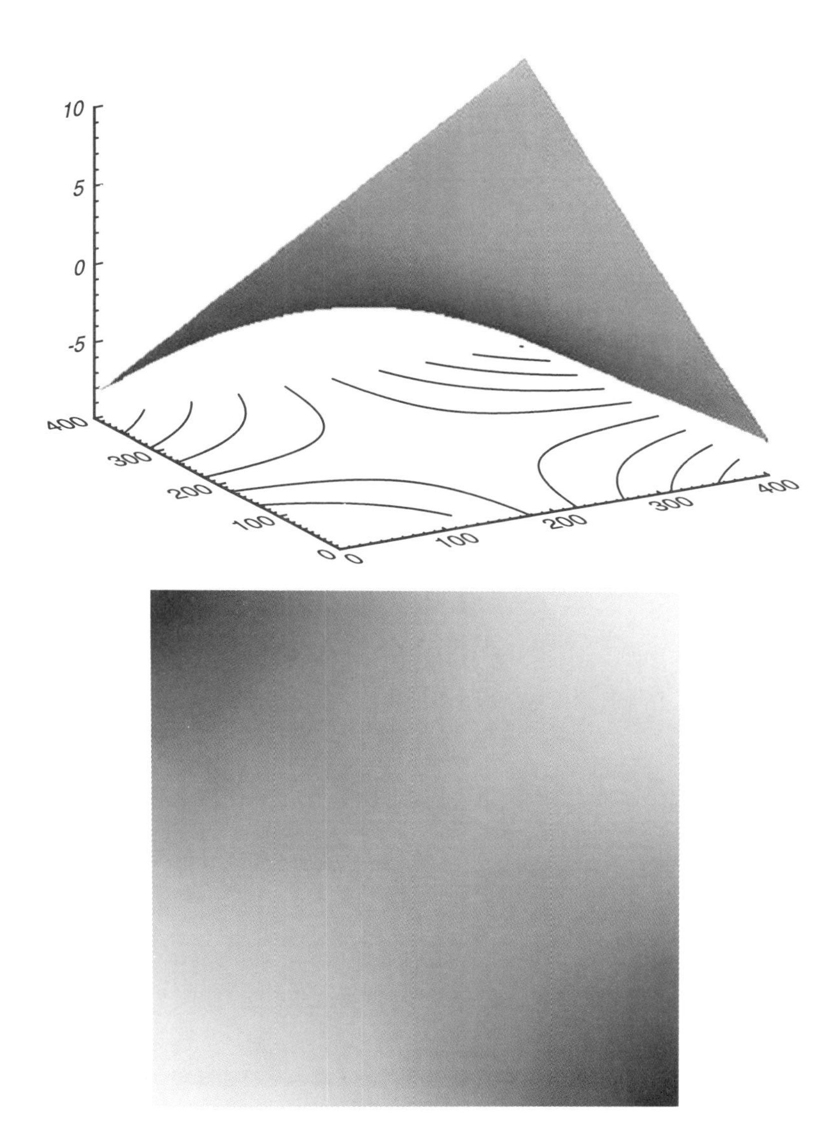

Figure 3.11 A hyperbolic $\mathcal{XY}$-type phase error shown as a shaded surface plot (top), and a two-dimensional brightness function (bottom). This function, whose constant phase contours are hyperbolas, results in a two-dimensional defocusing of the subsequent point target response. (See Figure 3.12.)

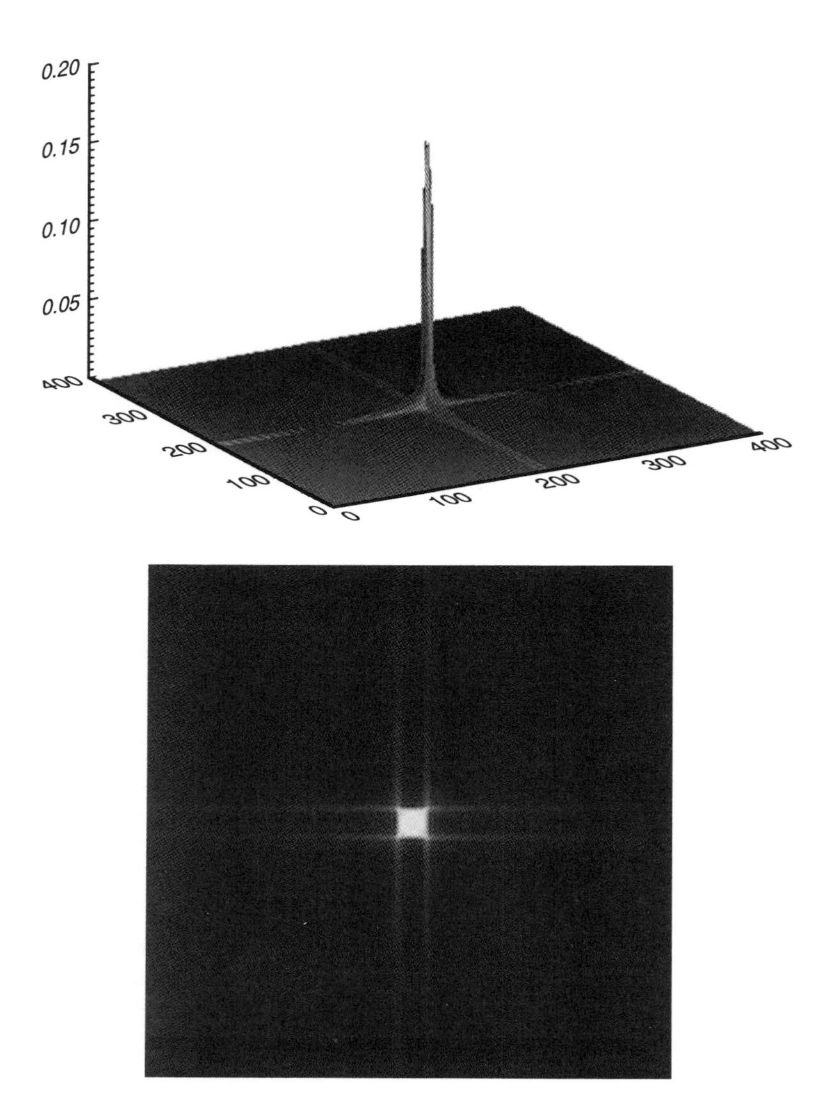

Figure 3.12 The two-dimensional defocusing of a point target by an $\mathcal{X}\mathcal{Y}$-type phase error shown as a shaded surface plot (top), and a magnitude image (magnified by 4X for clarity) at bottom.

and the relationship $k_0 = 4\pi/\lambda$ (where λ is the radar wavelength), this requirement becomes

$$\left| \frac{y_0'(\pi/\rho_{x'})^2}{2(4\pi/\lambda)} \right| \leq \frac{\pi}{4} \tag{3.21}$$

or simply

$$|y_0'| \leq 2\rho_{x'}^2/\lambda \; . \tag{3.22}$$

Next consider the $\mathcal{XY}$-product phase-error term in Equation 3.18. Using a criterion similar to that in Equation 3.20, we will require that

$$\left| \frac{x_0'(\Delta X'/2)(\Delta Y'/2)}{k_0} \right| \leq \frac{\pi}{2} \tag{3.23}$$

or

$$|x_0'| \leq 2\rho_{x'}\rho_{y'}/\lambda \; . \tag{3.24}$$

Equations 3.22 and 3.24 define the size of the scene that can be successfully imaged by Fourier transformation of the phase-history data directly on the polar grid. (Other authors arrive at the same conditions using somewhat different methods; see Reference [8].) As an example, consider a 10 Gigahertz SAR ($\lambda = 3$ cm) with enough bandwidth and angular diversity in the aperture to resolve 10 meters in range and cross range ($\rho_{y'} = \rho_{x'} = 10$m). Equations 3.22 and 3.24 imply that an image of dimensions *13,000* meters square can be formed using the direct Fourier transform method of processing. Now consider a SAR at the same center frequency, but with the capability of *1 meter* resolution in both dimensions. In this case, the maximum image dimensions shrink to *130* meters square. We conclude that, in general, low-resolution images can be successfully formed by Fourier transforming the polar data directly, without first performing polar-to-rectangular resampling. In high-resolution imaging situations for which spotlight-mode SAR is intended, however, this technique becomes impractical at best and in many cases totally inappropriate. The polar-to-rectangular resampling step is essential in any viable high-resolution spotlight SAR processor.

Figures 3.13 and 3.14 show a 1-meter resolution spotlight SAR image processed first without, then with, proper polar formatting. If processed without polar formatting, a gradual loss of focus is observed as a function of distance from the patch center. The type of space-variant defocus is a function of the phase error generated by the range and cross-range positions of the targets according to Equation 3.18. Isolated point targets exhibiting the space-variant defocus are extracted from the figures and shown in full resolution in the small sub-images. With proper polar formatting, focus is maintained throughout the image. See Appendix E for further examples of these types of phase errors and their effects on the phase history and synthetic target imagery.

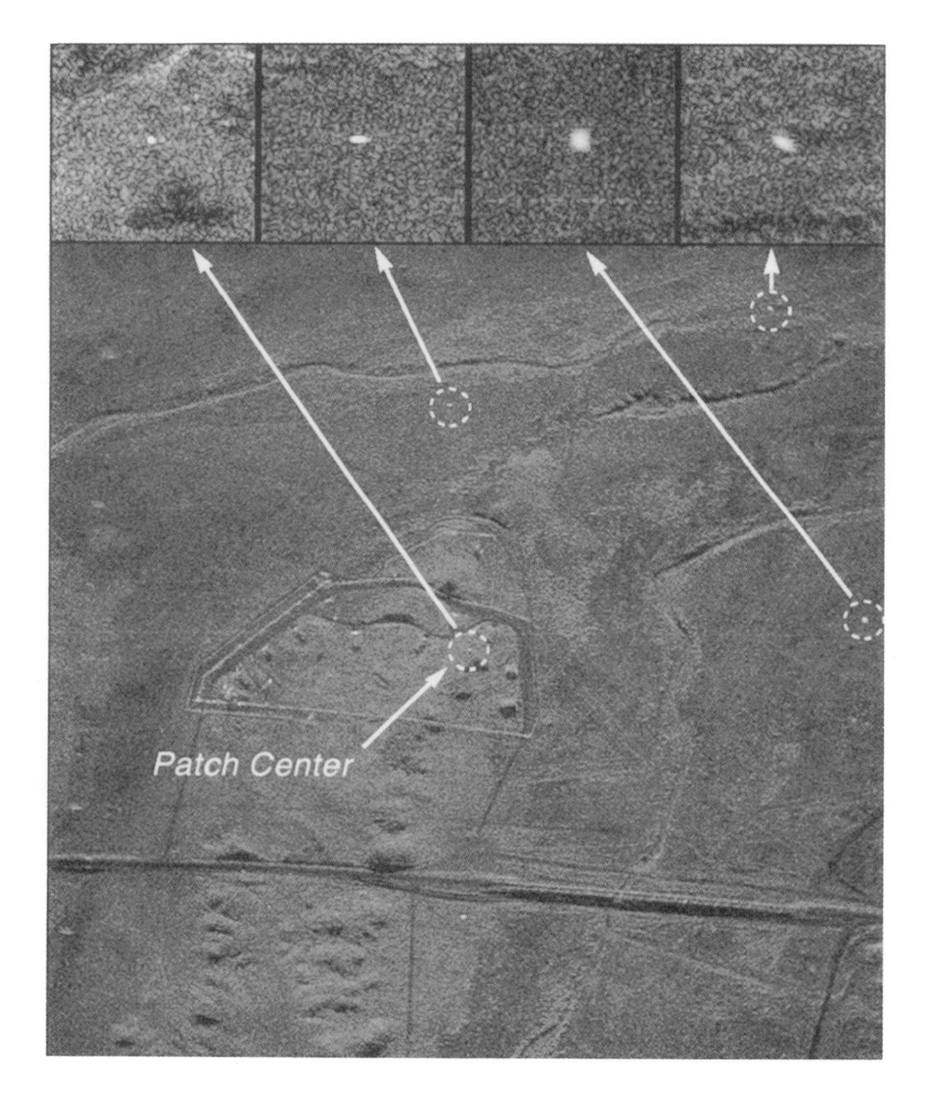

Figure 3.13 A high-resolution (1.0 m) spotlight SAR image processed *without* polar formatting. Note gradual loss of focus away from the patch center. Isolated point targets exhibiting the space-variant defocus are extracted from the main image and shown in full resolution in the small sub-images.

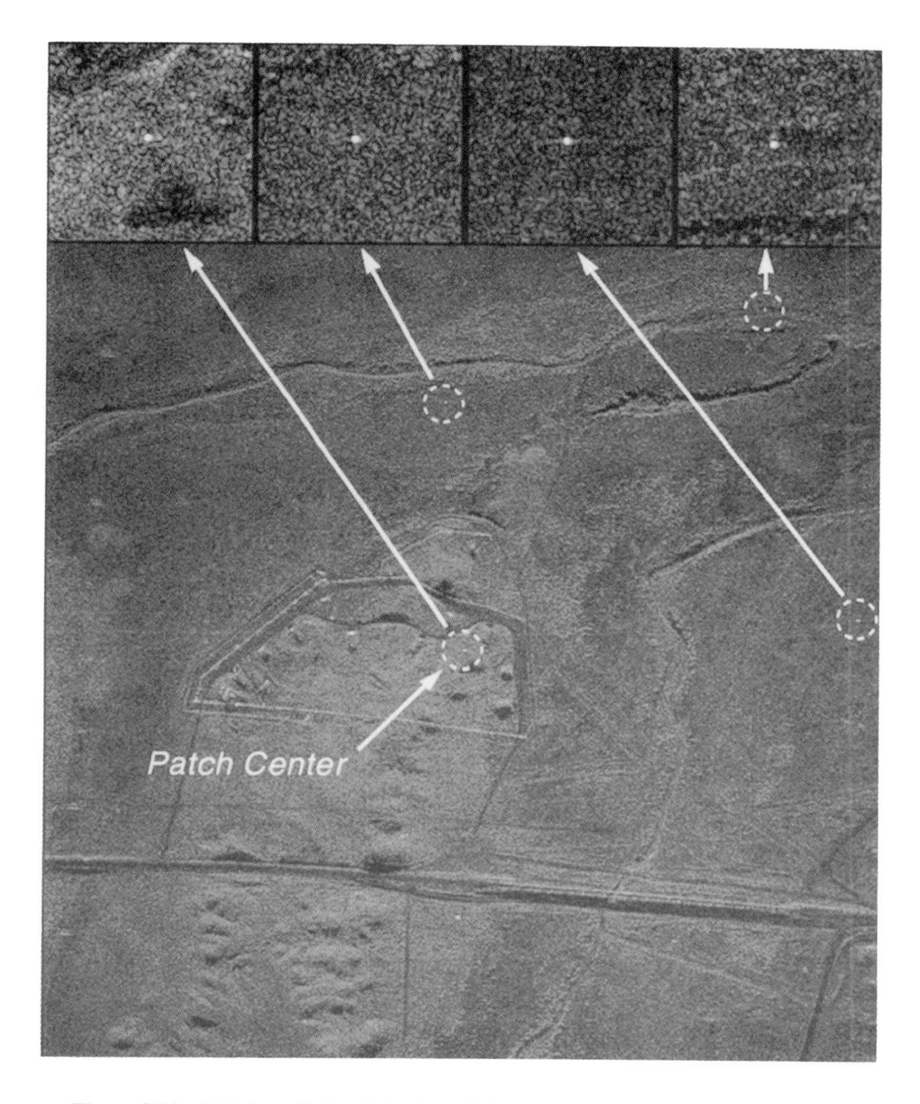

Figure 3.14 A high-resolution (1.0 m) spotlight SAR image processed *with* polar formatting. Note proper focus throughout the image as exemplified by the isolated targets shown in full resolution in the small sub-images.

Summary of Critical Ideas from Section 3.4

- **Computing the Fourier transform of the acquired data directly on the polar grid, assuming it was rectangular, constitutes a simple method of forming an image.**
- **The limits within which an acceptable image can be formed by this simple method depend heavily on the required resolution and on the desired image coverage.**
- **Two-dimensional series expansion of the phase errors resulting from not performing polar-to-rectangular resampling indicates that space-variant blurs are produced in the image.**
- **The magnitude of the blur depends on the target's distance from the imaged patch center.**
- **A target displaced from the patch center solely in the** *range* **dimension is defocused only in the** *cross-range* **dimension.**
- **A target displaced solely in the** *cross-range* **dimension is defocused in** *both* **the** *range and cross-range* **dimensions.**
- **In high-resolution imaging situations for which spotlight-mode SAR is intended, this simple image-formation technique becomes impractical at best. In many cases it is totally inappropriate.**
- **The polar-to-rectangular resampling step is essential in any viable high-resolution spotlight SAR processor.**

3.5 POLAR-TO-RECTANGULAR RESAMPLING

In the radar world, the polar-to-rectangular resampling problem (also referred to as polar resampling, polar reformatting, or polar processing) is unique to spotlight SAR processing. Why must we *interpolate* perfectly good samples acquired on a polar raster during collection to a rectangular grid? Why not just use our knowledge of the samples' polar geometry and process the data directly? Perhaps the best answers to these questions are obtained when we consider some of the practical constraints of image formation and processing.

We have already shown that the data acquired in spotlight-mode SAR, called the phase history, are essentially a slice of the Fourier transform of the terrain reflectivity and must be inverse transformed to form an image.[8] When dealing with sampled data, the 2-D fast Fourier transform (FFT) algorithm is the method of choice for efficient Fourier transformation. The 2-D FFT requires uniformly spaced samples on a rectangular grid because there is no known fast transform algorithm for samples on a polar grid [9]. Thus, the polar reformatting problem becomes one of interpolating the 2-D polar-gridded samples to a rectangular grid for ultimate use in a 2-D FFT image-formation process. (For further discussion on Fourier transforms in image formation, see Section 3.6.)

The interpolation process must maintain the underlying signal integrity and not introduce significant distortions. Specifically, the interpolation filter must be linear in its phase response (so as to not introduce phase errors into the complex data) and must meet certain bandwidth and aliased energy conditions. More will be said about these issues in a later section.

There are a variety of ways one might interpolate from a polar grid to a rectangular grid. The ideal solution is to perform 2-D interpolations, but this approach is computationally intensive. A significantly more efficient technique exploits separability, whereby one-dimensional interpolations are performed in the range (or fast time) dimension for each radar pulse, followed by one-dimensional interpolations in the cross-range (or slow time) dimension for each range line [10, 11, 12, 13].

This separable interpolation method that we adopt is called the *keystone*, or trapezoidal-grid technique [14]. It obtains its name from the geometry of an intermediate result obtained after range interpolation, as we shall soon see. The phase-history samples ideally lie on a polar grid depicted by the black dots in Figure 3.15. *Range interpolation* to the horizontal lines of a rectangular grid results in samples still residing

[8] The reader is encouraged to read Appendix C. This appendix includes a discussion on SAR imaging geometry and tells how the inherent geometry allows the establishment of a practical coordinate system from which phase-history data samples can be located in three-dimensional Fourier space.

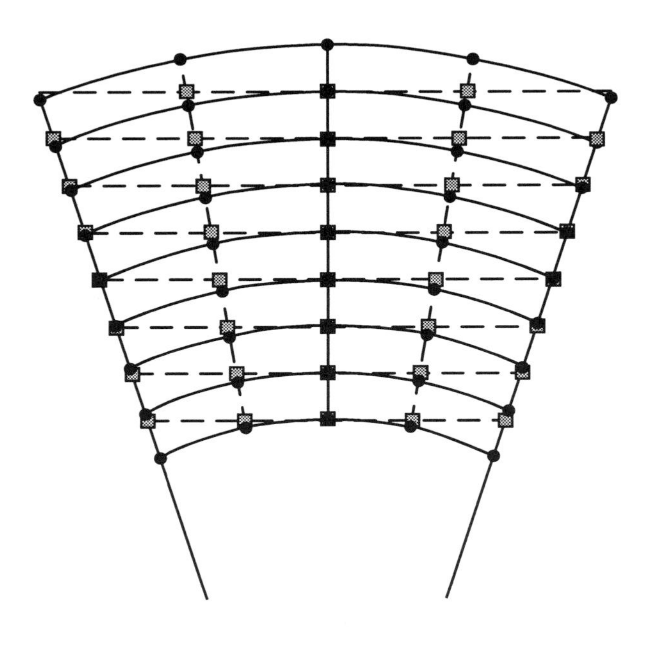

Figure 3.15 Range interpolation to keystone (trapezoidal) grid. The object is to use samples on the radial lines (depicted by the black dots) to obtain interpolated samples on the keystone grid (depicted by the grey squares). This interpolation can be done efficiently on a range-line-by-range-line basis by using one-dimensional digital filtering and resampling methods.

on the radial lines of the polar grid and depicted as the grey squares in Figure 3.15. Drawing a closed polygon connecting any two interpolated points on one horizontal line with two corresponding points on another horizontal line (bound by the radial lines) outlines a structure that resembles the keystone of an arch.

The second step in polar reformatting is *cross-range interpolation.* Range records processed as independent, one-dimensional signals and interpolated to a two-dimensional keystone grid become the input signals for cross-range interpolation. Each cross-range record on the keystone grid (i.e., the samples on each horizontal line) is treated as a one-dimensional signal to be interpolated to the vertical lines of the rectangular grid. After cross-range interpolation, the output samples lie on a rectangular grid, ready for 2-D FFT image formation. Figure 3.16 shows typical locations of interme-

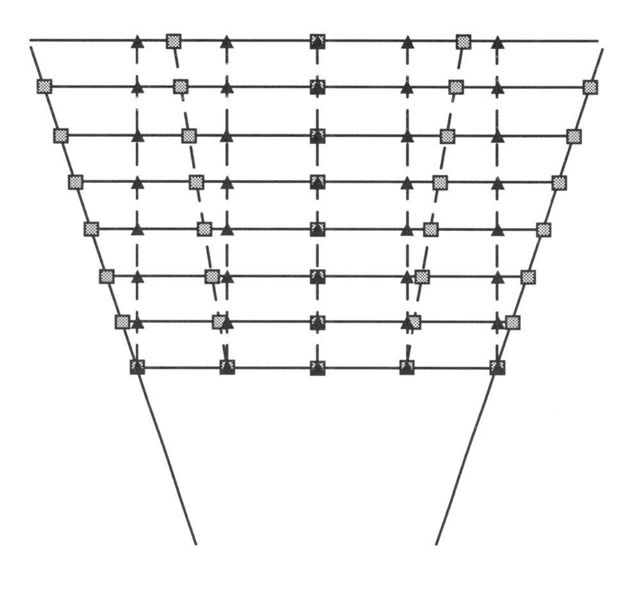

Figure 3.16 Cross-range interpolation. Similar in concept to range interpolation, the keystone samples (depicted by the grey squares) are used to obtain interpolated samples on the rectangular grid (depicted by the black triangles). This interpolation also can be done efficiently on a cross-range-line-by-cross-range-line basis by using one-dimensional digital filtering and resampling methods. Except for the varying sample spacings (which the filters must accommodate), range and cross-range interpolation are fundamentally identical.

diate (keystone) and final output sample locations shown as grey squares and black triangles, respectively.

Close scrutiny of Figures 3.15 and 3.16 illustrates two phenomena associated with cross-range interpolation. First, the keystone-grid samples on any cross-range line are unequally spaced (unlike the range interpolation data). These keystone samples are more closely spaced near the center and increase in separation towards the outer edges of the phase history. Second, the sample separation increases with distance from the polar origin. The interpolation filters must accommodate this space-varying sample rate. In practice, the variation in sample spacing along an individual cross-range record is small because the angular extent of the phase history is only a few degrees. Thus on an individual cross-range record, the interpolation filters can act as if the sample spacing is uniform without introducing intolerable errors. However, there is substantial change in sample separation with distance from the polar origin that the filters must accommodate.

Except for sample spacing issues, range and cross-range interpolation are fundamentally identical. In the following sections, we discuss some of the pertinent theory of interpolation and resampling, filter response and its connection to image properties, and the location of phase-history data samples in Fourier space to facilitate filter design. Simulations illustrating the concepts developed here are discussed in Appendix E.

3.5.1 Resampling Theory

As we have stated, the most practical method of implementing the two-dimensional interpolation required in polar reformatting is by separate one-dimensional interpolations in the orthogonal range and cross-range dimensions. In this section, we will develop the theory governing one-dimensional interpolation or *resampling* of an existing sampled data sequence to a new set of samples. We will present this discussion in the familiar context of time series analysis, even though we intend to apply the results in the SAR spatial-frequency domain. We will limit our discussion to uniform sampling for tractability and because, in practice, any deviation from uniformity in the sampling is typically small.

Let us assume that the data sequence we are given is a set of uniformly spaced samples of a bandlimited continuous signal. Let $x(t)$ denote such a signal and let $X(f)$ denote its Fourier transform:[9]

$$X(f) = \int_{-\infty}^{\infty} e^{-j2\pi ft} x(t) dt \; . \tag{3.25}$$

The presumed bandlimited nature of $x(t)$ means that $X(f)$ is identically zero for $|f| \geq B/2$, where B denotes the bandwidth of the signal. Assume further that $x(t)$ has been sampled at equally spaced points in time denoted by $t_n = nT$, $(-\infty < n < \infty)$. Then, the sampled data consist of the sequence $\{x_n | -\infty < n < \infty\}$ where

$$x_n = x(t_n) \; . \tag{3.26}$$

A well-known result from sampling theory [15] states that the function $x(t)$ can be reconstructed exactly (in principle at least) from the sequence x_n if its sample rate

[9] We find it convenient in this chapter to express the Fourier transform in a different form than we used in Chapter 2. We trust that this slight inconsistency will not create any confusion.

$f_s = 1/T$ is such that

$$f_s > B \ . \tag{3.27}$$

This condition is referred to as the Nyquist criterion, and the critical sample rate, equal to the bandwidth, is called the Nyquist frequency.[10] Sampling theory further states that the formula for reconstructing the continuous signal is

$$x(t) = \sum_{n=-\infty}^{\infty} x_n \frac{\sin \pi(t - t_n)/T}{\pi(t - t_n)/T} \tag{3.28}$$

A consequence of this sampling-theory result is that when the Nyquist criterion (Equation 3.27) is satisfied, the sequence x_n is a complete representation of the continuous signal $x(t)$, containing exactly the same information as $x(t)$. It should be clear that the sampled data sequence representing a particular signal is not at all unique. The sample locations and their spacing can be changed as long as the Nyquist criterion is satisfied. The idea of resampling a particular data sequence is to construct from it a second sequence that accurately represents the same underlying continuous signal, at least for part of its bandwidth. In order for the new sequence to represent the underlying signal in its entirety, the new sample rate must satisfy the Nyquist criterion. When this criterion is satisfied, the resampling process consists simply of evaluating Equation 3.28 at the new set of sample locations. If the new sample times are denoted by $t_m = mT'$, then each sample of the new sequence $\{x'_m | -\infty < m < \infty\}$ can be evaluated from the original data by the formula

$$x'_m = \sum_{n=-\infty}^{\infty} x_n \frac{\sin \pi(t_m - t_n)/T}{\pi(t_m - t_n)/T} \qquad \textbf{(full bandwidth case)}. \tag{3.29}$$

We will return to this full bandwidth resampling formula later, but first let us consider what is involved in resampling at a rate that is lower than the original sample rate.

From the sampling-theory result stated above, it is clear that if the new sample rate $1/T'$ is less than the signal bandwidth, then the resampled sequence cannot preserve the full integrity of the underlying continuous signal. If, however, we are willing to sacrifice some of the bandwidth of the original signal, we will at least be able to preserve that portion of the signal that corresponds to a new restricted bandwidth, provided that this new bandwidth is not larger than the new sample rate. This situation represents a *downsampling* of the data and requires a filtering operation that would not otherwise be needed. The resampling formula for this important case is derived in the following paragraph.

[10] Some authors define the Nyquist frequency as one half the critical sample rate, which is what we refer to as the *folding* frequency in Section 3.5.2.

The process of downsampling a data sequence can be thought of conceptually as: (1) reconstruction of the underlying continuous signal, (2) low-pass filtering to restrict the bandwidth, and (3) resampling at the new sample rate. The first of these steps is described by Equation 3.28, which we have written here in abbreviated form

$$x(t) = \sum_{n=-\infty}^{\infty} x_n s(t - t_n) \tag{3.30}$$

where

$$s(t) = \frac{\sin \pi t/T}{\pi t/T} . \tag{3.31}$$

The filtering step can be expressed by a convolution of the form

$$x'(t) = x(t) \otimes h(t) \tag{3.32}$$

where $h(t)$ is the impulse response of an appropriate low-pass filter. Let $H(f)$ denote the corresponding frequency response as given by the Fourier transform of $h(t)$. The purpose of this filter is to remove all frequencies greater in magnitude than half the sample frequency of the resampled data. An ideal frequency response for the filter is

$$H(f) = \begin{cases} 1 , & |f| \leq 1/2T' \\ 0 , & |f| > 1/2T' \end{cases} \tag{3.33}$$

where T' is used to denote the resample interval. The corresponding impulse response is

$$h(t) = \frac{\sin \pi t/T'}{\pi t} . \tag{3.34}$$

To see how this filter affects the resampling process, let us combine Equation 3.32 and Equation 3.30 to write

$$x'(t) = \sum_{n=-\infty}^{\infty} x_n s(t - t_n) \otimes h(t) . \tag{3.35}$$

Taking the Fourier transform of both sides of this equation gives

$$X'(f) = \sum_{n=-\infty}^{\infty} x_n e^{-j2\pi f t_n} \mathcal{R}(f) H(f) . \tag{3.36}$$

Here $\mathcal{R}(f)$ is the Fourier transform of $s(t)$, which (from Equation 3.31) has the form

$$\mathcal{R}(f) = \begin{cases} T , & |f| \leq 1/2T \\ 0 , & |f| > 1/2T \end{cases} . \tag{3.37}$$

With this result (and recognizing that $\mathcal{R}(f)H(f) = TH(f)$ when $T' > T$ in the downsampling case), Equation 3.36 becomes

$$X'(f) = T \sum_{n=-\infty}^{\infty} x_n e^{-j2\pi f t_n} H(f) \ . \tag{3.38}$$

By inverse transforming this expression, we obtain the filtered signal

$$\begin{aligned} x'(t) &= T \sum_{n=-\infty}^{\infty} x_n h(t - t_n) \\ &= \frac{T}{T'} \sum_{n=-\infty}^{\infty} x_n \frac{\sin \pi(t - t_n)/T'}{\pi(t - t_n)/T'} \ . \end{aligned} \tag{3.39}$$

The final step in constructing the downsampled data sequence $\{x'_m\}$ is simply to evaluate the filtered signal $x'(t)$ at the new sample times $t_m = mT'$, $(-\infty < m < \infty)$:

$$x'_m = \frac{T}{T'} \sum_{n=-\infty}^{\infty} x_n \frac{\sin \pi(t_m - t_n)/T'}{\pi(t_m - t_n)/T'} \qquad \textbf{(downsampled case)}. \tag{3.40}$$

Note that this result is identical in form to Equation 3.29 for the full bandwidth resampling case, except that it involves the new sample interval T' instead of the original T, and a scale factor consisting of the ratio of the two sample intervals.

Equations 3.29 and 3.40 represent the theoretical resampling formulas for the full bandwidth resampling case ($T' \leq T$) and the reduced bandwidth downsampling case ($T' > T$), respectively. Because these expressions both involve infinite sums, however, they are ill-suited for practical implementation. Still, these formulas do establish a framework from which we can derive practical resampling formulas. In particular, they cast the resampling process as a filter design problem, which is the subject of the Section 3.5.2.

3.5.2 Interpolation Filters

We have suggested that the problem of interpolation and resampling can be reduced to a digital filtering operation. The *sinc* function embodied in Equation 3.40 can be thought of as the impulse response of an ideal digital filter. We need to approximate this impulse response with one that is realizable. It remains for us to find a suitable filter function that does not require infinite sums, preserves phase, and provides suitable passband and stopband characteristics.

We wish to filter and resample a two-dimensional piece of Fourier space originally sampled on a polar grid to a rectangular grid for subsequent image formation by 2-D FFT. The piece of Fourier data, or phase history, that we possess contains complex sinusoids whose amplitudes are proportional to the magnitudes of terrain scatterers. Ideally, any filtering and resampling should not alter the spatial frequency amplitudes because doing so would inject artificial terrain reflectivity variations in the subsequent image. For example, a filter that has large ripples in its passband will induce these same ripples across the SAR image. It is important to design an interpolation filter with a flat passband to maintain spatial uniformity in image brightness.

Perhaps more important from a practical viewpoint is the filter's stopband characteristics on the resulting imagery. Because we are dealing with sampled data, any frequencies beyond half the Nyquist rate are folded back into the data in proportion to the filter stopband amplitudes. The resulting image then includes some of the aliased energy coming from outside the passband, or equivalently, from outside the processed patch of terrain desired. Thus, strong scatterers from just outside the processed patch of interest can be folded back into the image and be confused with scatterers in the patch.[11]

It is desirable to design interpolation filters with sufficiently large stopband attenuation in order to minimize aliased energy. Aliased energy contaminates the image with spurious target signatures and with increased clutter that reduces contrast.

The filter must also have linear-phase characteristics so as not to introduce phase distortion. In other words, the filter's frequency response must be such that the phase varies linearly with frequency. (The shift property of the Fourier transform dictates that a linear phase response in the frequency domain introduces only a shift of the function in the time domain.) Typically, only finite length (i.e., finite impulse response, FIR) filter functions can be designed with exactly linear phase. Recursive filters, on the other hand, have infinite length impulse responses (IIR) and cannot in general be designed with precisely linear phase [16]. Special two way (i.e., forward/backward) processing is required with recursive filters to produce exactly linear phase [17], representing an additional computational burden. For simplicity, we restrict ourselves to FIR filters for interpolation.

If an interpolation filter is designed to reduce the bandwidth of the signals it filters, this reduced bandwidth will produce a physically smaller image (i.e., the image will cover less physical terrain). Because we are independently performing range

[11] This describes the downsampling case only. In the full bandwidth case, where the antenna pattern spatially limits the field of view, aliased energy added by the interpolation filters is usually negligible.

and cross-range interpolations, range and cross-range filter bandwidths influence the extent of the range and cross-range dimensions in the subsequent image.

Specifically, if a large array of polar samples spanning the desired range and cross-range phase-history extent is downsampled (after corresponding bandwidth reduction) to a similar-sized rectangular grid with fewer samples in each dimension, the resulting image will have the desired resolution but will cover a physically smaller region of terrain. The size of the piece of the Fourier space spanned by the data determines the subsequent image resolution, whereas the number of samples on the rectangular grid (relative to the original polar grid) determines the image coverage (see Section 3.2 for details). These important points are summarized diagrammatically in Figure 3.17, and in the SAR imagery examples of Figures 3.18 and 3.19.

It should now be clear that ideally we would like to design an implementable interpolation filter that has a nearly rectangular passband, wherein the bandwidth controls final image coverage. Because a perfectly rectangular passband is unrealizable in practical filter design, we must make some compromises. It is not our intent to delve into the wealth of digital filter design methods that are thoroughly covered in the literature. Suffice it to say that any practical filter impulse response can be used for interpolation if it has linear phase, acceptable passband and stopband characteristics, and is easily implementable in an efficient manner for digital processing.

We have chosen to start our filter design from the ideal rectangular passband characteristics of the infinite-length *sinc* sampling function. Our approach is to observe the filter characteristics obtainable by truncating the *sinc* function to a reasonable number of samples and then to weight the truncated *sinc* samples by an appropriate window function to reduce passband ripple and stopband sidelobes. This is a simple methodology that captures the spirit of more elegant frequency sampling methods for filter design [18] or for generalized interpolation [19].

We recall from Fourier transform theory that multiplication of functions in the time domain is equivalent to convolving their transforms in the frequency domain. Thus, the process of truncating or windowing an infinite-length time function is equivalent to convolving the ideal frequency response by the transform of the window function. The window function selected has a direct bearing on the filter frequency response as we shall soon see.

The truncation and windowing method of filter design is best explained by using several examples. Let us begin with the ideal filter function

$$h(t) = B\frac{\sin \pi Bt}{\pi Bt} \tag{3.41}$$

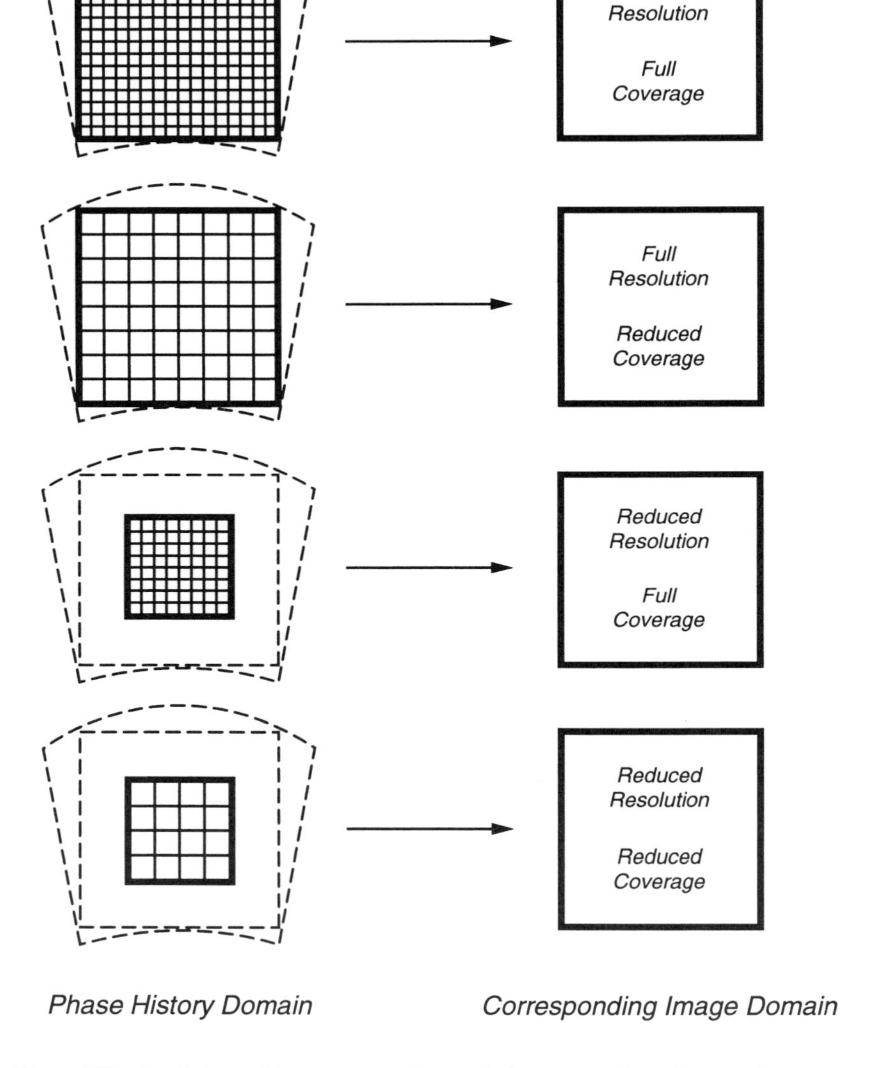

Figure 3.17 Resolution and image coverage issues. A dense array of samples spanning the phase history supports higher spatial frequencies over a large Fourier-domain aperture, thereby allowing the formation of a full coverage full resolution image (shown in upper portion of the figure). At the other extreme (lower portion of the figure), a less dense sampling of the phase history (appropriately filtered) over a smaller Fourier-domain aperture cannot support high spatial frequencies as before. Consequently, the formed image covers less area at reduced resolution. In practice, phase-history filtering and downsampling occur simultaneously (see Section 3.5.3).

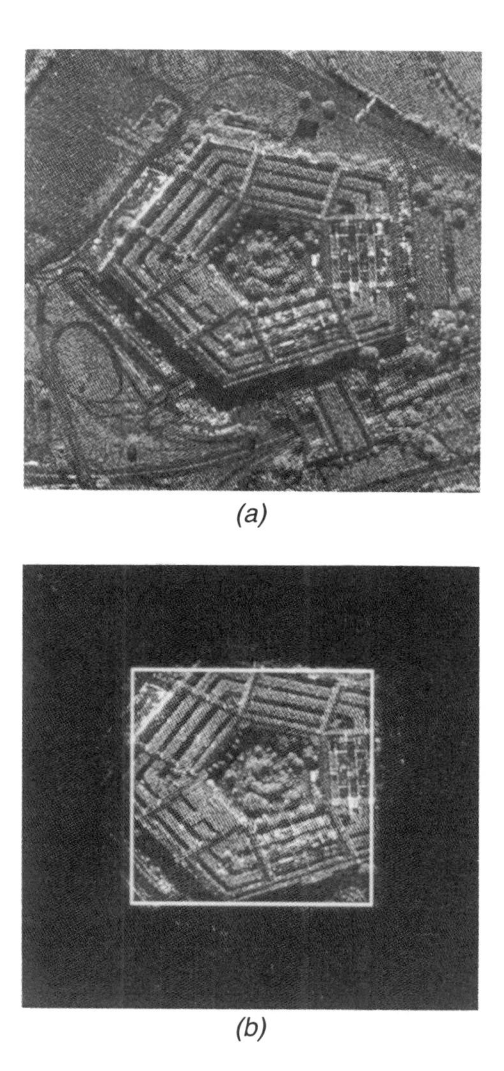

Figure 3.18 Actual SAR imagery showing tradeoffs of resolution versus area coverage. (a) Full resolution, full area coverage. (b) Full resolution, reduced area coverage. This reduced-area-coverage image was produced by using a five zero-crossing truncated *sinc* function interpolation filter (see Figure 3.21a) to reduce the bandwidth of the phase-history data by a factor of two with subsequent corresponding reduction in image coverage. The original sampling rate was maintained (i.e., not downsampled) to better illustrate the filter characteristics. Specifically, the white square indicates the image coverage supported by the Nyquist sampling rate of the interpolating filters. Any image content outside the white box would fold back into the box if the phase-history sampling rate were reduced to the Nyquist rate. All phase-history data were processed so that full resolution was maintained.

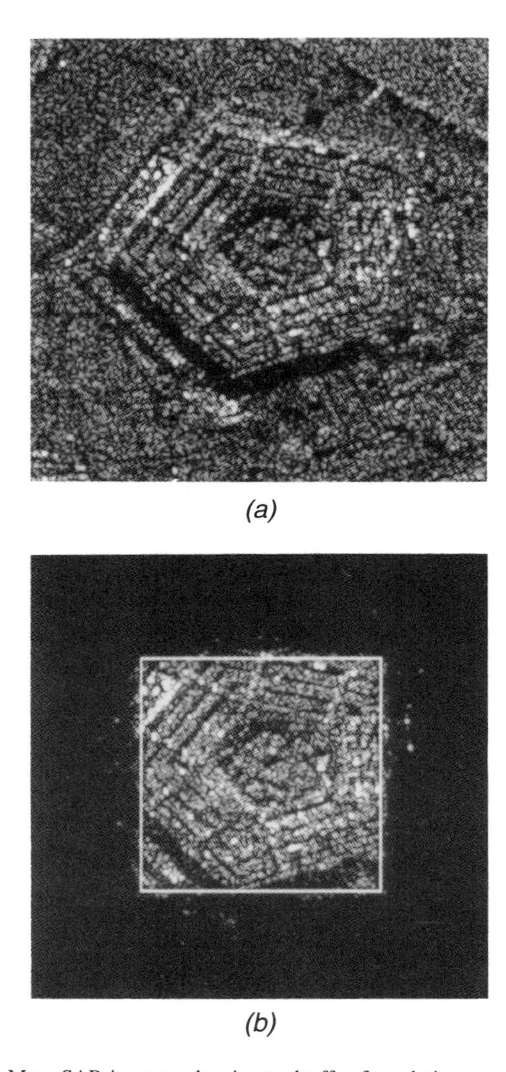

Figure 3.19 More SAR imagery showing tradeoffs of resolution versus area coverage. (a) Reduced resolution, full area coverage. (b) Reduced resolution, reduced area coverage. This reduced-area-coverage image is similar to the image in Figure 3.18(b), except that not all phase-history data were processed and this resulted in reduced resolution.

and its associated Fourier transform

$$H(\omega) = \begin{cases} 1, & |\omega| \leq \pi B \\ 0, & \text{otherwise} \end{cases} . \tag{3.42}$$

We wish to truncate $h(t)$ to some time interval $|t| \leq \tau_L/2$ and observe the frequency response. Mathematically,

$$h_L(t) = \begin{cases} h(t) & |t| \leq \tau_L/2 \\ 0 & \text{otherwise} \end{cases} . \tag{3.43}$$

A convenient way to parameterize the length of the truncated *sinc* is by specifying the number of one-sided zero crossings of the *sinc* function. We see that $h(t)$ has zero crossings at $t_k = k/B$ for $k = \pm 1, \pm 2, \ldots$. Therefore, by specifying the number of one-sided zero crossings N_z we specify τ_L as

$$\tau_L = 2N_z/B . \tag{3.44}$$

For sampled data systems with the sampling interval $T = 1/B$, we define the generic filter length L as

$$L = 2N_z + 1 \tag{3.45}$$

samples. The desired sample rate (Nyquist rate) is determined by the sample interval T. The length of the truncated *sinc* is specified by the number of samples, spaced T apart, that it spans.

Now let us examine the resulting filter response of various truncated *sinc* functions. The bandwidth and corresponding zero crossing spacing are set to unity for normalization purposes and only one-sided signals are plotted for simplicity. The Nyquist frequency is thus 1.0. Remember, any frequency content beyond the *folding frequency* (one half the Nyquist rate) aliases back into the filtered signal.

The first example, shown in Figure 3.20(a), shows both the one-sided *sinc* function truncated after sixteen zero crossings as well as the corresponding frequency response. Interpolation with this filter results in an image with rather large passband ripples. On the other hand, the frequency cutoff (transition) region is fairly sharp. Aliased energy just beyond the folding frequency is primarily due to stopband sidelobes and not to transition band energy folding.

If the *sinc* function is truncated to eight zero crossings as shown in Figure 3.20(b), we see identical passband and stopband ripple magnitudes as was true in the previous case, but the transition band rolloff is not as sharp. It is also important to note that the ripples are broader (which leads to more aliased energy). Therefore, terrain scene content just beyond the folding frequency appears in the image with more energy than it does in the sixteen-zero case.

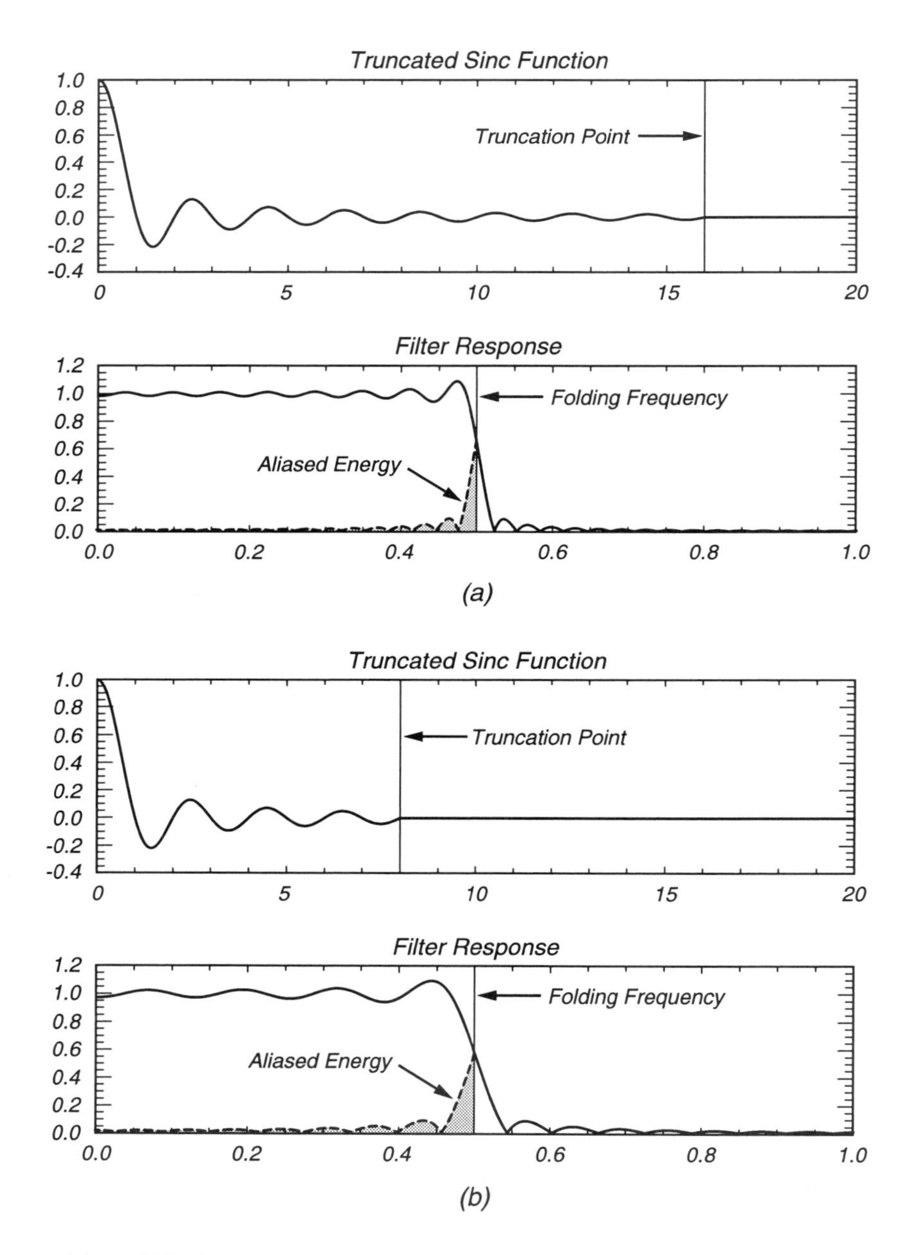

Figure 3.20 One-sided *sinc* function (a) truncated after sixteen zero crossings and the corresponding frequency response, (b) truncated after eight zero crossings (top) and the corresponding frequency response (bottom).

Carrying this process a couple of steps further results in Figure 3.21(a) and (b), showing the responses with five zero crossings and three zero crossings, respectively. It is obvious from these examples that the ripple magnitudes are independent of the filter truncation length. The ripple magnitudes are strictly the result of truncation with a rectangular window for which the frequency response is a *sinc* function with fixed sidelobe amplitude behavior. The rate of falloff of the stopband ripples is, however, a function of the truncation length as is the falloff of the passband ripples from the filter response edge. In addition, the transition band steepness is proportional to the filter length. In theory, the ideal rectangular passband is approached as the filter length becomes infinite. In practice, however, windowing always introduces a tradeoff between passband/stopband ripples and transition band steepness.

It is interesting to note that passband ripple effects can be removed from the resulting imagery by compensating the image amplitudes as a function of pixel position. Aliased energy, however, is in the image to stay and cannot be removed by a post-processing step.

In order to alleviate the potentially detrimental effects of aliased energy, it is necessary to reduce the sidelobe level of the truncated *sinc* filter response. A common technique is to weight the truncated *sinc* function with an appropriate window function whose frequency response has an acceptable sidelobe level.

The subject of weighting or window functions is addressed in some detail in Section 3.6.4. In this section we will consider only the so-called von Hann window, or more commonly the Hann[12] or raised cosine window [17]. The Hann window is used because it is simple and because it adequately illustrates the concepts at hand. It has the form

$$w(x) = (1/2)[1 + \cos(2\pi x)] \qquad |x| \leq 1/2 \, . \tag{3.46}$$

The window function is adjusted to match the length of the truncated *sinc* function as shown in Figure 3.22.

Applying the Hann window to our set of truncated *sinc* functions produces the series of frequency responses shown in Figures 3.23 and 3.24. It is obvious that the passband and stopband performance has been significantly improved over the unweighted cases. A more subtle effect occurs in the transition band. It is a fact of weighting that sidelobes are reduced at the expense of reducing the steepness of the filter rolloff. This effect can be seen by comparing Figures 3.20, 3.21, 3.23, and 3.24 with the tradeoff summarized in Figure 3.25. The effect of this broadened transition band on the imagery is explained as follows. Let us assume that the filter

[12] It is sometimes referred to in the literature as a Hanning window.

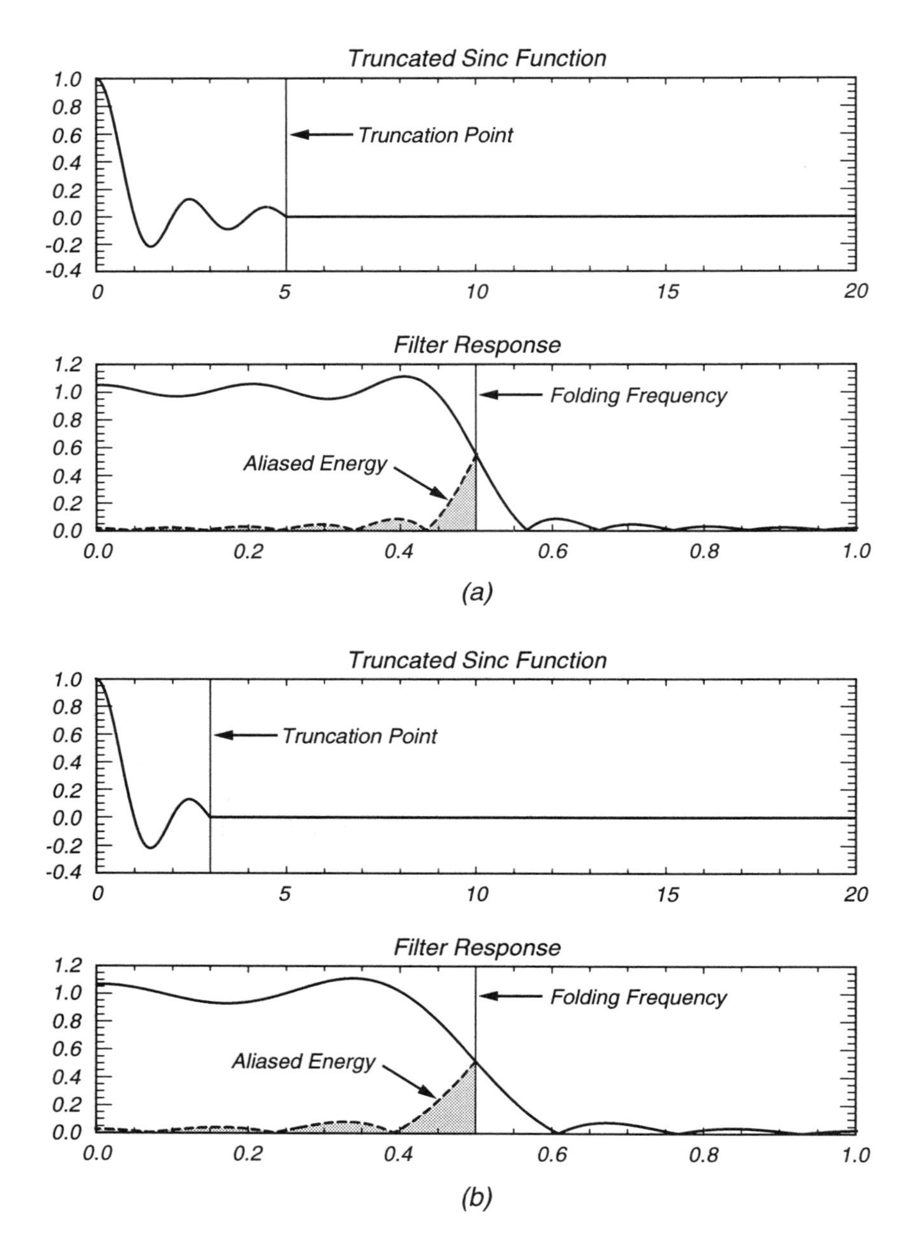

Figure 3.21 One-sided *sinc* function (a) truncated after five zero crossings (top) and the corresponding frequency response (bottom), (b) truncated after three zero crossings (top) and the corresponding frequency response (bottom).

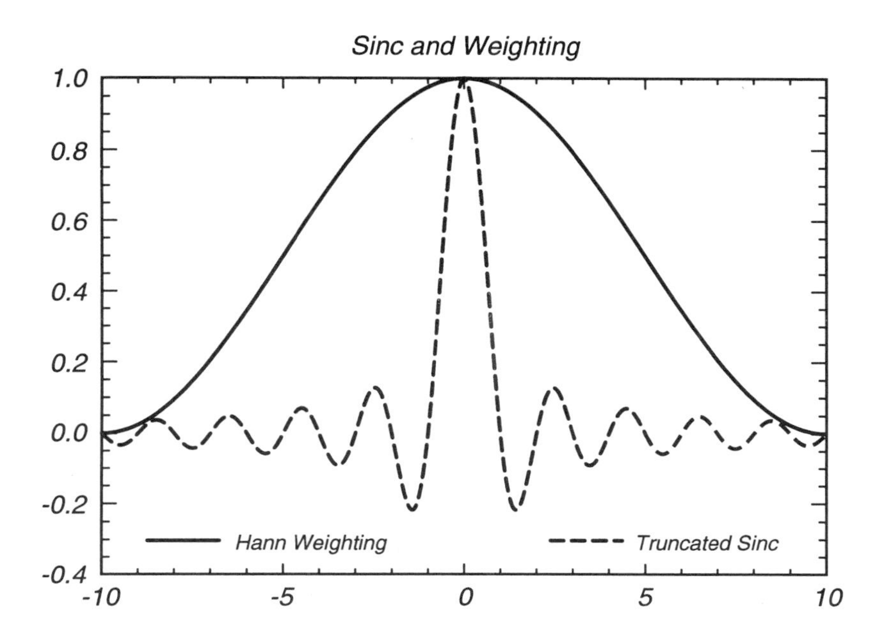

Figure 3.22 Alignment of the desired weighting with the *sinc* function. The product of the weighting with the *sinc* function produces the desired filter function used for interpolation.

response sidelobe level is adequately reduced. The broadened transition band now has the potential for folding back into the image edges more aliased energy than do the sidelobes. This is simply a result of a finite rate of frequency rolloff beyond the folding frequency.

In order to reduce the folded energy to some acceptable level at some specified point near (but not at) the image edge, it is necessary to reduce the bandwidth of the interpolating filter relative to the Nyquist sampling rate. One cannot expect the same performance of a filter all the way out to the bandwidth supported by the Nyquist sampling rate. Some tradeoff must be made. It is common practice to reduce the filter bandwidth by a few (e.g., 5 to 10) percent relative to Nyquist so that the highest frequency passed by the filter is slightly oversampled. In the resulting image, this means that the scene coverage will be slightly reduced relative to the number of samples spanning the scene. This slight oversampling helps prevent energy from just outside the patch of interest from folding in toward the patch edges.[13]

[13] It is also common practice to retain only a centered set of image samples (i.e., to discard a narrow band of image samples around the edge) so that the remaining image meets desired amplitude flatness and aliased energy constraints at the truncated image boundaries.

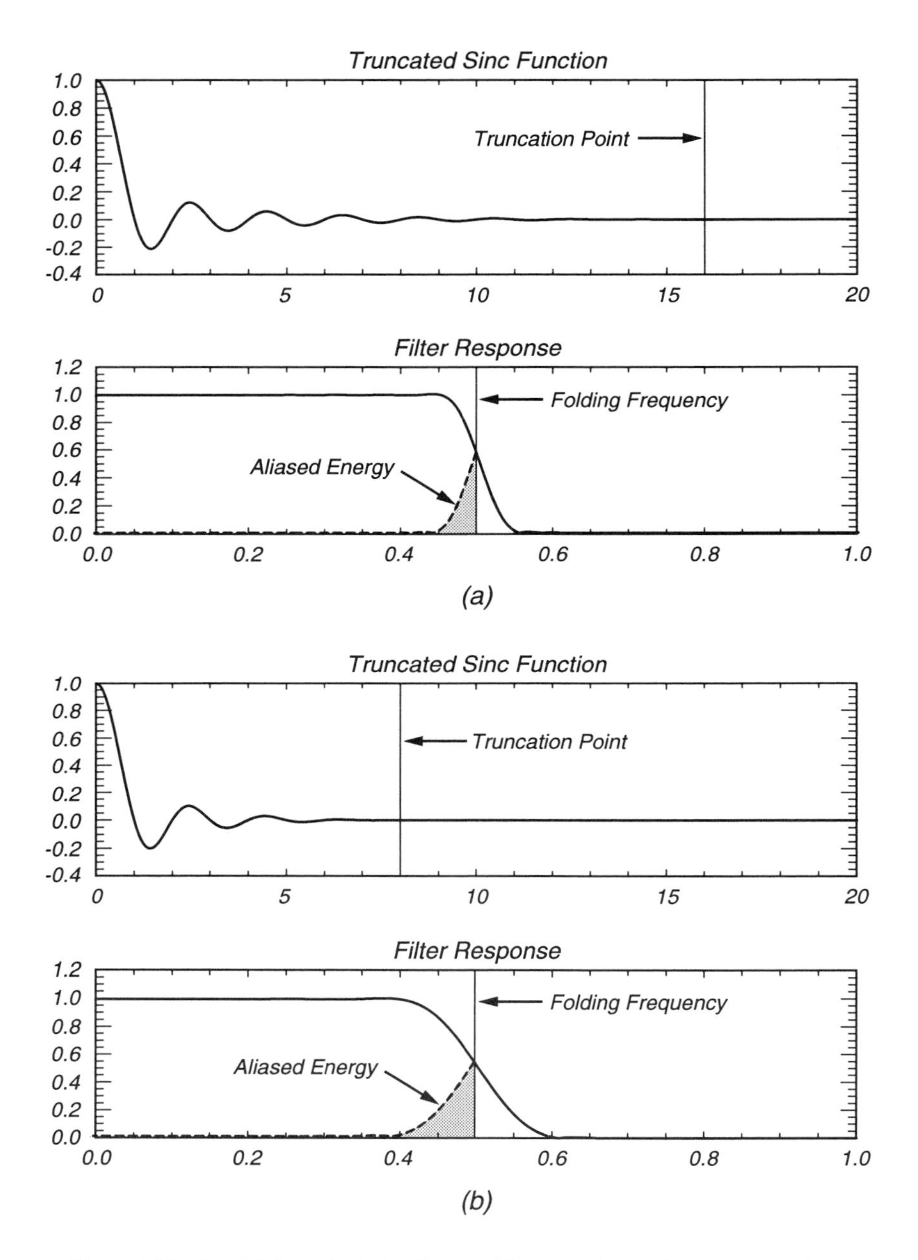

Figure 3.23 One-sided weighted *sinc* function (a) truncated after sixteen zero crossings (top) and the corresponding frequency response (bottom), (b) truncated after eight zero crossings (top) and the corresponding frequency response (bottom).

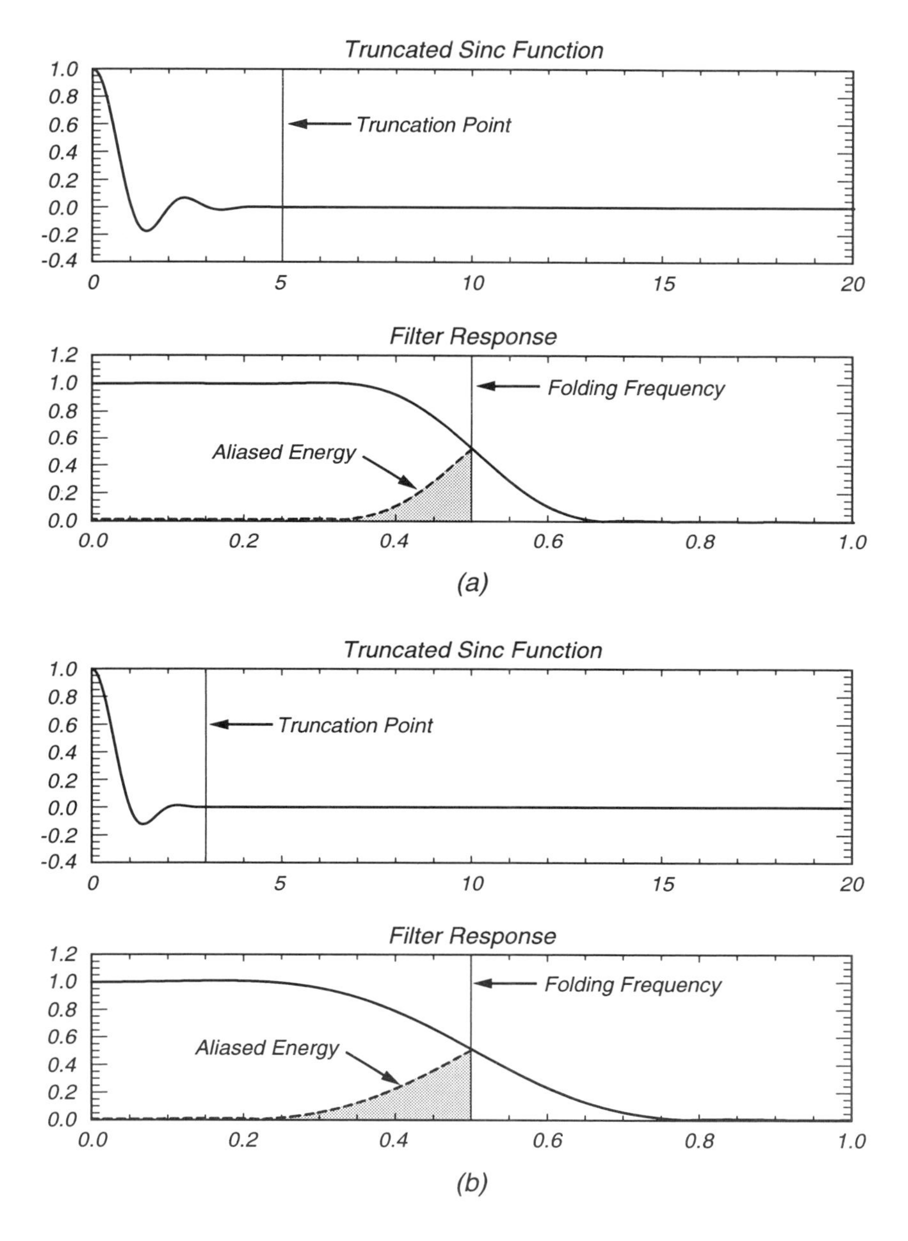

Figure 3.24 One-sided weighted *sinc* function (a) truncated after five zero crossings (top) and the corresponding frequency response (bottom), (b) truncated after three zero crossings (top) and the corresponding frequency response (bottom).

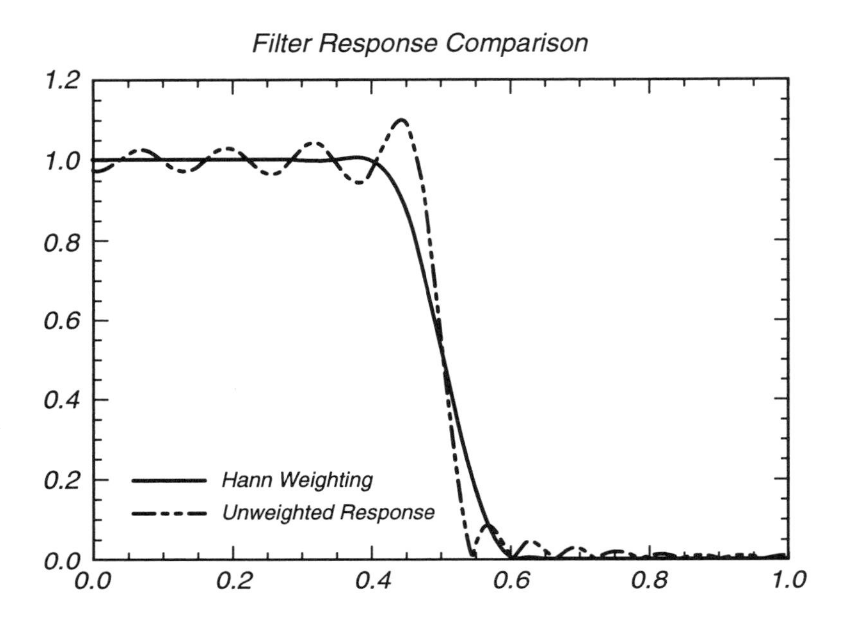

Figure 3.25 Comparison of weighted and unweighted filter response functions. Weighting reduces passband ripples and stopband sidelobes at the expense of broadening the transition band.

Reducing the filter bandwidth broadens the filter impulse response (shown as the dotted line in Figure 3.26). This broadened impulse response would be used for the interpolation. Figure 3.27 shows a comparison of filter functions and frequency responses that have undergone a bandwidth reduction.

Mathematically, the bandwidth reduction factor β is incorporated into the weighted filter function as

$$h'(t) = B[0.5 + 0.5\cos(\frac{2\pi t}{\beta\tau_L})]\frac{\sin \pi\frac{B}{\beta}t}{\pi\frac{B}{\beta}t} \tag{3.47}$$

for $|t| \leq \beta\tau_L/2$, where $\beta \geq 1.0$ and is typically chosen empirically to meet specific peak aliased amplitude limits at the desired image edges. We have found that setting $\beta = 1.04$, for an eight-zero-crossing filter, produces acceptable results in most practical applications. Keep in mind that no matter how the interpolation filters are designed, filter transition-band sharpness and stopband sidelobes determine how closely the folding frequency can be approached while keeping the aliased energy to some acceptable level.

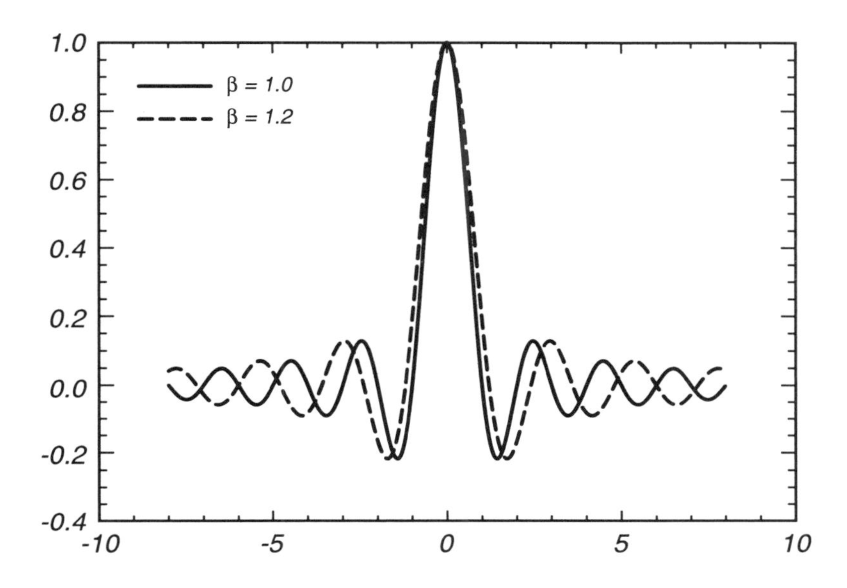

Figure 3.26 Effect of reducing the filter bandwidth on the filter impulse-response function. Reducing the filter bandwidth implies broadening the filter impulse response as shown by the dotted line. This broadened filter impulse-response function would be used for the interpolation.

3.5.3 Physical Mechanics of Interpolation and Resampling

There are many ways in which to *implement* the digital interpolation process. These depend on various practical constraints that do not concern us here. Figure 3.28 shows, on a conceptual level, the mechanics of interpolation and resampling. An interpolated sample at a desired position is obtained by centering the filter function on that sample location so the appropriate weighted summation of input samples can be computed (Equation 3.47). Once the desired interpolated value is obtained, the filter function is moved to the next desired location and the process is repeated. This process simultaneously accomplishes the tasks of bandwidth reduction and downsampling. The filter is designed to block passage of frequencies higher than those that can be supported by the output sample rate. This is accomplished by specifying the nominal zero-crossing spacings and by selecting an appropriate bandwidth reduction factor β.

These filter parameters can be modified on a range-line-by-range-line basis to accommodate slight changes in output sample spacings required in the keystone inter-

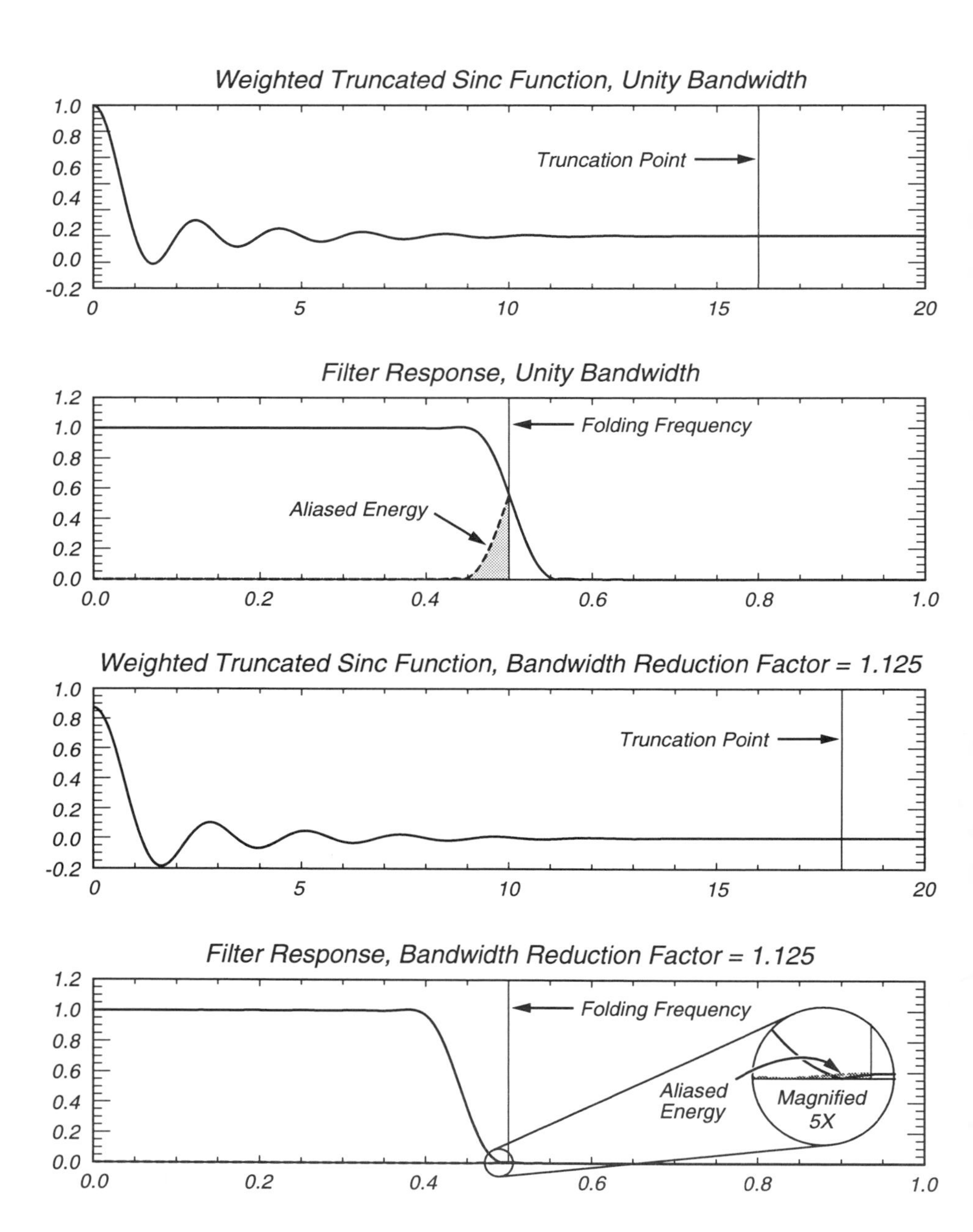

Figure 3.27 Comparison of filter impulse-response functions and frequency responses that have undergone a bandwidth reduction. The broadened transition band of the weighted filter frequency response can fold (alias) more energy into the passband than the unweighted filter. However, by slightly reducing the bandwidth relative to the Nyquist sampling rate, aliased energy from the transition band can be significantly reduced.

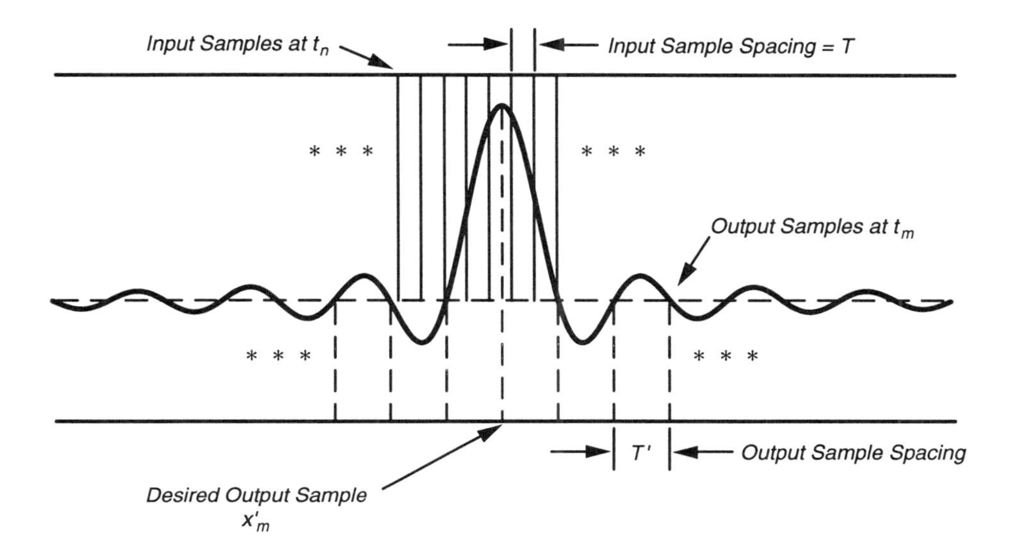

$$x'_m = \frac{T}{T'} \sum_n x_n \frac{\sin \pi (t_m - t_n)/T'}{\pi (t_m - t_n)/T'}$$

Figure 3.28 The mechanics of interpolation. To obtain an interpolated sample at a desired position, the filter function is centered on that sample location so the appropriate weighted summation of input samples can be computed. Once the desired interpolated value is obtained, the filter function is moved to the next desired location and the process repeated. This process simultaneously accomplishes the tasks of bandwidth reduction and downsampling. The filter is designed to block passage of frequencies higher than those that can be supported by the output sample rate. This is accomplished by specifying the nominal zero crossing spacings and by selecting an appropriate bandwidth reduction factor.

polation step. Similar line-by-line filter parameter changes can be made in the cross-range interpolation process to accommodate output sample spacings that change as a function of range. Any other required bandwidth or sample spacings can be easily accommodated during this process. This simple and robust filtering process allows us to accommodate interpolations required in more complex situations such as: out-of-plane correction, ground-plane projections, squinted imaging, bistatic spotlight SAR, and interferometry.

Filter lengths influence performance and computational load. Longer filters provide better performance at the expense of increased computation. In practice, polar-to-rectangular resampling often requires more processing time than the actual Fourier inversion using a 2-D FFT.

Summary of Critical Ideas from Section 3.5

- **The acquired polar samples of the spotlight SAR phase history must be resampled to a rectangular grid for FFT-based image formation.**
- **The generic 2-D interpolation is accomplished by a separable process known as range and cross-range interpolation.**
- **Interpolation can be thought of as a digital filtering operation.**
- **The interpolation filter frequency response directly influences image properties.**
- **Practical and easily implementable filters always require tradeoffs between passband flatness, transition-band sharpness, and stopband rejection.**
- **A Hann-weighted truncated *sinc* function (with, say, eight zero crossings) and a 1.04 bandwidth reduction factor provides a viable interpolation filter.**
- **The filter length influences performance and computational load. Longer filters provide better performance at the expense of increased computation.**
- **In practice, polar-to-rectangular resampling requires somewhat more processing time than the 2-D FFT.**

3.6 FOURIER TRANSFORMS, APERTURE WEIGHTING, AND OVERSAMPLING

We have stated several times that the central and essential step of forming an image from spotlight-mode SAR data is a two-dimensional Fourier transform. We showed in Chapter 2 that the demodulated radar return data consist of samples of the spatial Fourier transform of the radar reflectivity scene function. This implied that a two-dimensional transform would produce an image of the scene function. In Section 3.5, we motivated the polar-to-rectangular reformatting process by the desire to put the data in a form appropriate for processing with a Fast Fourier Transform (FFT) algorithm. In this section, we are prepared to address this critical step in greater detail. We will show how the properties of the discrete Fourier transform and the related issues of aperture weighting and oversampling affect the formed SAR image. Because the two-dimensional Fourier transform in Cartesian coordinates is separable into two orthogonal components, most of our discussion will be presented in the context of one-dimensional Fourier analysis. Furthermore, we will use the notation and language of time series analysis, because this will make many of the concepts easier to state and, we believe, easier to understand.

3.6.1 The Continuous Fourier Transform

The one-dimensional (continuous) Fourier transform is defined by the following pair of relationships:

$$H(f) = \int_{-\infty}^{\infty} h(t)e^{-j2\pi ft}dt \tag{3.48}$$

$$h(t) = \int_{-\infty}^{\infty} H(f)e^{j2\pi ft}df \tag{3.49}$$

where t typically represents time and f denotes frequency (in Hertz). Mathematically, of course, these two transform expressions are identical except for the signs of the exponential phase terms (which in fact are reversed in some definitions); nonetheless, it is customary and instructive to make a distinction between them. A common way of differentiating between them is to call the first one the *forward* transform of $h(t)$ and the second the *inverse* transform of $H(f)$. A more descriptive distinction is to refer to Equation 3.48 as the *analysis* transform and to refer to Equation 3.49 as the *synthesis* transform. We think of the analysis transform decomposing the function $h(t)$ into its frequency components, or spectrum, expressed by $H(f)$, while the synthesis transform reconstructs $h(t)$ from its spectrum $H(f)$. In the context of SAR image formation, we often think of synthesizing the image in space from its spatial frequency components. For the purposes of this section, however, it is more convenient to think of the image-formation transform in terms of analysis of the SAR data to determine their frequency content in the two image dimensions. This analysis interpretation will associate the spatial variables of the image with two orthogonal frequency variables.

3.6.2 The Discrete Fourier Transform

We have used continuous functions to describe the radar signals and the image-formation process in our discussion up to this point. In practice, of course, we have only a certain set of discrete samples that cover a limited time interval to represent one of these continuous signals. It follows that we should be talking about a discrete version of the Fourier transform. The Discrete Fourier Transform (DFT) is represented by

$$H_k = \sum_{n=0}^{N-1} h_n e^{-j2\pi nk/N} \,, \qquad k = 0, 1, ..., N-1 \tag{3.50}$$

$$h_n = \sum_{k=0}^{N-1} H_k e^{j2\pi nk/N} \,, \qquad n = 0, 1, ..., N-1 \,. \tag{3.51}$$

We will concentrate on Equation 3.50, which represents the forward or analysis transform analogous to Equation 3.48. The quantities n and k here denote discrete sample indices in the time and frequency domains, respectively, while N denotes the number of samples in both domains. The quantity h_n will be thought of a sample of the continuous function $h(t)$ at time $t = nT$, and H_k will be interpreted as an approximation of the continuous spectrum $H(f)$ at the frequency $f = f_k = k/(NT)$. The discrete Fourier transform (Equation 3.50) evaluates N frequency domain samples from N time domain samples. The corresponding inverse DFT (Equation 3.51) reconstructs the original time series from the frequency samples.

It is important to recognize some fundamental relationships between the discrete and the continuous versions of the Fourier transform. In SAR processing, as in most situations where the DFT is used, the intent of the sampled data is to represent a particular continuous signal. But as we have mentioned, the sampled data constitute an incomplete representation of the continuous signal in two important respects: (1) the sampled data measure the signal at discrete points in time, and (2) they span a limited extent of the total duration of the signal. The implications of both these aspects of sampled data were mentioned in Section 3.2, but we will restate them here in relation to the DFT. Consider the discrete sampling aspect. Recall that one of the implications of sampling is that the resulting data sequence can accurately represent a continuous signal only within a bandwidth equal to the sample rate $1/T$. The range of frequencies evaluated by the DFT coincides exactly with this bandwidth. Because there is no information in the sampled data sequence outside this range, it is not appropriate for the DFT to attempt to evaluate the spectrum there. As we stated in Section 3.2, the implication of this frequency limit in SAR is that the dimensions of the formed image will be limited to the reciprocal of the phase-history sample spacing in each dimension.

The fact that the sampled data sequence h_n is of finite extent was described in Section 3.2 as the cause of limited resolution in the image (or spectrum). To relate this effect to the DFT of the sampled data sequence, suppose $h(t)$ denotes a continuous signal of unlimited duration. A truncated version of this signal, of duration NT, can be represented by

$$h_s(t) = r_T(t)h(t) \tag{3.52}$$

where

$$r_T(t) = \begin{cases} 1\,, & |t| \leq NT/2 \\ 0\,, & |t| > NT/2 \end{cases} . \tag{3.53}$$

The spectrum (Fourier transform) of the truncated signal is

$$H_s(f) = \mathcal{S}_T(f) \otimes H(f) \tag{3.54}$$

where

$$\mathcal{S}_T(f) = \frac{\sin \pi f NT}{\pi f} = NT \operatorname{sinc} fNT \tag{3.55}$$

is the Fourier transform of $r_T(t)$. The DFT effectively samples the smoothed spectrum $H_s(f)$ instead of the original $H(f)$. This smoothing effect removes the fine structure in the original spectrum, thereby limiting the *resolution* in the observed spectrum. Quantitatively, the resolution is limited roughly to $1/NT$, the nominal width of the *sinc* function kernel, which in turn is equal to the reciprocal of the signal duration in time. In the context of SAR, the image resolution is limited to the reciprocal of the signal extent in the phase-history domain.

The relationship between a continuous signal and its discrete counterpart is shown in Figure 3.29. Part (a) shows a certain continuous infinite-duration signal (left side) and its bandlimited spectrum (right side). This spectrum consists of a small number of discrete components within its bandwidth B. In Part (b), we show the effect of truncating (windowing) the time-domain signal, which as we have said is to smooth the spectrum by convolving it with a *sinc* function. Note that this signal is not bandlimited, even though the continuous spectrum from which it was derived was bandlimited. This illustrates a fundamental principle that a signal cannot simultaneously be time limited and frequency limited. The act of truncating in time

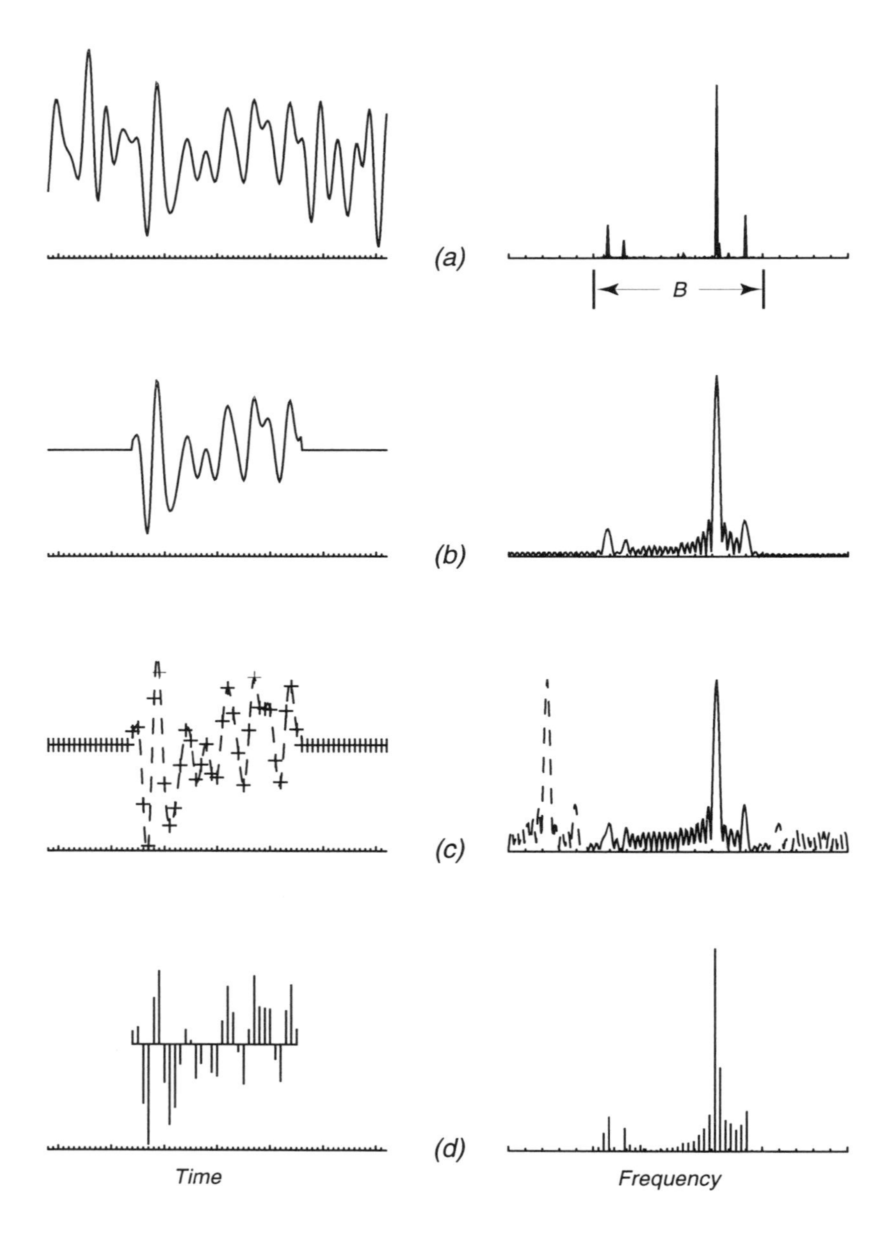

Figure 3.29 Discrete signal analysis showing various continuous and discrete-time signals (left) and their corresponding spectra (right). (a) A continuous infinite-duration signal and its bandlimited spectrum. (b) A truncated signal and its smoothed spectrum. (c) The effect of sampling the signal. (d) A finite discrete-time sequence and its DFT.

precludes bandlimiting in frequency. Figure 3.29(c) shows the result of sampling the truncated signal over its infinite extent (effectively multiplying by an endless series of Dirac delta functions). The frequency-domain effect is to replicate the spectrum from (b) at multiples of the sample rate. Because the spectrum being replicated is not bandlimited, the shifted replicas cause energy from outside the nominal signal bandwidth to be overlaid onto the baseband region $[-B/2, +B/2]$. This phenomenon, referred to as *aliasing*, is a source of distortion that must be dealt with in the sampled-data spectrum. Finally, Figure 3.29(d) illustrates the use of a finite number of time-domain samples (spanning only the extent of the truncated signal) and application of a DFT to compute the spectrum at an equal number of discrete points in frequency. This discrete frequency-domain representation is complete in the sense that the time-domain samples can be exactly reconstructed (synthesized) from them by applying an inverse DFT. The extent to which the discrete analysis produces a rendition of the continuous spectrum can be seen, however, to be somewhat limited. The techniques of oversampling and aperture weighting, described below, are attempts to mitigate some of these limitations.

3.6.3 Image (Spectral) Oversampling

This section describes a technique for producing an image that is oversampled, meaning simply that it is evaluated (sampled) more densely than the Nyquist criterion would dictate. The reason for oversampling an image is both to make it more pleasing visually and to facilitate certain image processing procedures that may be applied to the image. The following description is presented in the language of spectral oversampling, with the understanding, again, that the spectrum is analogous to the image.

Note that the frequency sampling interval ($\Delta f = 1/NT$) dictated by the DFT is identical to the peak-to-null width of the *sinc* function in Equation 3.55. This interval coincides exactly with the minimum sample rate (in the frequency domain) needed to represent the spectrum $H_s(f)$ that is being sampled. Thus, sampling at a rate lower than $1/NT$ in the frequency domain is inadequate to represent the signal; however, we are free to sample at a higher rate if we so choose. An oversampled spectrum is one sampled at a rate somewhat greater than the critical DFT rate of $1/NT$. It should be emphasized that oversampling in the frequency domain does not provide additional information about the detailed structure of the original spectrum $H(f)$, as this information has already been limited by the act of truncating the function $h(t)$ in time. What oversampling does accomplish is simply to *interpolate* values of $H_s(f)$ at a different set of discrete frequencies.

Oversampling is accomplished in practice by *zero padding* the data in the time (phase-history) domain, extending the sequence length to a value $N' > N$. A DFT of length N' is then computed on this extended sequence with the result

$$H'_k = \sum_{n=0}^{N'-1} h_n e^{-j2\pi nk/N'} = \sum_{n=0}^{N-1} h_n e^{-j2\pi nk/N'}, \quad k = 0, 1, \ldots, N' - 1 . \quad (3.56)$$

The new frequency-domain samples are at $f'_k = k/N'T$, $(k = 0, 1, \ldots, N' - 1)$. An example of spectral oversampling is illustrated in Figure 3.30.

3.6.4 Aperture Weighting

In Equation 3.54, we showed that the effect of truncation of the time series was to convolve the original spectrum $H(f)$ with a *sinc* function. In this section, we will see how *aperture weighting* can alter this effect to produce a somewhat improved estimate of $H(f)$. Aperture weighting, as its name implies, involves multiplying the time-domain sampled data sequence by an appropriate function. This effectively replaces the *sinc* function convolution kernel in Equation 3.54 with the Fourier transform of the weighting function. To the extent that we can find weighting functions whose Fourier transform have desirable characteristics, we can improve our spectral estimate by this technique.

Figure 3.31 illustrates the effects of aperture weighing. It shows a hypothetical high-resolution spectrum (Part a) followed by the result of convolving this spectrum with a *sinc* function (Part b). The result in Figure 3.31(b) is representative of an unweighted spectral analysis of a finite-length sequence of data. Note that in addition to the smoothing effect we mentioned earlier, the convolution with the *sinc* function has all but totally obscured the low-amplitude features in the original spectrum under the "sidelobes" of the dominant peaks. Figure 3.31(c) shows the spectrum obtained from the same set of data when weighing has been applied. The reduction in sidelobe level in this spectrum has allowed the low-amplitude components in the original signal to be clearly visible. The beneficial effect of aperture weighting has not been achieved without a price, however. This is illustrated in Figure 3.31(c) by the fact that two closely spaced frequency components that are distinct in the unweighted spectrum are fused together in the weighted spectrum. The inevitable loss of resolution accompanying aperture weighting is generally considered to be an acceptable cost for the increased dynamic range that is achieved by reducing the sidelobes. A typical SAR imaging situation can involve targets that span several decades of dynamic range. Therefore, it is essential that the interfering effect of sidelobes associated with large targets be minimized even if resolution must be sacrificed in the process.

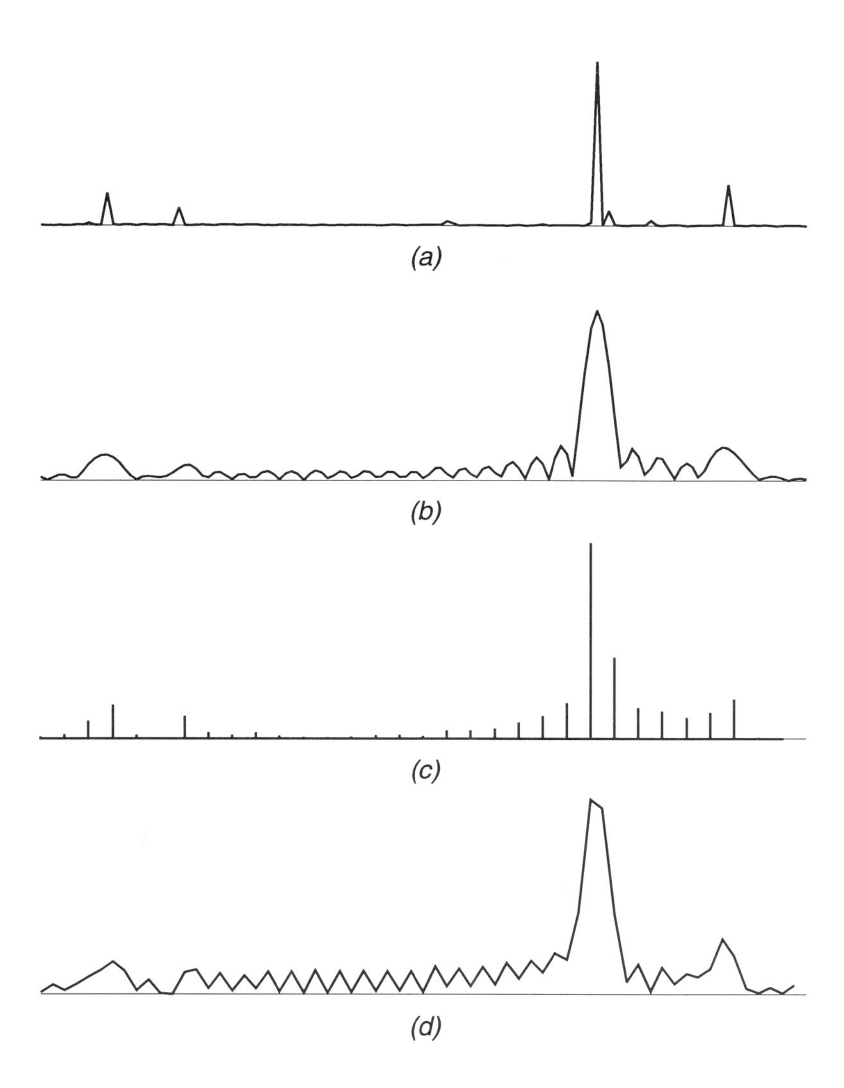

Figure 3.30 The effect of spectral oversampling. (a) A continuous hypothetical spectrum (image). (b) Smoothed spectrum by virtue of signal truncation. (c) Very nearly Nyquist-sampled version of the smoothed spectrum in (b). (d) Oversampled spectrum more nearly approximates smoothed spectrum in (b). In practice, oversampling a spectrum produces a more visually pleasing result.

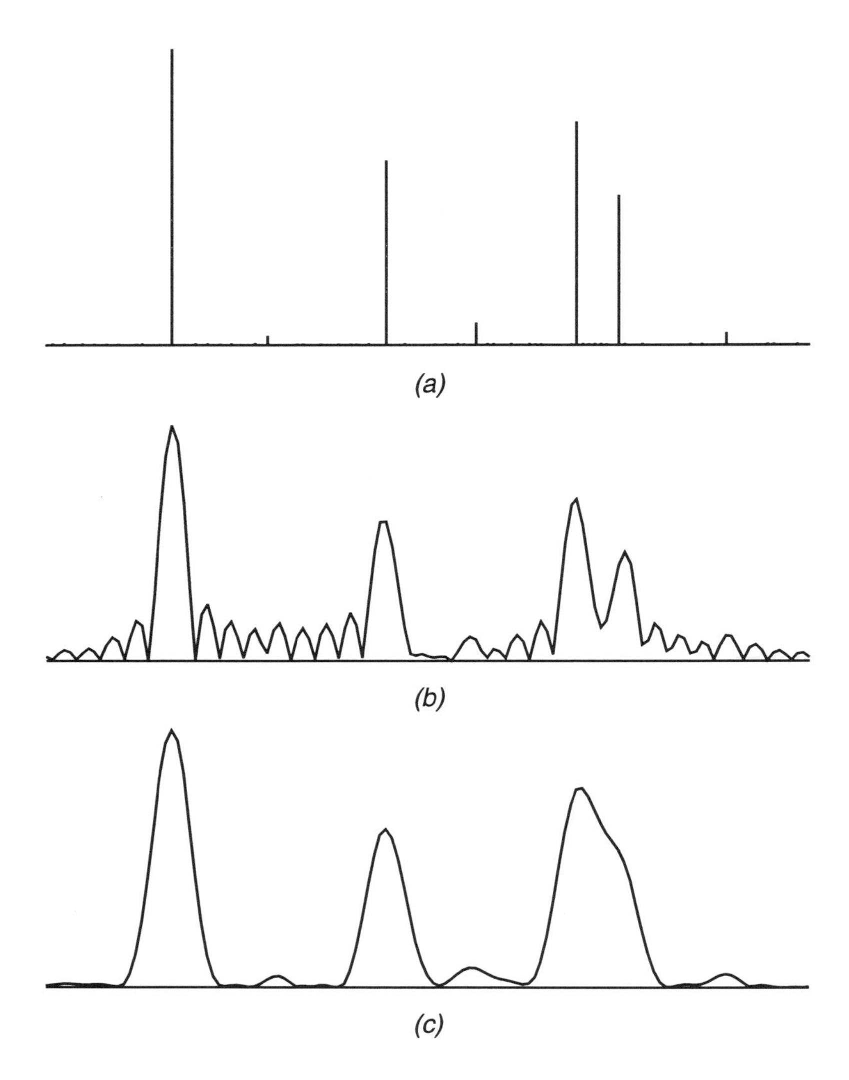

Figure 3.31 Aperture weighting. (a) A hypothetical high-resolution spectrum (image). (b) The result of convolving a *sinc* function with the same spectrum. (c) Aperture weighting leading to the reduction in sidelobe level in this spectrum has allowed the low-amplitude components in the original signal to be clearly visible. Two closely spaced frequency components that are distinct in the unweighted spectrum are fused together in the weighted spectrum. The inevitable loss of resolution accompanying aperture weighting is generally considered to be an acceptable cost for the increased dynamic range that is achieved by reducing the sidelobes.

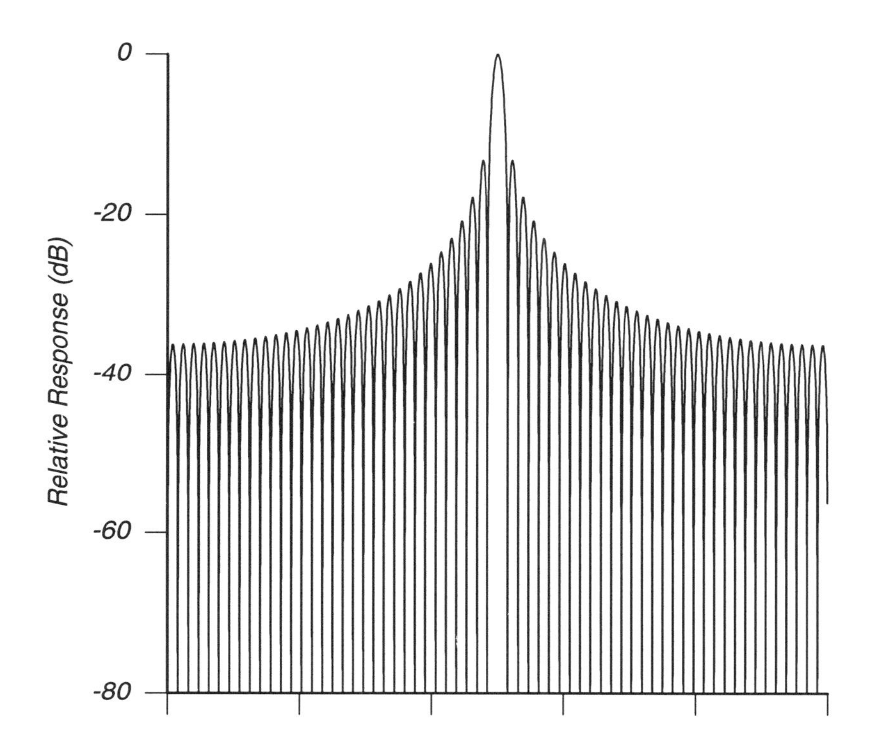

Figure 3.32 The *sinc* function response for an unweighted aperture.

A desirable weighting function is, in general, one whose frequency response has lower sidelobes and a wider mainlobe than the *sinc* function response shown in Figure 3.32. A variety of candidate weighting functions, or windows, have been proposed for suppressing sidelobes. They range from purely heuristic forms to solutions of complicated optimization criteria. A few of the most commonly used signal processing window functions are shown in Figure 3.33. In Figures 3.34-3.35, we show the frequency response functions for these windows, along with the idealized Dolph-Chebyshev response. The Taylor window actually encompasses a family of functions that approximate the Dolph-Chebyshev response to varying degrees by trading off sidelobe level against mainlobe width. (In Figure 3.35(b) we have specified a sidelobe attenuation of 43 dB to match the maximum sidelobe of the Hamming response.) A comparison of mainlobe and near sidelobe characteristics for these various responses is plotted in Figure 3.36. The Taylor window has been used extensively in the SAR community and is described in some detail in the following

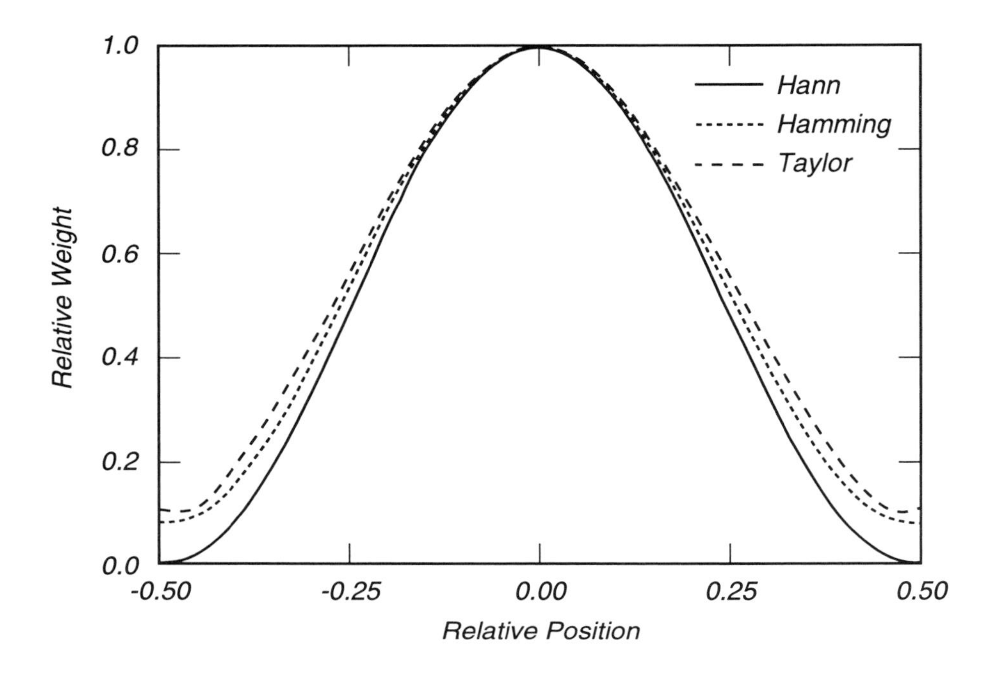

Figure 3.33 Some candidate aperture-weighting (window) functions

paragraphs. It should be noted, however, that there are countless other options that may be considered to meet specific requirements. John Adams [20] presents a thorough overview of alternative optimum forms.

The Dolph-Chebyshev function is an optimum frequency response in the sense that it has a minimum mainlobe width for a specified peak sidelobe level. (Here again, lower sidelobes imply a wider mainlobe.) The response function of the Taylor window approximates the Dolph-Chebyshev response and is expressed in terms of a finite series of the form [21]

$$w(\xi) = 1 + \sum_{m=1}^{\bar{n}-1} F_m \cos(2\pi m \xi) \ , \quad |\xi| \leq 1/2 \ . \tag{3.57}$$

A discrete set of weighting coefficients is obtained by evaluating Equation 3.57 at N equally spaced points spanning the interval $[-1/2, 1/2]$:

$$w_n = w(\xi_n) \tag{3.58}$$

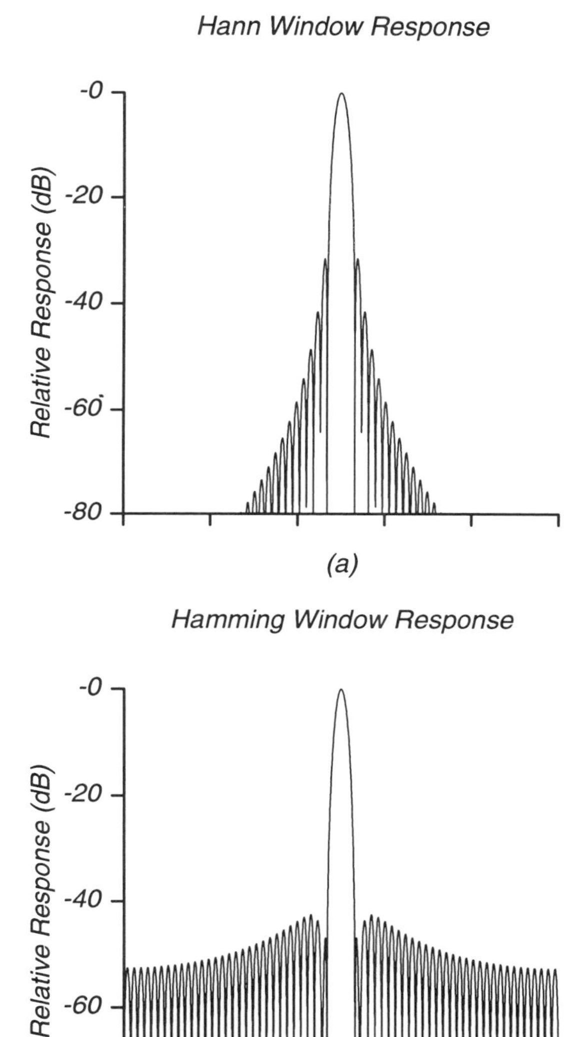

Figure 3.34 Impulse response functions for some commonly used windows. (a) Hann window response. (b) Hamming window response.

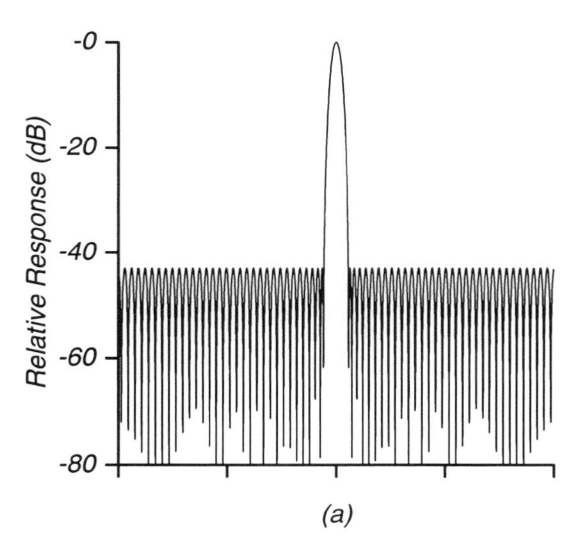

(a)

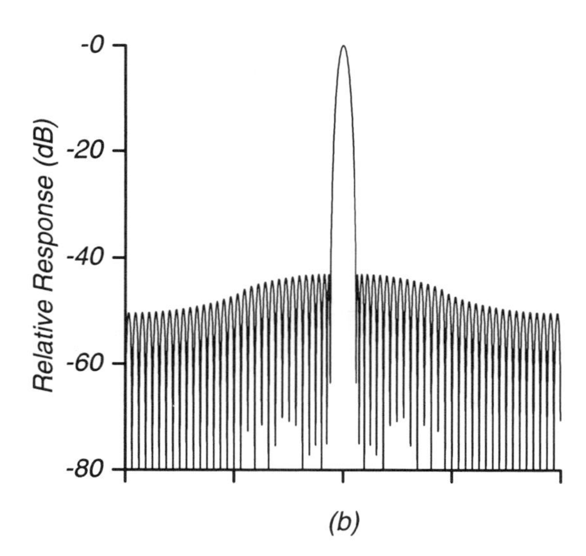

(b)

Figure 3.35 Impulse response functions for some commonly used windows. (a) 43-dB Dolph-Chebyshev window response. (b) 43-dB Taylor window response.

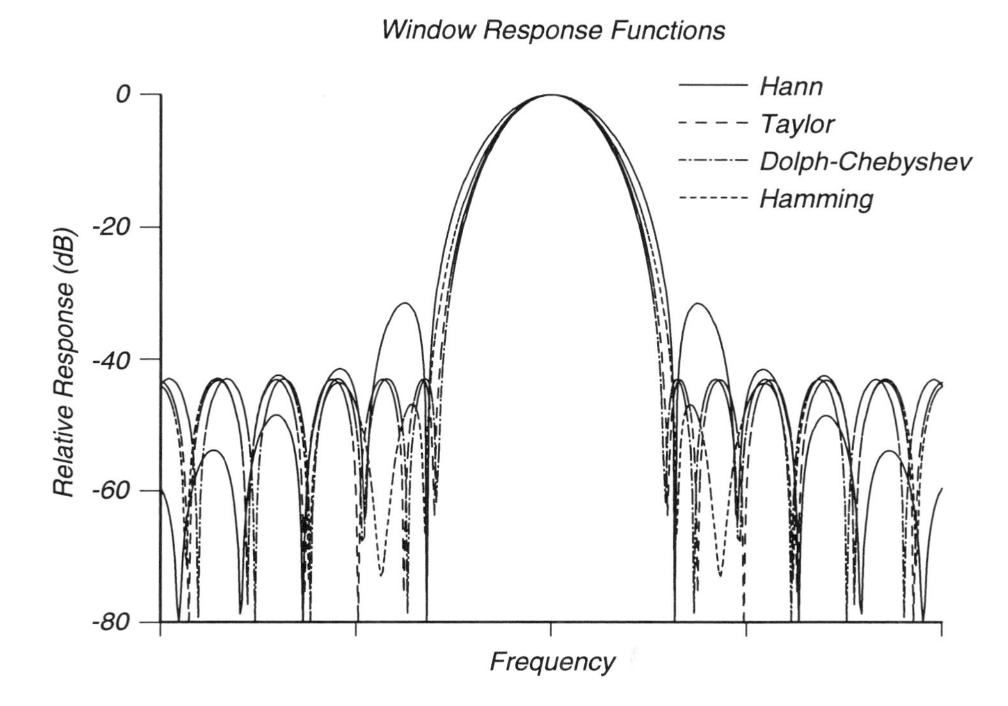

Figure 3.36 A comparison of the frequency response of various aperture-weighting (window) functions.

where

$$\xi_n = \frac{n + 1/2}{N} - \frac{1}{2} \qquad n = 0, 1, ..., N-1 \,. \tag{3.59}$$

The coefficients of the series in Equation 3.57 are given by

$$F_m = \frac{(-1)^{m+1} \prod_{n=1}^{\bar{n}-1} \left[1 - \frac{m^2/\sigma_p^2}{A^2 + (n-1/2)^2} \right]}{\prod_{\substack{n=1 \\ n \neq m}}^{\bar{n}-1} \left[1 - \frac{m^2}{n^2} \right]} \tag{3.60}$$

where the parameter A is determined by the specified sidelobe attenuation S_L (in decibels) according to

$$A = \frac{1}{\pi} \cosh^{-1}(10^{S_L/20}) \tag{3.61}$$

and σ_p is the ratio of mainlobe width at 3 dB (half amplitude) to the width of the ideal Dolph-Chebyshev response at 3 dB. This mainlobe broadening factor is related to the attenuation factor A and the number of terms $\bar{n}$ in the sum by the expression

$$\sigma_p = \frac{\bar{n}}{\sqrt{A^2 + (\bar{n} - \frac{1}{2})^2}} > 1 . \tag{3.62}$$

The maximum amount of mainlobe broadening relative to Dolph-Chebyshev is

$$\sigma_{pm} = \frac{\sqrt{4A^2 + 1}}{2A} \tag{3.63}$$

that occurs when

$$\bar{n} = \bar{n}_0 \simeq 2A^2 + 1/2 . \tag{3.64}$$

As $\bar{n}$ is made increasingly large we have

$$\lim_{\bar{n} \to \infty} \sigma_p = 1 . \tag{3.65}$$

The Taylor window approaches the ideal Dolph-Chebyshev characteristics of minimum mainlobe width and uniform sidelobe levels in the limit as $\bar{n}$ approaches infinity. For $\bar{n}$ finite (but larger than $\bar{n}_0$ above), it represents a good approximation in the vicinity of the mainlobe. In this case, the sidelobes are approximately uniform at the specified level out to a certain frequency, beyond which they decay inversely with frequency. In practice, it is not necessarily desirable to choose a very large $\bar{n}$ in an effort to approximate Dolph-Chebyshev. Because the approximation for finite $\bar{n}$ already has lower sidelobes than the "ideal," it is prudent to choose $\bar{n}$ only slightly larger than $\bar{n}_0$ (perhaps twice as large). This results in a window with lower integrated sidelobe energy than a Dolph-Chebyshev function at the expense of a small increase in mainlobe width. (The tradeoffs between peak and integrated sidelobe levels are discussed in [20].) As an example of the mainlobe broadening, consider the 43-dB Taylor window shown in Figure 3.35(b). It has $\bar{n} = 15$ and $A = 1.8$, which imply $\sigma_p = 1.027$. Thus, this response is only 2.7% wider in the mainlobe than the Dolph-Chebyshev response.

Summary of Critical Ideas from Section 3.6

- **Because the two-dimensional Fourier transform in Cartesian coordinates is separable into two orthogonal components, the description and analysis of this image-formation step can be understood in the context of one-dimensional signal analysis.**

- **Forward and inverse Fourier transforms represent** *analysis* **and** *synthesis* **transforms, respectively.**

- **The image-formation transform can be thought of in terms of analysis of the SAR data to determine the frequency content of the SAR data in two dimensions.**

- **The discrete Fourier transform (DFT) and its efficient implementation as a fast Fourier transform (FFT) constitute the means by which discrete Fourier estimates are obtained from a set of discrete signal values.**

- **The DFT (FFT) produces samples of a smoothed spectrum instead of samples of the underlying continuous spectrum.**

- **The extent to which the discrete analysis produces a rendition of the continuous spectrum is limited. The techniques of oversampling and aperture weighting are attempts to mitigate some of these limitations.**

- **Oversampling a spectrum (image) produces a visually more pleasing result and is accomplished by zero padding the data prior to Fourier transformation.**

- **Aperture weighting is a technique used to suppress sidelobe levels and thereby reduce the interference of large amplitude signals with neighboring low-level components.**

- **Aperture weighting involves multiplying the sampled data by a function whose Fourier transform has certain desirable characteristics.**

- **The Taylor window is an approximation of the Dolph-Chebyshev function, which is optimum in the sense that it has a minimum mainlobe width for a specified peak sidelobe level. Taylor windows are commonly used in SAR image formation.**

3.7 SLANT-PLANE IMAGE FORMATION AND IMAGE PROPERTIES

3.7.1 Introduction

We will now describe a simple spotlight SAR collection geometry and will examine properties of the resulting phase history and subsequent imagery. Let us consider an ideal collection with the SAR platform undergoing straight and level flight. The SAR transmits, receives, and demodulates the returns as it travels the straight and level flight path. Each demodulated signal represents a line in Fourier space that evaluates a segment of the three-dimensional Fourier transform of the complex terrain reflectivity. A series of such Fourier lines acquired over the collection time for the synthetic aperture forms the two-dimensional phase history over a perfectly planar surface. Samples along each Fourier line form a polar record. The totality of records define samples of the 2-D phase history on a polar raster in the *slant plane*. The convex hull of the polar samples form the polar annulus that defines the support of the phase-history-domain aperture.

3.7.2 Slant Plane Description

For our ideal straight and level collection, the notion of *slant plane* is unambiguous because all polar records (and samples) lie exactly in the same plane. Polar-to-rectangular resampling followed by aperture weighting for sidelobe control, and 2-D Fourier transformation yields a *slant-plane image*.

Recall that tomographic principles indicate that such a slant-plane image is an orthogonal projection of the three-dimensional scene into the slant plane. This image is also analogous to an optical image obtained by viewing the scene normal to the slant plane (i.e., the same orthogonal projection applies).

Straight-line SAR flight paths always lead to perfectly planar Fourier acquisitions. The notion of a slant plane and subsequent slant-plane images are a natural consequence of this operational mode and allow a simplified look at the resulting phase history and image properties.

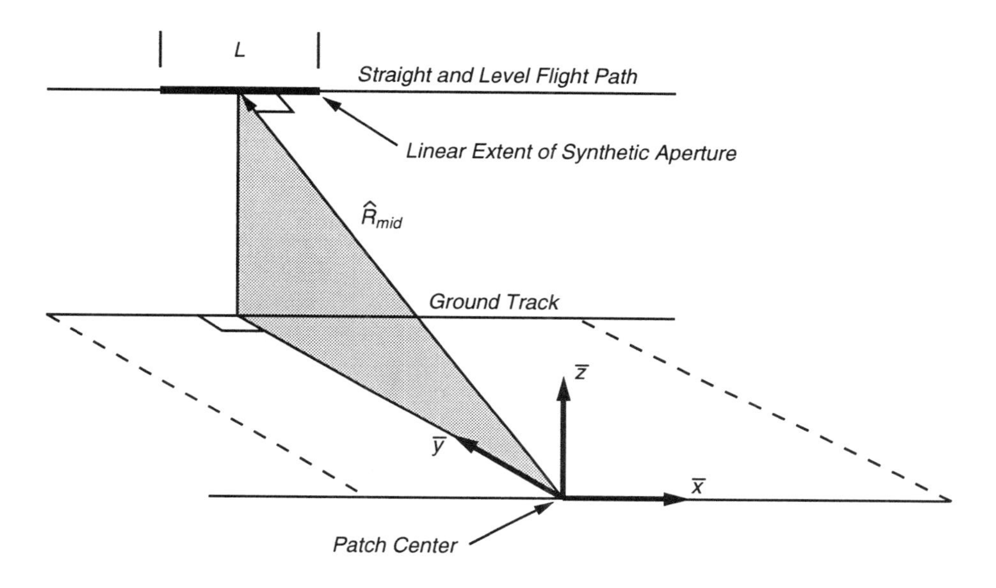

Figure 3.37 An ideal *broadside* collection with the SAR platform undergoing straight and level flight.

3.7.3 Straight and Level Broadside Acquisition

Let us consider an ideal *broadside* collection with the SAR platform undergoing straight and level flight as shown in Figure 3.37.[14] We will now examine some of the properties of the slant-plane phase history and resulting imagery with simulated but realistic signals that capture all the features of actual SAR signals without the uncertainties of a real collection. We will do this by generating demodulated SAR signals from an arrangement of point targets using our *synthetic target generator.*[15]

We begin by selecting a set of target generator parameters for a broadside spotlight collection that provides approximately 0.375 meters resolution in both range and cross range. The patch diameter, A/D sample rate, and PRF are selected to provide a nominal dataset size. The parameters chosen for our particular test case are summarized below.

[14] See Appendix C for further quantitative development of imaging geometry issues and definitions of relevant terms.

[15] See Appendix D for a thorough treatment of the synthetic target generator philosophy and the significance of the relevant parameters associated with dataset generation.

- Radar platform velocity, $V_p = 500$ m/s.
- Radar platform height above ground plane, $h_p = 15000$ m.
- Depression (or grazing) angle at broadside, $\psi_b = 30$ deg.
- Initial slant-plane squint angle, $\theta_{s_i} = -1.17$ deg.
- Final slant-plane squint angle, $\theta_{s_f} = +1.17$ deg.
- Radar center frequency, $f_0 = 1.0 \times 10^{10}$ Hz
- Transmitted bandwidth, $B = 4.015 \times 10^8$ Hz
- Linear FM chirp rate, $\dot{f} = 1.0 \times 10^{12}$ Hz/s
- Pulse-repetition frequency, $PRF = 163.5$ Hz
- Slant-plane patch diameter, $D = 150$ m
- A/D sample rate, $f_s = 1.0 \times 10^6$ Hz

Some computed parameters:

- Number of samples per pulse = 400
- Number of pulses = 400
- Maximum valid pulse-time interval, $T_{eff} = 4.005 \times 10^{-4}$ sec.
- Effective transmitted bandwidth,[16] $B_{eff} = 4.005 \times 10^8$ Hz.
- Effective system Q, $Q_{eff} = f_0/B_{eff} = 24.9688$

It is noted, in this example, that the target generator produces 400 polar records; each record contains 400 complex samples. The totality of records spans a polar annulus of ± 1.17 degrees in the slant plane. Some of the relevant features of this acquisition are summarized in Figure 3.38. In particular, note that the Fourier data are offset from the polar origin by an amount proportional to the effective system Q. Knowledge of the polar sample locations from knowledge of their position along the polar radius and the instantaneous angle that a given polar record makes with respect to the established coordinate system provides all the information needed to begin the polar-to-rectangular resampling process.

[16] See Equation D.6.

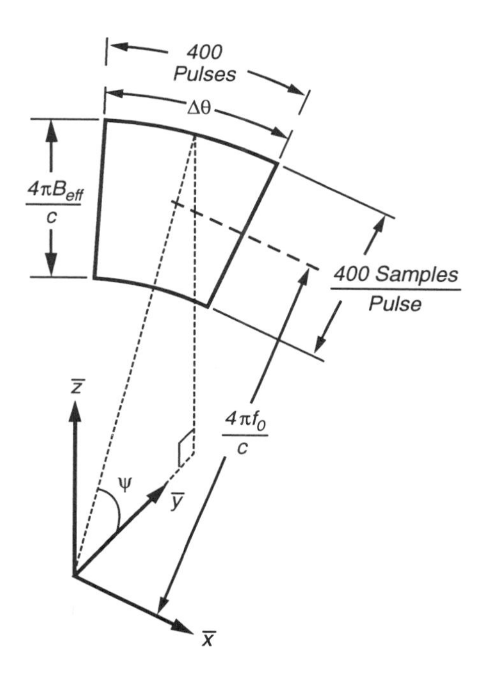

Figure 3.38 Some relevant features of a particular broadside acquisition. The synthetic target generator produced a ± 1.17 degree polar annulus encompassing 400 pulses; each pulse contained 400 samples. In terms of samples, the annulus center is offset from the origin by 400 x $Q_{eff} = 9987.52$ samples. The established ground-plane unit vectors are designated by $\bar{x}$, $\bar{y}$, and $\bar{z}$.

The system Q factors into the resolution equations in the following way. We know that the range and cross-range resolutions are given by $\rho_{y'} = c/2B_{eff}$ and $\rho_{x'} = \lambda/2\Delta\theta$, respectively. Setting the range and cross-range resolutions equal (desirable in most practical situations), and manipulating the above equations yields

$$\Delta\theta = \frac{1}{Q_{eff}} . \tag{3.66}$$

With $Q_{eff} = 24.9688$, we find that $\Delta\theta$ spans ± 1.147 degrees. The synthetic target data, in this example, were gathered over a slightly larger angular span (± 1.17 degrees) to allow some flexibility in forming imagery (from sampled data) with equal resolutions and equal (range and cross-range) scale factors (i.e., a square aspect ratio). Resolution and scale-factor issues will be covered in additional detail in Section 3.10.

3.7.4 Inscribed vs. Circumscribed Rectangular Grid

The polar-to-rectangular resampling process computes samples on a rectangular grid from knowledge of the polar sample locations with respect to the desired locations of output samples. It is necessary to place a rectangular grid over the acquired polar data so that interpolation can follow. There are many ways for the rectangular grid to be defined given the polar geometry just described. Figure 3.39 illustrates some possibilities. The rectangle can be oriented to *circumscribe* the polar region of support, as shown in Figure 3.39(a and c); it can be *inscribed* entirely within the region of support, as in Figure 3.39(b and d); or it can be defined as a compromise between the two extremes, like that shown in Figure 3.39(e)

Inscribing a rectangle naturally sacrifices some of the data, with subsequent loss of resolution, but eliminates the odd-shaped region of support that ultimately affects the sidelobe structure in the final image IPR if a circumscribed rectangle were used. In practice, a rectangle is usually inscribed within the data for just such reasons. However, in the examples that follow (Section 3.7.5), we chose to circumscribe a rectangle so that the polar region of support is visible when viewing the phase-history data. Keeping the region of support visible helps to convey the nature of the resampling process and graphically illustrates some of the not-so-obvious distortions that arise from squinted imaging and ground-plane projections to be discussed in Section 3.9.

3.7.5 Synthetic Target Phase Histories and Imagery

We now wish to place ideal synthetic point targets in a physical region of space that will help to illustrate phase-history properties and image projection and height-dependent layover effects (see Figure D.1 in Appendix D, for imaging geometry). The synthetic-target arrangement and the broadside imaging geometry are shown in Figure 3.40.

We now collect the demodulated complex returns through our synthetic target generator, circumscribe a rectangular array of 400-by-400 samples, perform the polar-to-rectangular resampling (interpolation), and display the real part of the complex phase history in Figure 3.41.

Because our point targets are symmetrically arranged, it is no surprise that the superposition of the various sinusoids also produces a symmetrical pattern in the phase history. It is also easy to see the polar annulus region of support.

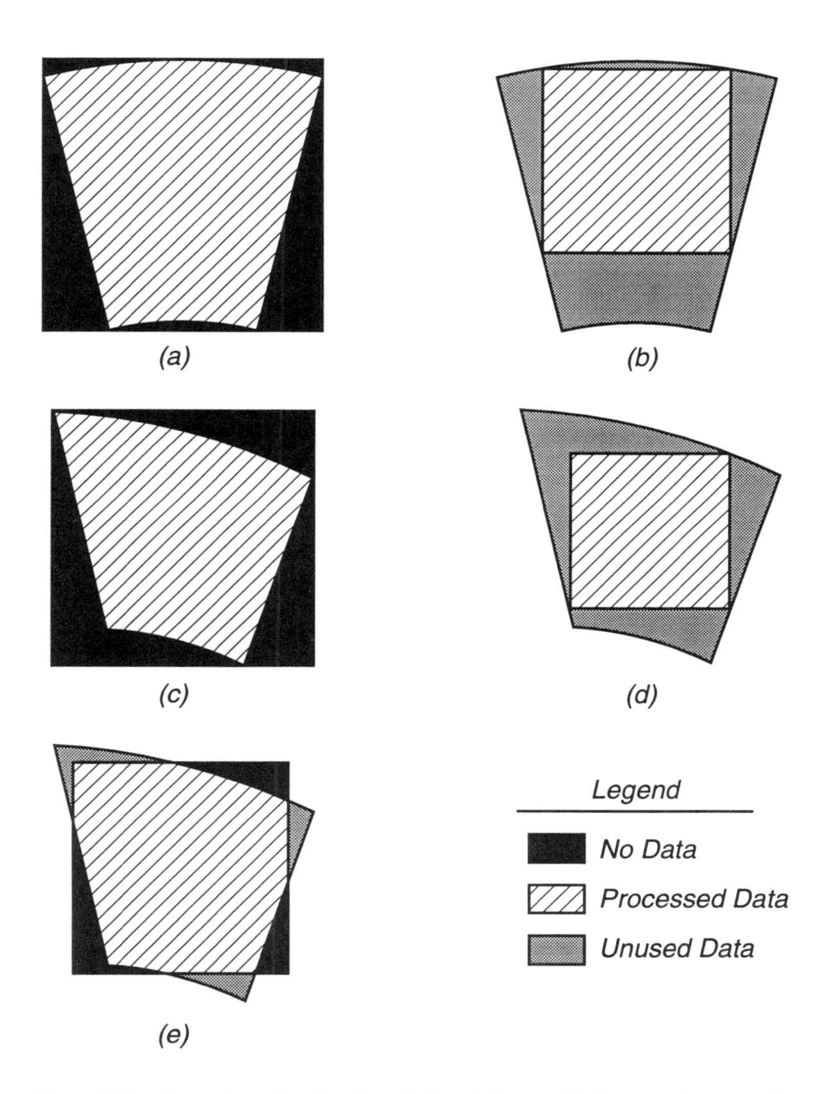

Figure 3.39 Comparison of various inscribed and circumscribed rectangular arrays (exaggerated geometry). (a) Rectangular array circumscribing the Fourier annulus. All data are used, but the rectangular array contains regions of missing data which influence the impulse-response function. (b) Rectangular array is inscribed within the Fourier annulus, which leaves some of the data unused (sacrifices resolution). (c) and (d) Possible circumscribed and inscribed arrays when the Fourier data are distorted by squinted acquisitions and phase-history projections into the ground plane. (e) A possible hybrid approach that makes a compromise between missing and unused data.

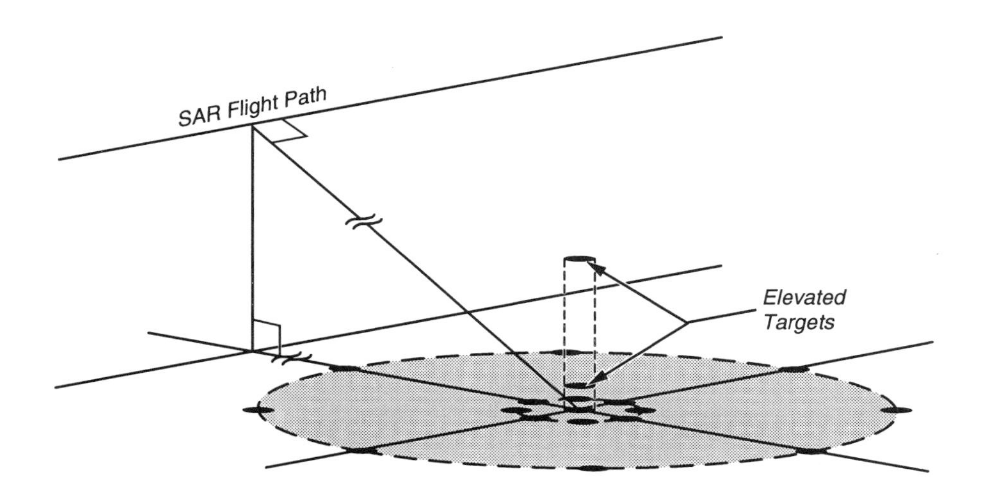

Figure 3.40 A nineteen target geometry consisting of one target of unity reflectance amplitude at the imaged patch center for geometric reference, eight unity-amplitude targets placed symmetrically around a circle of 10 meters radius at 45 degree increments, an additional eight double-amplitude targets placed on a circle of 50 meters radius at 45 degree increments, and two double-amplitude targets elevated above the ground plane directly over the patch center by 10 meters and 50 meters, respectively.

The 400-by-400 rectangular array is processed by weighting in both dimensions with a 35 dB Taylor window for sidelobe control, zero padding to 512-by-512 samples, and Fourier transforming via a 2-D FFT. The magnitude of the resultant image is linearly scaled for display and is shown in Figure 3.42. Keep in mind that in all phase-history and image examples the vertical y axis always refers to the range dimension with positive y pointing toward the SAR. The x axis refers to cross range. The origin of the x-y system is always at the image center or the phase-history center, depending on the applicable discussion.

The symmetric array of point targets in Figure 3.42 shows two predictable projection effects. The first effect is the distortion of the circular array to an elliptical shape. This is a direct result of projecting the circular array of targets on the ground into the slant plane in a direction normal to the slant plane. In this case, the range dimension is compressed by the cosine of the nominal grazing or depression angle of the acquisition, as expected.

The second projection effect noted in Figure 3.42 involves the resulting positions of the two elevated targets. The elevated targets also project orthogonally into the slant plane and appear in the image in exactly the same location as certain ground-plane

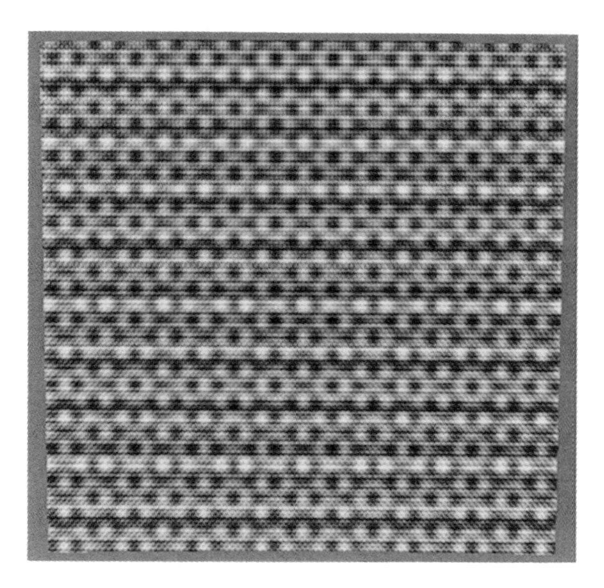

Figure 3.41 Real part of the nineteen-target phase history after polar reformatting. Note that the superposition of the various sinusoids produces a somewhat symmetrical pattern in the phase history. The polar-annulus region of support is also evident.

point targets would. This is the projection phenomena of "layover," where elevated targets project into the slant plane and appear to lay over in a direction normal to the nominal ground track of the SAR flight path.[17] In our broadside imaging example, the elevated targets lay over in the range dimension in the direction of "near range" (i.e., toward the SAR). In the early years of SAR, virtually all imaging was conducted in a broadside mode. Consequently, the layover phenomena was termed "range layover" because elevated targets laid over purely in the range dimension.

This well-entrenched range layover notion led to some conceptual confusion when more advanced SAR systems began collecting imagery in non-broadside or squinted modes. In these non-broadside acquisitions, elevated targets do not lay over strictly in the range dimension as one might initially expect. Correct understanding of projection principles, such as those embodied in the tomographic paradigm, help make layover phenomena clear and quantitative.

[17]More correctly, elevated targets lay over in a direction orthogonal to the line-of-intersection between the slant and ground planes. See Appendix C for further details.

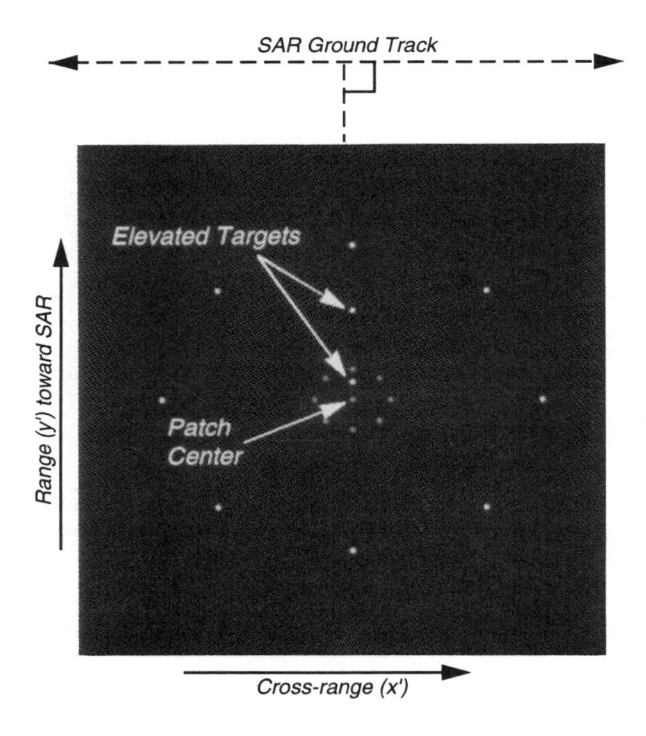

Figure 3.42 Linearly scaled magnitude of the nineteen-target slant-plane image. The symmetric array of point targets as well as the two expected projection effects are easily seen. The distortion of the circular array to an elliptical shape is a direct result of projecting the circular array of targets on the ground into the slant plane, in a direction normal to the slant plane. In this case, the range dimension is compressed by the cosine of the nominal grazing or depression angle of the acquisition, as expected. The second projection effect noted involves the resulting positions of the two elevated targets. This is the projection phenomena of "layover," where elevated targets project into the slant plane and appear to lay over in a direction normal to the nominal ground track of the SAR flight path. The origin of the $x' - y'$ coordinate system is at the patch center.

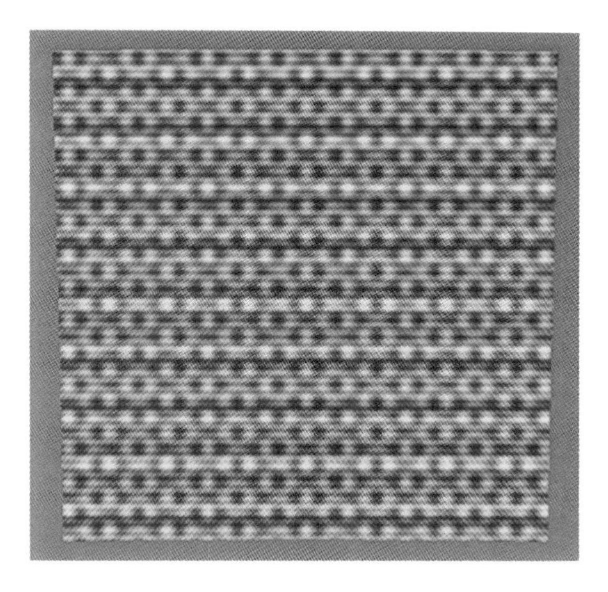

Figure 3.43 Real part of the phase history of the nineteen-target case processed with a bandwidth reduction factor of $\beta = 2.0$. Compared to Figure 3.41, this figure shows a reduction in spatial-frequency content. Some target signatures have been reduced or eliminated.

This is an appropriate point to illustrate pictorially what happens to the phase history and subsequent imagery if the interpolation filters are specified to have a bandwidth reduction of $\beta = 2.0$ instead of the typical $\beta = 1.04$. If the output sample spacing in the reduced bandwidth case is kept the same as before, the resampled phase history will still contain 400 by 400 samples, but only spatial frequencies up to *one half* of the maximum previously supported will be passed by the interpolation filters. This subtle loss in spatial frequency is evident in Figure 3.43, which shows the real part of the phase history for the 19-target case processed with $\beta = 2.0$ (compared to Figure 3.41).

The associated formed image is shown in Figure 3.44 from which it is evident that the spatial frequencies from the extreme range and cross-range targets did not pass the interpolation filters and consequently the point responses are missing in the image. (The image now covers *one half* of the original spatial extent.) Perhaps more interesting in this example is the fact that the four remaining distant targets have been reduced in brightness, but not eliminated. These targets happen to produce spatial frequencies that fall in the transition band of the filters. Consequently, the

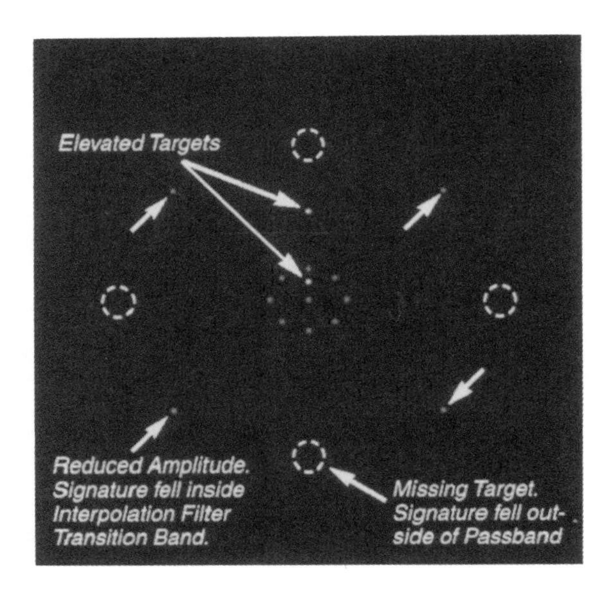

Figure 3.44 Linearly scaled magnitude of the nineteen-target slant-plane image processed with a bandwidth reduction factor of $\beta = 2.0$. It is evident that the spatial frequencies from the extreme range and cross-range targets do not pass the interpolation filters. Consequently, the point responses are missing in the image (dashed circles). In addition, the four remaining distant targets (arrows) have been reduced in brightness but are not eliminated because they produce spatial frequencies that fall in the transition band of the filters and are only reduced in amplitude. (Compare this figure to Figure 3.42.)

spatial frequencies are only reduced in amplitude. The location of the transition band defines the maximum possible image dimensions. Even though all extreme targets are equidistant from the patch center, the spatial frequencies (produced during the broadside acquisition at a 30 degree grazing angle) fall just within the filter transition band and are not entirely eliminated. The elevated targets are unaffected because their spatial frequencies fall well within the filter passband. These simple examples illustrate the connection between spatial frequencies passed by the interpolation filters and the subsequent extent of the image coverage.

We will now examine some *non-broadside* acquisition examples to further illustrate various aspects of image formation and projection issues. Consider the slightly more complex *squinted* imaging geometry shown in Figure 3.45. Here, we still consider straight and level flight, but now the radar is squinted forward (off broadside) by

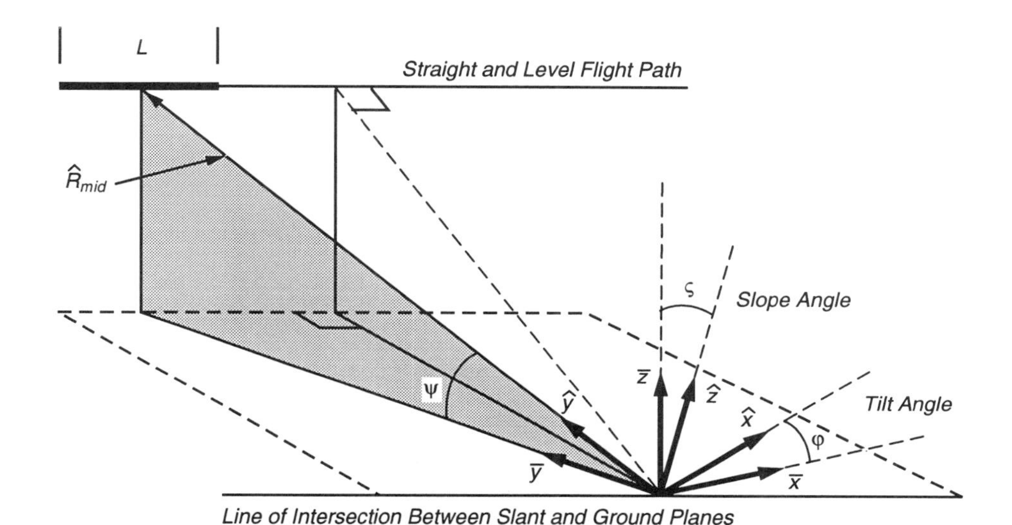

Figure 3.45 Squinted imaging geometry with straight and level flight. The radar is squinted forward (off broadside) by several degrees. Here again, $\bar{x}$, $\bar{y}$, and $\bar{z}$ refer to the *ground-plane* unit vectors whereas, $\hat{x}$, $\hat{y}$, and $\hat{z}$ refer to the *slant-plane* unit vectors. The imaging geometry details are covered in greater detail in Appendix C.

several degrees.[18] Phase-history data are acquired in exactly the same slant plane as for the broadside case. The synthetic targets are distributed in exactly the same way as before. The radar flight path has not changed, but the data are acquired at a non-broadside position. In order to maintain the same cross-range resolution as in the broadside case, we must sweep out the same polar angle in the slant plane. Our squinted case (i.e., 30 degree squint forward of broadside as measured in the slant plane) requires that data be acquired over ± 1.17 degrees from the nominal squint angle. All other radar parameters remain the same.

Because our processing geometry is established with the mid-aperture pointing vector (i.e., the vector pointing from the patch center to the SAR at the middle of the synthetic aperture) defining the Y' axis in the phase history and the subsequent imagery, we expect to see a *rotation* of the phase history and imagery compared to the broadside case. This rotation is reflected in the phase-history data shown in Figure 3.46. The polar region of support remains the same as it is in the broadside case because the aperture dimensions in the slant plane have not changed. The

[18] See Appendix C for a more thorough discussion of these imaging geometry issues.

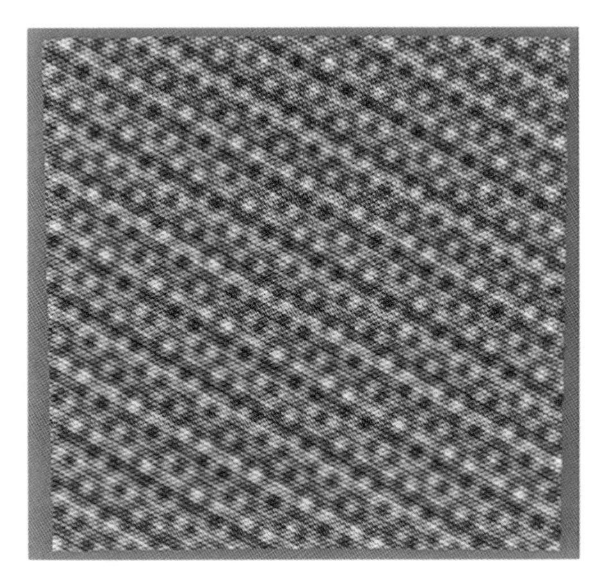

Figure 3.46 Real part of the phase history of the nineteen-target case with squinted imaging. A rotation of the constant phase lines (compared to the broadside case) is evident. The polar region of support remains unchanged from the broadside case because the same size phase-history annulus is acquired.

image formed from the polar processed data under squinted conditions is shown in Figure 3.47. Note the apparent rotation of the imagery compared to the broadside case as explained previously.

Also evident are the projection effects of the ground plane and elevated targets. Now, the elevated targets *do not* appear to lay over strictly in the range (i.e., y') direction, but rather in a direction *orthogonal to the ground track* of the SAR. Because the vertical y' axis points toward the SAR at the effective imaging point and the layover direction is orthogonal to the ground track (for straight-and-level flight), it is easy to verify from this figure that the SAR was indeed in a nominal 30-degree forward-squint mode (i.e., the angle between the y' axis and the layover direction vector is 30 degrees in this slant-plane image). The circular array of targets is compressed into an elliptical shape by the *same* orthogonal projection into the slant plane, even though the nominal grazing angles are different between the broadside and squinted case. While the resultant image appears rotated and compressed in an unusual way compared to the broadside case, clear depictions of the imaging geometry and the projection effects make the resulting image understandable in quantitative terms.

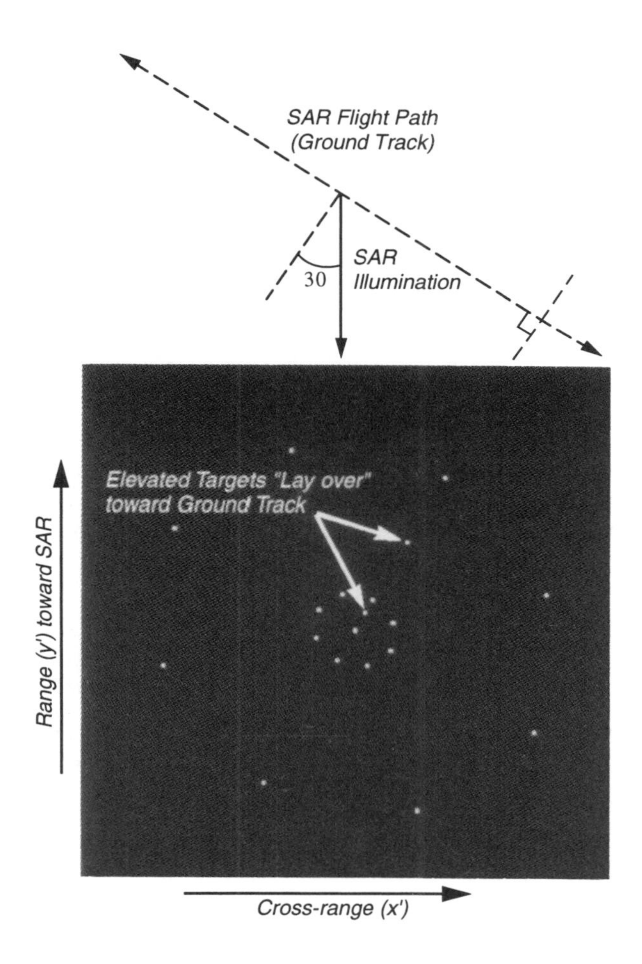

Figure 3.47 The nineteen target *slant*-plane image with *squinted* imaging. A rotation of the image (compared to the broadside case) is evident. The circular array of targets is compressed to an elliptical shape by the *same* orthogonal projection into the slant plane, even though the nominal grazing angles are different between the broadside and squinted case (see text). Relevant angles for this example are: (1) slope angle $\varsigma = 30.0$ deg., (2) slant-plane squint $\theta_s = 30.0$ deg., (3) nominal grazing angle $\psi = 25.66$ deg., and (4) nominal tilt $\varphi = 16.1$ deg.

Summary of Critical Ideas from Section 3.7

- **For ideal straight and level collection the notion of *slant plane* is unambiguous because all polar records (and samples) lie exactly in the same plane. Polar-to-rectangular resampling followed by aperture weighting (for sidelobe control) and 2-D Fourier transformation yields a *slant-plane image*.**

- **The slant plane is a logical image processing plane. Either inscribed or circumscribed rectangular grids are easily defined to encompass the phase-history data.**

- **Tomographic principles indicate that such a slant-plane image is an orthogonal projection of the three-dimensional scene into the slant plane. This image is also analogous to an optical image obtained by viewing the scene normal to the slant plane (i.e., the same orthogonal projection applies).**

- **Straight-line SAR flight paths always lead to perfectly planar Fourier acquisitions and allow a simplified look at the resulting phase history and image properties.**

- **Squinted imaging results in a rotation of the imaging geometry and distortion of the imagery by now well-understood projection effects.**

- **"Layover" phenomena are easily explained by orthogonal projections. Layover refers to height-dependent displacement in the final image, whether the laid-over objects superimpose on other structure or are merely displaced.**

3.8 OUT-OF-PLANE CORRECTION AND PROJECTION EFFECTS

For spotlight acquisitions where the SAR platform does not follow a straight flight path, phase-history data are collected on a non-planar ribbon in 3-D Fourier space. Treating the non-planar Fourier ribbon as if it were a planar surface induces phase errors by distorting Fourier space. Subsequent image formation using 2-D Fourier transformation produces defocused imagery. An out-of-plane correction is needed to produce a well focused image in this situation.

3.8.1 Out-of-Plane Correction

The Fourier-domain description of spotlight SAR indicates that a collection of scatterers in 3-D physical space produce a collection of 3-D complex sinusoids in Fourier space. Any planar slice of Fourier space results in a collection of 2-D complex sinusoids. This collection forms the phase history. Subsequent Fourier analysis (image formation) yields focused imagery with 2-D *sinc* impulse responses for all scatterers that constitute the terrain. In addition, all points on the 3-D object are completely focused and appear at locations in the 2-D image corresponding to the projection of the 3-D object normal to the Fourier slice taken (typically the slant plane). Therefore, planar slices of Fourier space produce focused imagery for all points on a 3-D object.

In contrast, non-planar slices of 3-D sinusoids do not yield 2-D sinusoids. Subsequent Fourier analysis of these distorted signals does not yield focused (*sinc* function) responses. Therefore, it is not possible to treat a curved collection surface as if it were planar and to obtain focused imagery as a result.

However, scatterers that lie in some *object plane* in the scene produce sinusoids of *zero* spatial frequency in Fourier space in a direction *orthogonal* to this object plane. This direction of zero spatial frequency defines a direction along which the data do not vary. Thus, projection of the 3-D phase-history data onto a processing plane in a direction *orthogonal* to the object plane preserves the 2-D sinusoidal phase variations (parallel to the object plane). A 2-D Fourier transform of data projected onto the processing plane in this way will result in well focused imagery [10, 11]. The phase history projection involved here is illustrated in Figure 3.48, where the object plane is identified as the ground plane, or focus plane, and the processing plane has been chosen to be the slant plane.

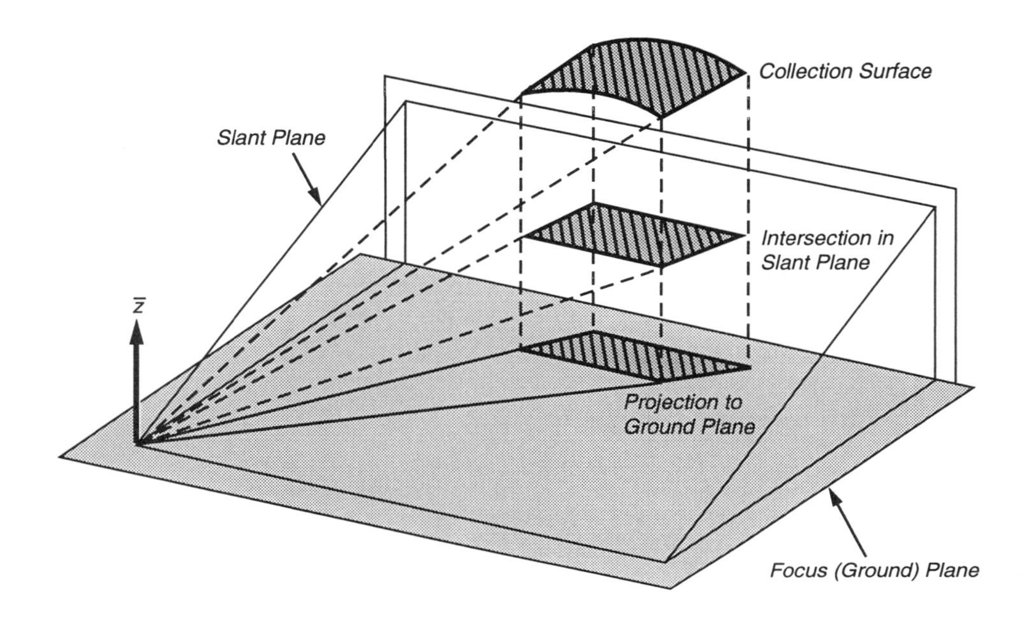

Figure 3.48 A schematic of phase-history projections. In this example, the portion of phase history to be processed happens to be way above the nominal slant plane. (In practice, the nominal slant plane would be selected to match more closely the collection surface.) Projection of these data (orthogonal to the focus plane) into the slant plane (or any other plane) compensates for out-of-plane motion *only* for targets that lie in the focus plane.

The term *focus plane* is commonly used to identify the plane whose normal specifies the direction of the out-of-plane data projection. It usually coincides with the object plane described above, but it can be chosen at will. Actually, the focus plane *defines* the plane in which targets are in focus for non-planar collections, as a result of the projection.

While the above technique allows objects in the focus plane to be focused, it causes targets outside the focus plane to be defocused when the collection surface is non-planar. The amount and nature of the defocus depends on the amount and nature of the out-of-plane motion and on the distance above or below the focus plane the target lies. These "depth-of-focus" issues are discussed further in Section 3.8.2.

In summary, a non-planar collection must be compensated in the phase-history domain prior to image formation. This compensation is referred to as out-of-plane correction and amounts to projecting the phase-history data into a desired processing plane (typically the nominal slant plane or the ground plane) in a direction

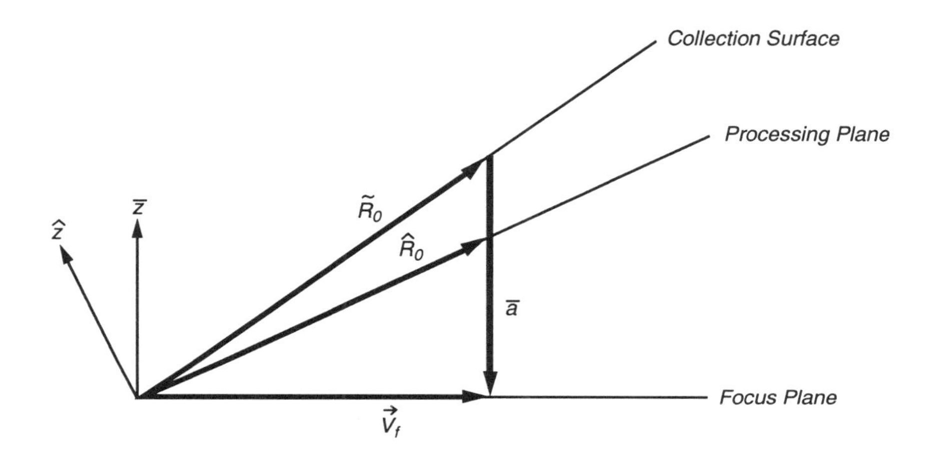

Figure 3.49 Out-of-plane projections on a pulse-by-pulse basis. The established coordinate system geometry permits the location of samples (indicated by $\tilde{R}_0$) on the instantaneous collection vector to be projected orthogonal to the focus plane. The intersection of the projection vector $\bar{a}$ with the desired processing plane establishes the location of the projected sample ($\hat{R}_0$). The interpolation filter is then centered on the desired output sample point (the rectangular grid is established in the desired processing plane) where it produces a weighted sum of the projected input samples to form the desired output sample. See Appendix C for information on how the imaging geometry is established.

orthogonal to the focus plane.[19] Figure 3.49 illustrates the conceptual nature of out-of-plane compensation on a pulse-by-pulse basis. The established coordinate system geometry permits the locations of samples on the instantaneous collection (i.e., pointing) vector to be projected orthogonal to the focus plane. The intersection of the projection vector $\bar{a}$ with the desired processing plane establishes the location of the projected sample in the processing plane. This projected location of a particular sample on the rectangular grid (defined in the processing plane) determines the contribution of that sample to the interpolated output array.

In practice, out-of-plane correction affects *only* the range interpolation step of polar reformatting. Range interpolation places the out-of-plane polar samples onto equally spaced range samples (through the proper geometric projections as discussed) but onto unequally spaced cross-range samples whose structure resembles the keystone of an arch (see Section 3.5.1 on the resampling process). Once the keystone grid

[19] See Appendix C for further discussions on how the imaging geometry allows development of a coordinate system to accomplish these projections.

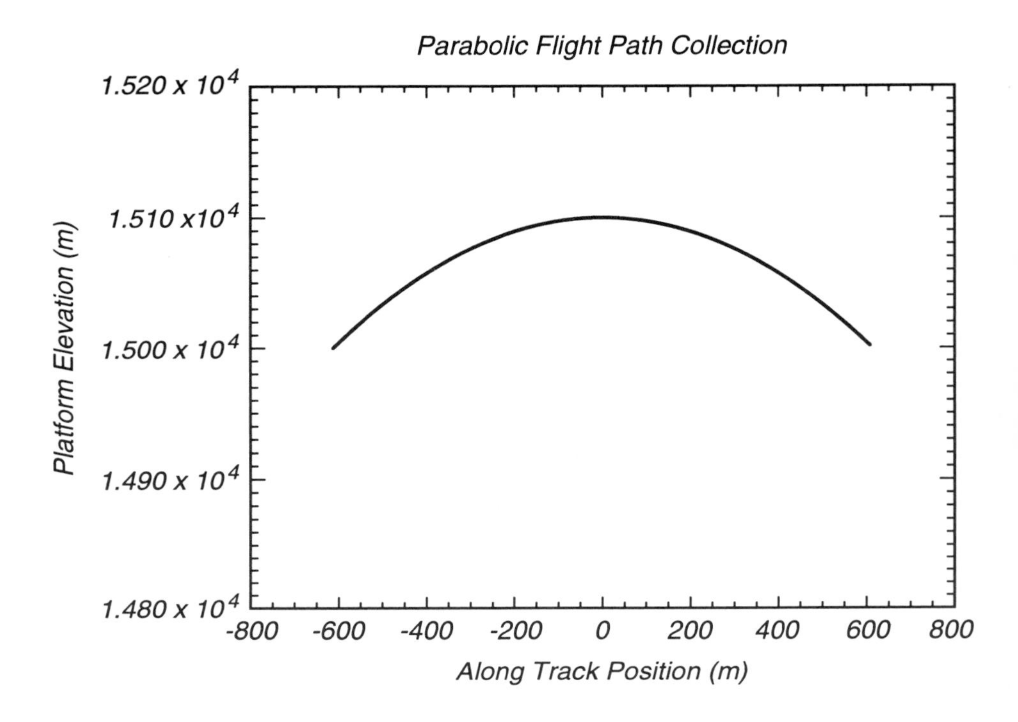

Figure 3.50 Flight path elevation profile for a curved collection surface. The maximum out-of-plane motion is 100 meters over a 1200+ meter synthetic aperture length.

has been constructed, the data have been placed in the chosen processing plane and no further correction is needed in the cross-range interpolation step.

The results of curved flight paths on the resulting imagery (with and without out-of-plane correction) are best illustrated with several examples. Figure 3.50 represents the elevation profile that our synthetic SAR will follow to generate data on a curved collection surface (100m maximum out-of-plane motion). Other than the curved elevation profile, the synthetically generated data have been produced with exactly the same radar and imaging geometry parameters as in our previous broadside example in Section 3.7.3. We now polar process these data as if the collection surface were indeed planar. In fact, we use the pointing vectors from the straight and level broadside flight to establish the geometry and to guarantee that the processing takes place in the established slant plane.

By purposely not performing any out-of-plane compensation, the curved collection surface induces phase errors on the target signals by virtue of Fourier space distortion. The subsequently formed image is shown in Figure 3.51(a). Notice that no defocusing of the central target occurs, as expected, because it alone produces a zero-spatial-frequency signal. Somewhat surprisingly, for those cross-range targets that have the same range as the patch center, there is no significant defocusing. Apparently, this particular curved slicing of Fourier space imparts very little phase error to cross-range-only targets. There is considerable defocusing, however, of the other ground-plane targets and the two elevated targets. The spatial frequencies of the defocused targets suffer distortions from the improperly corrected curved collection surface.

When the data are properly compensated for out-of-plane errors, focused imagery is produced for all targets that lie in the focus plane (see Figure 3.51b). An additional example is illustrated in Figure 3.52. Elevated targets are defocused in an amount proportional to their distance out of the ground plane. This is discussed in Section 3.8.2.

3.8.2 Depth-of-Focus Issues

Assuming that out-of-plane compensation is always performed in practice (sufficiently planar collections are difficult to achieve, especially for high-resolution SARs), we can consider the issue of *depth-of-focus.* The mathematics behind how much phase error is induced by out-of-plane motion is identical to that derived in the context of SAR interferometry. We refer the reader to Chapter 5 for a thorough discussion, but in the meantime we assume that the connection between the interferometric scale factor and the depth-of-focus calculation is valid.

The scale factor describing the change in phase per unit change in height of a scatterer as seen by the SAR is given by the following equation (see Equation 5.83 in Chapter 5):

$$\frac{\Delta\phi}{\Delta h} = S = \frac{4\pi}{\lambda}\frac{\Delta\psi}{\cos\psi} \qquad \text{radians/meter} \tag{3.67}$$

where λ is the radar wavelength in meters, ψ is the nominal grazing (or depression) angle, and $\Delta\psi$ is the change in grazing angle. In interferometric applications, $\Delta\psi$ is the difference in nominal grazing angles between two collections. Here, for depth-of-focus analysis, we can ascribe $\Delta\psi$ to the maximum change in grazing angle due to out-of-plane motion of the SAR during collection. More correctly, $\Delta\psi$ is a continuously varying function over the synthetic aperture and describes the nature of the out-of-plane motion, which ultimately quantifies the nature of the phase error. However, for smooth but curved collections (e.g., parabolic), we assume that the maximum flight path altitude deviation occurs at the aperture center (as was the

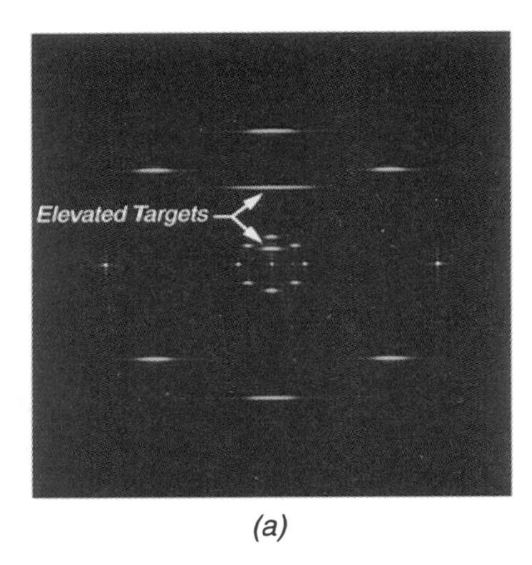

(a)

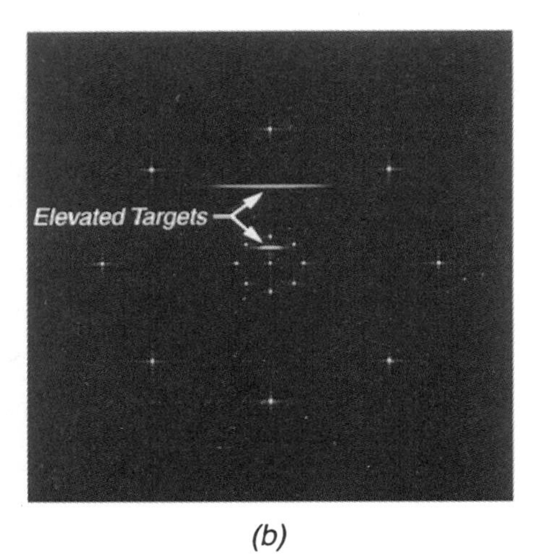

(b)

Figure 3.51 Out-of-plane motion examples. (a) Image formed from curved collection surface and processed as if the collection surface were planar. Notice that no defocusing of the central target occurs, as expected. Somewhat surprisingly, for those cross-range targets that have the same range as the patch center, there is no significant defocusing. There is considerable defocusing of the other ground-plane targets and the two elevated targets. The spatial frequencies of the defocused targets suffer distortions from the improperly corrected curved collection surface. (b) Image formed from curved collection surface and processed by projection of the phase history into the slant plane. With the proper phase-history projection, all ground-plane (focus-plane) targets are properly focused. The two elevated targets are defocused in an amount proportional to their distance out of the ground plane. Phase-history projection can correct non-planar collection distortions *only* for targets that lie in the focus plane.

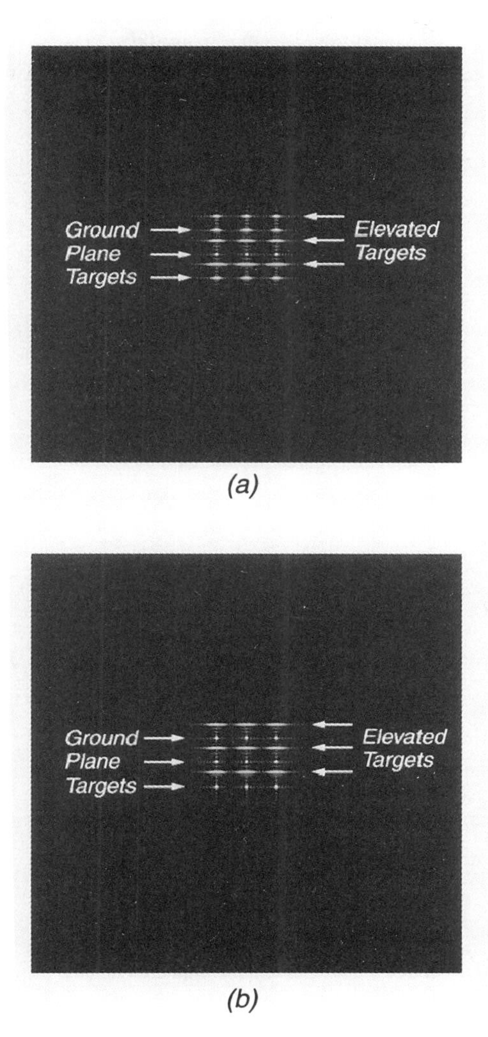

Figure 3.52 Additional out-of-plane motion examples. (a) Image formed from curved collection surface and processed as if the collection surface were planar. The image consists of 9 point targets on 10 meter centers arranged in a square pattern on the ground. An additional 9 targets are placed 10 meters above the ground targets. Notice the defocus of the ground targets and the elevated targets. (b) Image formed from curved collection surface and processed by *projection* of the phase history into the slant plane. The ground targets are properly focused while the elevated targets (although "laid over") defocus in proportion to their elevation out of the focus (ground) plane. In this example, all elevated targets defocus by the same amount because they are all at the same height above the ground.

case for the synthetic flight path shown previously in Figure 3.50) and that this flight path imparts a parabolic (quadratic) phase error on elevated targets due to the way Fourier space is sliced.

Assuming a simple geometry, it is easily verified that

$$\Delta\psi \approx \frac{\Delta z_{max}}{R_0} \cos\psi \tag{3.68}$$

where Δz_{max} is the maximum altitude deviation from planar and R_0 is the nominal slant range at aperture center. Substituting Equation 3.68 into Equation 3.67 yields the depth-of-focus scale factor

$$S_D = \frac{4\pi}{\lambda}\frac{\Delta z_{max}}{R_0} . \tag{3.69}$$

Thus, targets that lie outside the focus plane by an amount Δh will be injected with an amount of (predominantly) quadratic phase error given by

$$\Delta\phi = S_D \Delta h = \frac{4\pi}{\lambda}\frac{\Delta z_{max}}{R_0}\Delta h . \tag{3.70}$$

Substituting our simulated imaging parameters from Section 3.7.3 into Equation 3.70 with $\lambda = 0.03$, $\Delta z_{max} = 100$, $R_0 = 30,000$, and $\Delta h = 10$ yields

$$\Delta\phi = \frac{40\pi}{9} \qquad \text{radians.} \tag{3.71}$$

As a result, even though out-of-plane correction is applied, a target elevated by 10 meters above the focus plane acquires $40\pi/9$ radians of quadratic phase error from the curved collection surface with the imaging and radar parameters given. For quadratic phase errors exceeding 2π radians (on unweighted data), the impulse response mainlobe width increases by approximately one mainlobe width for each $\pi/4$ radians of error.[20] Therefore, we expect to see an impulse broadening (defocus) by a factor of 17.78 for unweighted data. Similarly, the 50-meter-elevated target would be broadened by an additional factor of 5 for a total broadening of 88.88 times the nominal unweighted IPR width. Aperture weighting somewhat reduces the effects of phase errors on IPR broadening, especially for relatively small amounts of quadratic error. However, once the quadratic error exceeds several cycles (i.e., several times 2π radians) across the aperture, the weighted IPR width will broaden (defocus) in the same manner as does the unweighted IPR width. It is readily

[20] The IPR mainlobe width (from unweighted data) broadens by a factor of approximately $4\phi_{peak}/\pi$, where ϕ_{peak} is the peak quadratic phase error in radians.

verified that the 10-meter and 50-meter-elevated targets of Figure 3.51(b) have been broadened in proportion to their distance out of the focus plane.[21]

If we say that phase errors less than $\Delta\phi_{max} = \pm\pi/4$ radians yield acceptably focused imagery, then by rearranging Equation 3.70 and solving for Δh yields the expression for *depth-of-focus*

$$\Delta h_D = \frac{|\Delta\phi_{max}|}{S_D} = \frac{\lambda}{16}\frac{R_0}{\Delta z_{max}} . \tag{3.72}$$

It is easy to see that as the out-of-plane deviation, Δz_{max}, goes to zero, the depth-of-focus becomes infinite (i.e., all targets in 3-D space will be focused). If we use our previous parameters in Equation 3.72 we find that the depth-of-focus is $\Delta h_D = 0.5625$ meters. Therefore, only targets that lie within approximately ± 0.5625 meters of the focus plane will remain in focus (by our chosen phase-error tolerance). Tighter control on phase errors makes the depth-of-focus more restrictive. Such a small depth-of-focus for a mildly curved collection surface illustrates how sensitive image focus is for non-planar collections. Figure 3.53 illustrates some of these depth-of-focus issues with actual SAR imagery of the Solar Power Tower at Sandia National Laboratories, Albuquerque, New Mexico. The tower is a tall (60-meter) structure that houses a thermal receiver. A large group of heliostats (mirrors) mounted on the ground focus solar energy on the receiver. The radar scatterers at the top of the structure lie at elevations that are several times the depth-of-focus above the ground plane and exhibit height-dependent defocus for this curved collection (approximately 5 meters peak out-of-plane motion across the synthetic aperture). For visual reference, an aerial photo of the Solar Power Tower is shown in Figure 3.60(c).

In general, curved collection surfaces impart a height-dependent defocus that cannot be compensated globally (i.e., through phase-history projections). Interestingly though, local focus corrections can be used as a crude estimate of target height in a local area. The relationship between the target height and the amount of quadratic error required to focus targets at that height is given by Equation 3.70. Therefore, a single SAR image collected with a known curved collection surface encodes height in a regionally varying focus. However, terrain-height estimates obtained from this depth-of-focus technique tend to be less accurate, lower in resolution, and less robust than similar measurements derived from interferometry.

---

[21] Draw radial lines from the patch center to each edge of the blur of the 50-meter-elevated target. Notice that the blur of the 10-meter-elevated target also lies within these radial lines, confirming the linear relationship between blur width and target height.

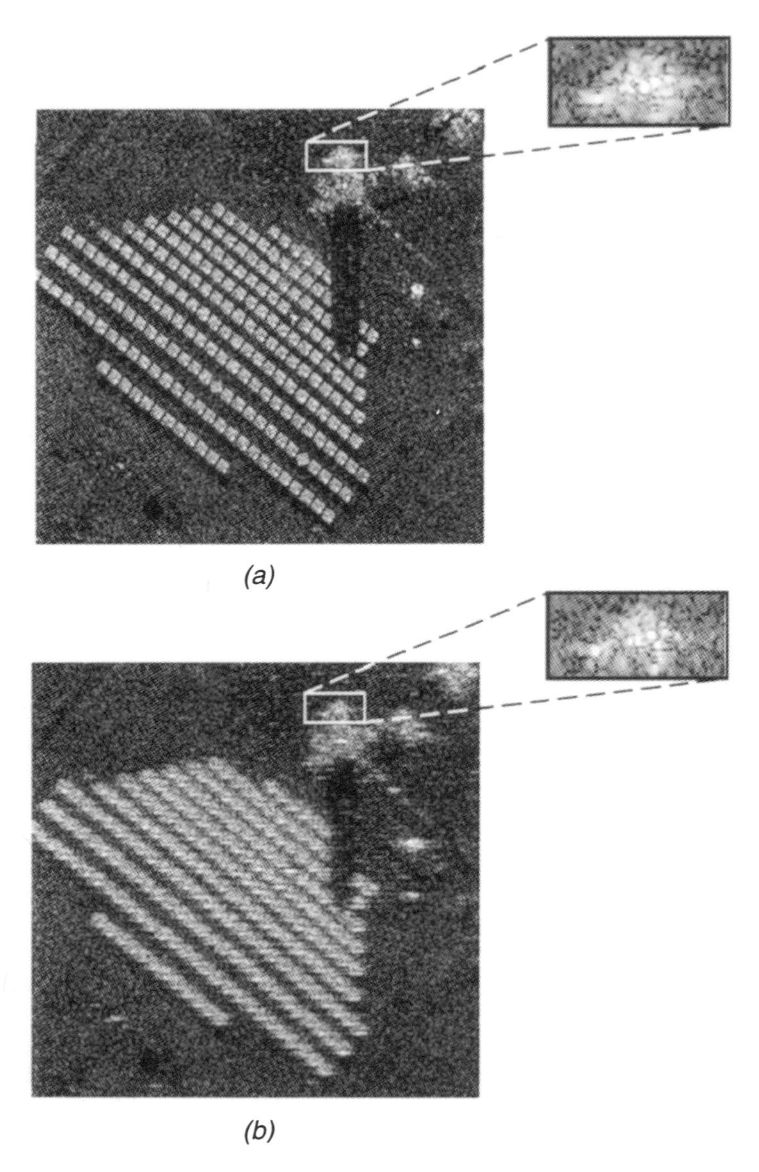

Figure 3.53 Actual 1-meter resolution SAR imagery of the Solar Power Tower, obtained from non-planar collection encompassing several meters of vertical relief. (a) The scatterers at the top of the tower (highlighted) exceed the tolerable depth-of-focus limit for this collection and therefore become blurred. Because of the lack of isolated scatterers in the highlighted region (and in the full-resolution accompanying piece), it may be difficult to tell that they are blurred. (b) However, if the data are reprocessed so that the top of the tower is the origin of the focus plane (i.e., the top of the tower becomes the new phase stabilized motion compensation point), all scatterers on the ground defocus by the same amount as did the top of the tower in (a). This reprocessing more clearly depicts the amount of defocusing resulting from this curved collection.

Summary of Critical Ideas from Section 3.8

- **Phase-history data are generally collected on a non-planar ribbon in 3-D Fourier space.**
- **Treating this Fourier ribbon as if it were planar induces phase errors by distorting Fourier space. Subsequent image formation by 2-D Fourier transformation produces defocused imagery.**
- **Out-of-plane Fourier collections can be compensated by projection of the data in a direction orthogonal to the focus plane onto a nominal processing plane.**
- **Only targets that lie in the focus plane will be fully focused in the subsequent imagery.**
- **After out-of-plane compensation, elevated targets will defocus in proportion to their distance above or below the established focus plane.**
- **The amount of defocus depends on the actual deviations from a planar collection as well as on the elevation changes of the imaged terrain.**
- **The tolerable out-of-plane defocus establishes the depth-of-focus for a particular SAR collection. Depth-of-focus requirements can put a very tight tolerance on allowable out-of-plane motion for high-resolution SARs.**
- **Height-dependent defocus allows a crude estimate of terrain elevation on a coarsely sampled grid. The grid spacings are always much coarser than the inherent SAR resolution.**

3.9 GROUND-PLANE IMAGE FORMATION AND IMAGE PROPERTIES

In many situations it is desirable to produce SAR imagery in the ground plane instead of in the slant plane as described above. By a *ground-plane* image we mean one that presents an orthographic view of the scene. That is, it is analogous (for flat terrain at least) to an optical view looking down on the scene from directly above. This is in contrast to a slant-plane image, which provides a view of the scene from an oblique angle (normal to the slant plane). One way of producing a ground-plane image is to warp the slant plane image to remove the geometric distortion associated with its oblique view. In this section we will describe a method of processing phase history data that produces ground-plane imagery directly, without the need for warping.

Ground-plane image formation is accomplished by the same phase-history projections used to compensate for out-of-plane motion. Instead of projecting the collection surface (phase history) into a nominal slant plane and forming a slant-plane image, the data are projected all the way into the nominal ground (focus) plane. The image formed from such a projection is indeed a ground-plane image and is orthographically correct for all targets that lie in the ground plane. Elevated targets still "lay over" according to their projection into the ground plane in a direction orthogonal to the slant plane. Ground-plane phase-history projection preserves the spatial-frequency relationships (just as out-of-plane compensation does) to produce correct ground-plane imagery.

There are definite advantages to ground-plane imagery produced in this manner, along with some disadvantages. These advantages and disadvantages will become clear as we explore some of the interesting projection effects of the phase-history annulus as seen in the focus plane. The distortion of the polar region of support in the focus plane impacts how the rectangular grid should be selected to encompass the projected data, which subsequently affects certain image properties.

Let us begin by using the same synthetically generated data as before, acquired at broadside in straight and level flight. When these data are projected into the ground plane and the rectangular array is established to circumscribe the data, the result is shown in Figure 3.54. There is only a subtle difference in the appearance of the phase-history structure compared to that shown previously in Figure 3.41; specifically, there is a slight compression of the vertical (Y) dimension of the polar region of support. This compression is in an amount equal to the cosine of the nominal grazing angle as expected. Not surprisingly, this phase-history compression increases the range component of the spatial frequency by the inverse of this amount. It is precisely that increase that causes range-displaced targets to appear farther from

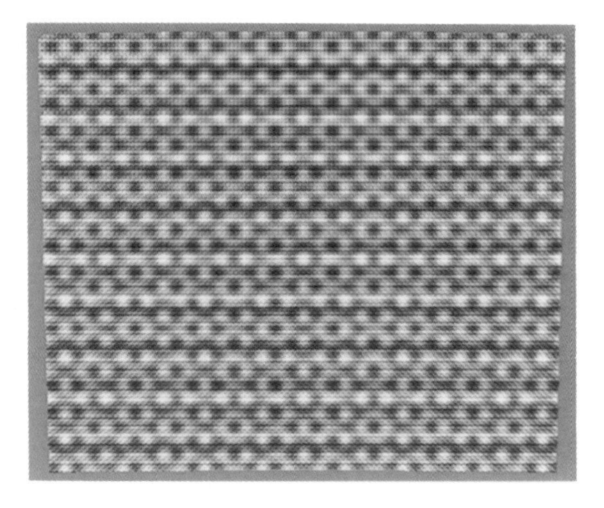

Figure 3.54 Real part of the phase history projected into the ground plane after acquisition with straight and level flight and broadside imaging. There is a predictable compression of the phase history in the vertical dimension (by the cosine of the grazing angle) that compensates for and undoes the elliptical arrangement of targets in the corresponding slant-plane image.

the patch center and appear in their orthographically correct circular arrangement as shown in Figure 3.55. Elevated targets lay over according to established projection principles.

Squinted imaging and ground-plane projections cause some interesting and more extreme phase-history distortions. (This distortion is similar to that encountered during bistatic spotlight SAR acquisition as discussed in Chapter 2.) If we use the same 30-degree squinted imaging data as before (with straight and level flight) and project these data into the ground plane, we find the interesting situation shown in Figure 3.56. Here we see not only a rotation of the data by virtue of the squinted geometry, but a rhombus-shaped region of support resulting from the projection into the ground plane. The region of support is distorted in this manner by virtue of rotation and compression in the Y dimension according to all the previously established projection principles. Specifically, the Y dimension is compressed by

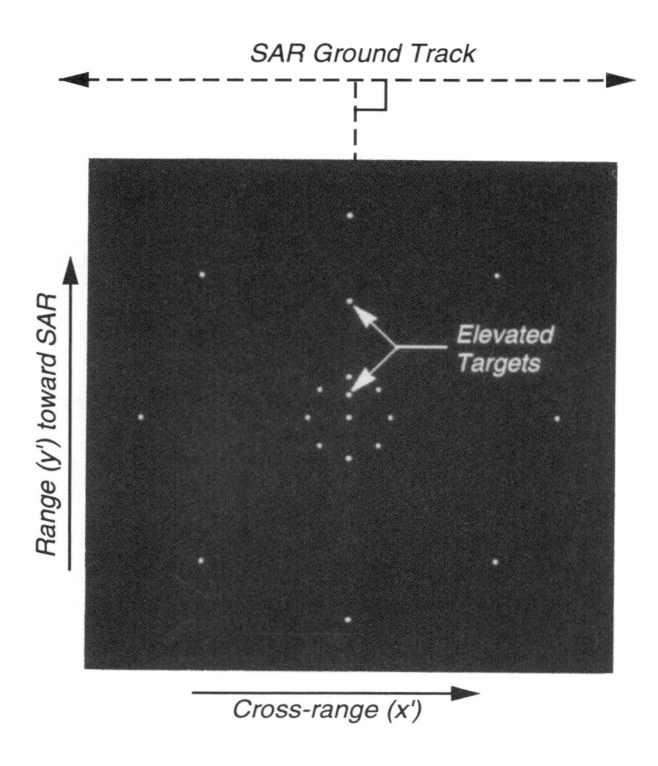

Figure 3.55 Image formed from ground-plane projection of phase-history data gathered in a straight and level broadside acquisition. Ground-plane targets appear in their orthographically correct circular arrangement and elevated targets lay over according to established projection principles.

the cosine of the grazing angle ψ and the X dimension is compressed by the cosine of the tilt angle φ.[22]

The ground-plane image is shown in Figure 3.57 and is orthographically correct for all ground-plane targets. The image is rotated, of course, from that obtained during an equivalent broadside imaging condition because the established processing coordinate system rotates with squint angle. Elevated targets lay over according to the previously discussed projections.

In squinted imaging, the polar region of support becomes distorted when projected into the ground plane (see Figure 3.58). We are free to establish the desired rectangular grid any way we desire. In the previous example, the grid circumscribed the data to show the region of support. In actual image-formation practice though,

[22] See Appendix C.

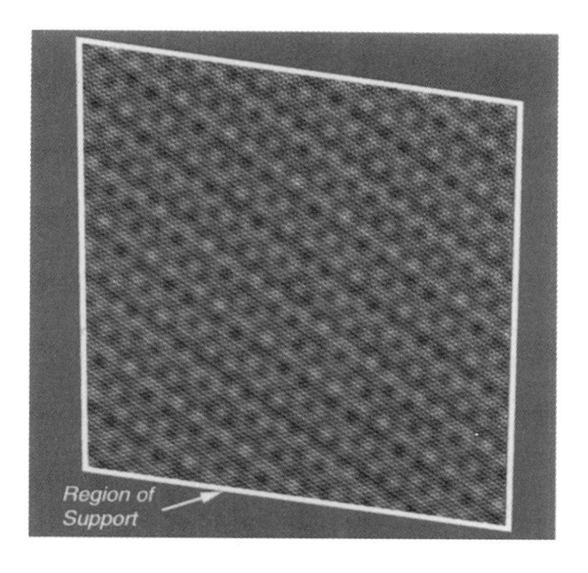

Figure 3.56 Real part of the phase history projected into the ground plane after acquisition with straight and level flight in a squinted mode. There is a rotation of the data by virtue of the squinted geometry. A rhombus-shaped region of support results from the projection of the phase-history annulus into the ground plane (the rectangular grid was chosen to circumscribe the data to allow the region of support to show).

the rectangular grid is typically inscribed within the region of support to produce a square phase-history aperture. Processing the phase history to lie on an inscribed rectangle regardless of the imaging geometry produces IPRs that resemble 2-D *sinc* functions, with orthogonal sidelobes. This is in contrast to the non-orthogonal sidelobes that typically result from processing an irregularly shaped region of the phase history. It also differs from the result obtained from warping an already formed slant-plane image into ground-plane coordinates. In that case an orthogonal sidelobe structure in the slant plane image becomes skewed when the image is warped. This effect is shown in Figure 3.59. The image warping shown here is accomplished by removing the slant-plane shear angle ξ[23] from the slant-plane image, stretching the y dimension by the reciprocal of the cosine of the grazing angle ψ and stretching the x dimension by the reciprocal of the cosine of the tilt angle φ.

[23] See Appendix C.

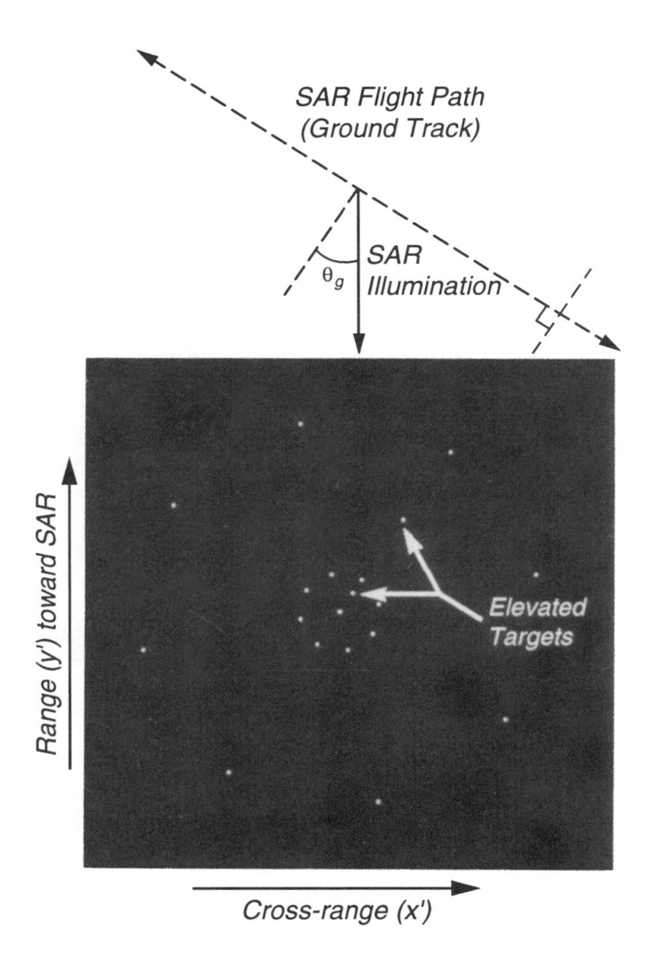

Figure 3.57 Image formed from ground-plane projection of phase-history data acquired with squinted imaging. Ground-plane targets appear in their orthographically correct circular arrangement (but are rotated because of the processing axes rotation with squint) and elevated targets lay over according to established projection principles. Relevant angles for this example are: (1) slope angle $\varsigma = 30.0$ deg., (2) slant-plane squint angle $\theta_s = 30.0$ deg., (3) ground-plane squint angle $\theta_g = 33.69$ deg., (4) nominal grazing angle $\psi = 25.66$ deg., and (5) nominal tilt angle $\varphi = 16.1$ deg.

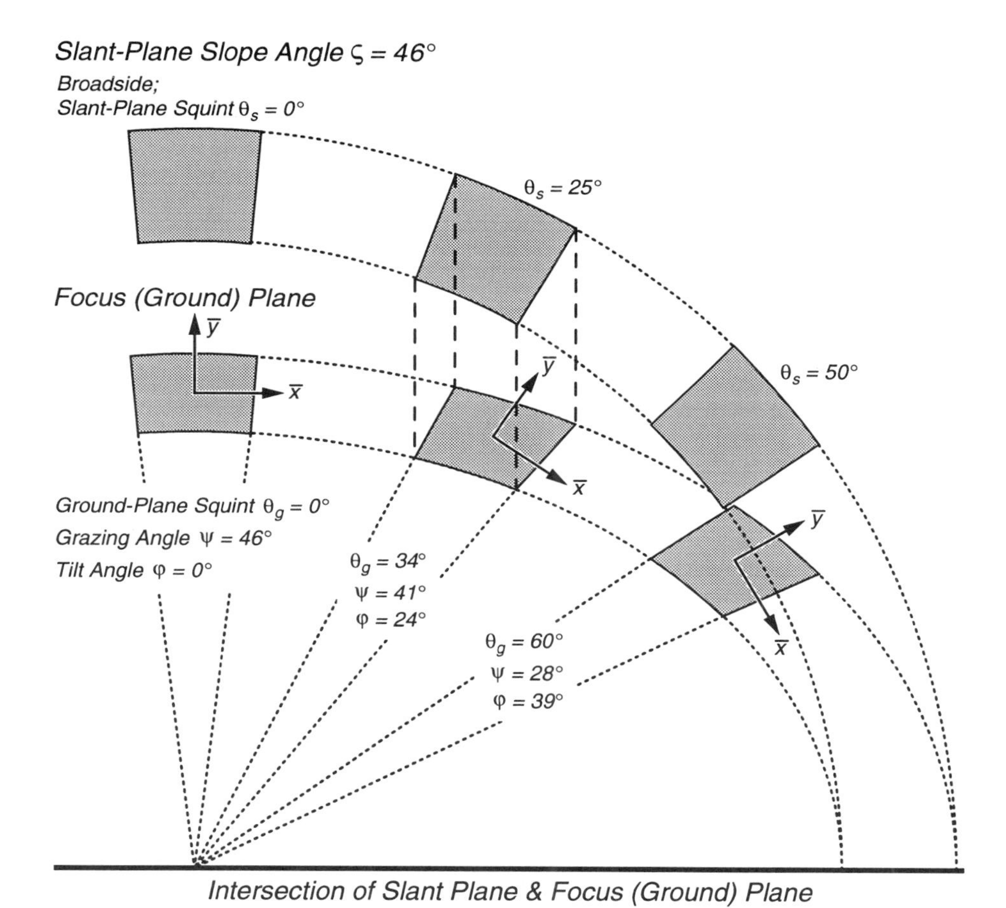

Figure 3.58 The Fourier-domain aperture acquired in the slant plane is distorted when projected into the ground plane. Three slant-plane apertures, at various squints, are shown in the upper circular arcs of the figure. Projection of these apertures into the ground plane result in the corresponding three distorted versions shown in the elliptical arcs. Processing coordinates are shown in the ground-plane apertures along with the resulting relevant angles that influence image properties.

Maintaining common image properties between squinted imaging geometries becomes important in interferometry, as is discussed in Chapter 5. Ground-plane projection of the phase history offers significant advantages for the processing of interferometric pairs.

In summary, ground-plane projection of phase-history data yields orthographically correct imagery for all targets lying in the ground plane. The impulse response structure remains invariant under acquisition geometry changes. Elevated targets project into the ground plane in a direction orthogonal to the nominal slant plane. Figure 3.60 shows actual SAR imagery collected in a squinted mode and processed in both the slant and ground planes for comparison.

Unlike slant-plane images that resemble optical images taken in a direction orthogonal to the slant plane, ground-plane images are not true projections of the scene in the same sense. That is, there is no location in space from which an optical image could produce the projections seen in a ground-plane image of non-planar terrain. This peculiar characteristic of ground-plane imagery (whether produced by image warping or phase-history projection) should be kept in mind when interpreting the result. A true orthographic projection would require an independent measure of terrain elevation (e.g.,through interferometry or existing topographic data).

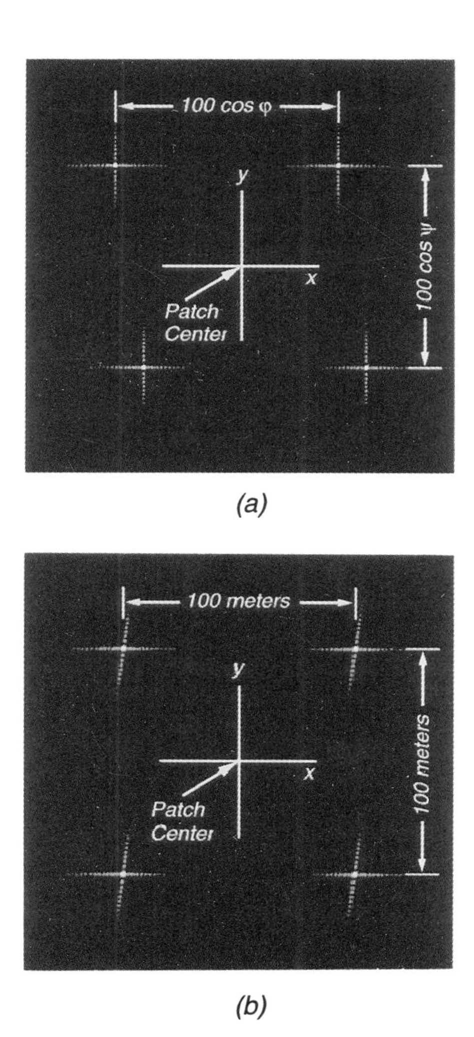

Figure 3.59 Result of warping a slant-plane image (acquired in a squinted mode) into a ground-plane image. (a) A square arrangement of point targets appears distorted in this slant-plane squinted image. (b) Warping the slant-plane image into a ground-plane image results in an orthographically correct rendition. However, note the skewed sidelobe structure of the point targets resulting from this warping. In contrast, ground-plane image formation through phase-history projections prevents skewed sidelobes and maintains common image properties between image pairs used in interferometry. The contrast has been enhanced to show the sidelobes more clearly.

Figure 3.60 Actual 1-meter resolution SAR imagery of the Solar Power Tower acquired with squinted geometry and processed in both the slant and ground planes using phase-history projections, (as opposed to image warping). (a) Slant-plane image. (b) Ground-plane image. (c) Aerial view of the Solar Power Tower. Additional properties of these images are examined in Appendix C.

Summary of Critical Ideas from Section 3.9

- **Phase-history projections for out-of-plane motion established a way to form ground-plane imagery by projecting the phase history into the nominal ground (or focus) plane.**
- **The image formed from such a projection is indeed a ground-plane image and is orthographically correct for all targets that lie in the ground plane.**
- **Elevated targets still "lay over" according to their projection into the ground plane in a direction orthogonal to the slant plane.**
- **Ground-plane phase-history projections preserve the correct spatial-frequency relationships (as is the case for out-of-plane compensation) to produce correct ground-plane imagery.**
- **The polar region of support of the phase-history data becomes distorted when projected into the focus plane (especially during squinted imaging).**
- **This distorted phase-history region impacts how the rectangular grid should be selected to encompass the projected data, which subsequently affects certain image properties.**
- **Ground-plane projections offer advantages for phase-history matching in interferometric applications.**
- **Ground-plane images have the (perhaps) undesirable property of containing both slant-plane and ground-plane properties. There is no location in space from which an optical image can be obtained to produce the projections seen in a ground-plane image of non-planar terrain.**

3.10 SPATIAL FREQUENCIES, IMAGE SCALE FACTORS, IMAGE DIMENSIONS, OVERSAMPLING, AND RESOLUTION SUMMARIZED

Thus far, we have covered much ground in discussing various aspects of spotlight-mode image formation. Throughout the previous sections, we have quantified some of the phase history and image properties resulting from various signal processing operations (e.,g., resolution, area coverage, sampling and interpolation, phase errors, etc.). At other times, we have only hinted that specific properties are affected by the particular way in which an operation is carried out (e.g., rectangular grid placements, phase-history projections, etc.). Details were omitted for the sake of concept clarity. It is worthwhile at this point, however, to revisit some of the previously discussed issues and quantify specific image properties in terms of known radar, geometry, and sampling parameters.

To this end, we now examine a typical phase-history annulus and rectangular grid placement with all the relevant parameters specified. The very important issues of spatial frequency, image scale factors, image dimensions, oversampling, and resolution will be made explicit in terms of known parameters.

Consider the diagram shown in Figure 3.61. A phase-history annulus has been acquired along some ribbon-like slice through Fourier space by our hypothetical spotlight SAR. If we project this ribbon of data into some desired processing plane (e.g., nominal slant or ground plane) by the methods described in Sections 3.8 and 3.9, we obtain a projected annulus that might look something like that shown in Figure 3.61. This annulus spans an angle θ_p as projected into the chosen processing plane. (By simple geometry, a *projected* angular span is always somewhat larger than the acquired angular span in the slant plane.) Physical dimensions in Fourier space are scaled in terms of *spatial frequency* with units of 2π cycles per meter (i.e., radians/meter). In the Y (range) dimension, the annulus spans a spatial frequency of

$$\Delta Y = \frac{4\pi}{c} B_{eff} P_0 \qquad \text{radians/meter} \tag{3.73}$$

where B_{eff} is the effective transmitted bandwidth in Hz,[24] c is the speed of light in meters per second, and P_0 is a projection scale factor.[25] P_0 can be near unity for projection into the nominal slant plane (assuming a nearly planar acquisition), or it can be as small as $\cos\psi_0$, when projecting data into the ground plane (ψ_0 is

[24] More correctly, B_{eff} is that portion of the bandwidth actually used in the polar-reformatting process.

[25] Specifically, P_0 is the nominal length of the polar radius at mid-aperture projected into the desired processing plane, relative to the unprojected mid-aperture polar radius. This projection is shown diagrammatically in Figures 3.48 and 3.49.

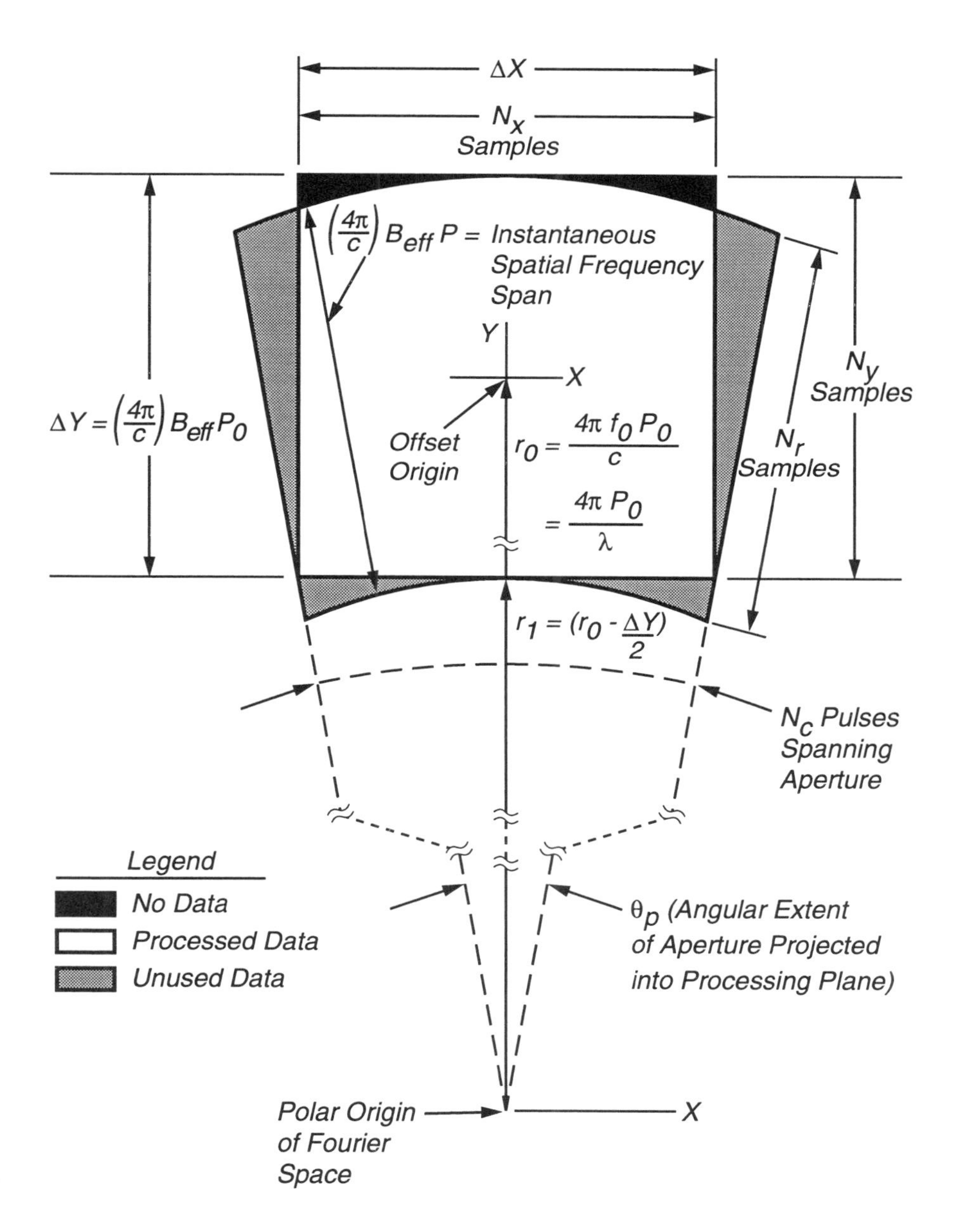

Figure 3.61 A typical phase-history annulus projected into a selected processing plane. The rectangular grid is established for polar-to-rectangular resampling. Spatial frequency dimensions, sample dimensions, and aperture span are shown.

the nominal grazing angle at aperture center). In general, the projection scale factor P changes on a pulse-by-pulse basis, but we shall use the mid-aperture value P_0 to establish the rectangular grid location, and spatial-frequency span ΔY.

Following the geometry shown in Figure 3.61 we compute the X-dimension spatial-frequency span, denoted ΔX, as

$$\Delta X = 2r_1 \tan \frac{\theta_p}{2} \tag{3.74}$$

where

$$r_1 = r_0 - \frac{\Delta Y}{2} = \frac{4\pi}{c}(f_0 - \frac{B_{eff}}{2})P_0 \tag{3.75}$$

is the spatial frequency measured from the polar origin to the beginning of the rectangular grid, and f_0 is the effective radar center frequency.[26] Each demodulated radar pulse is sampled at a rate that provides N_r samples over the span of data actually used (see Figure 3.61). There are N_c pulses (polar records) that compose the annulus of interest. When projected into the selected processing plane, these N_c pulses span θ_p radians.

Because we have all relevant dimensions in spatial frequency, it is now a simple matter to compute spatial-frequency spacings as

$$\begin{aligned} \delta X &= \frac{\Delta X}{N_x} \\ \delta Y &= \frac{\Delta Y}{N_y} \end{aligned} \tag{3.76}$$

where δX and δY are the cross-range and range spatial-frequency spacings, and N_x, N_y are the respective sample dimensions of the rectangular grid. Substituting Equations 3.73, 3.74, and 3.75 into Equation 3.76 and simplifying yields

$$\begin{aligned} \delta X &= \frac{8\pi}{c} \frac{(f_0 - \frac{B_{eff}}{2})P_0 \tan(\frac{\theta_p}{2})}{N_x} \\ \delta Y &= \frac{4\pi}{c} \frac{B_{eff} P_0}{N_y} \, . \end{aligned} \tag{3.77}$$

[26] The effective center frequency defines the middle frequency of the bandwidth portion actually used in the polar-reformatting process. For example, if the entire transmitted bandwidth is used (e.g., all samples in a demodulated pulse are used) then f_0 is the actual radar center frequency. If some subset of the samples in the demodulated pulse is selected for processing, f_0 is the middle frequency of that selected bandwidth.

Equation 3.77 represents the spatial-frequency spacing in terms of the relevant radar, geometry, and sampling parameters.

The maximum physical image dimensions possible are limited by the Nyquist sampling rate and are reciprocally related to the spatial-frequency spacings (see Equation 3.10). Therefore, the maximum cross-range and range image dimensions (in the selected processing plane) are

$$D_x = 2\pi N_x / \Delta X$$
$$D_y = 2\pi N_y / \Delta Y \ . \tag{3.78}$$

Equation 3.78, when used with Equations 3.77 and 3.76, agrees qualitatively with Equation 3.10, but more accurately accounts for the details of the processing.

Cross-range and range image scale factors are easily computed from Equation 3.78 as

$$S_x = \frac{D_x}{N'_x} \qquad \text{meters/pixel}$$
$$S_y = \frac{D_y}{N'_y} \qquad \text{meters/pixel} \tag{3.79}$$

where N'_x and N'_y are the cross-range and range FFT lengths used in image formation. Typically, the FFT lengths are somewhat longer than the rectangular grid dimensions, so that the resultant image is oversampled relative to the number of samples required to span the scene. The respective oversampling ratios are

$$OSR_x = \frac{N'_x}{N_x}$$
$$OSR_y = \frac{N'_y}{N_y} \ . \tag{3.80}$$

Finally, the image resolutions (defined by, say, the -3 dB width of the impulse-response function) are given by (see Equation 3.7 for qualitative comparison)

$$\rho_x = \frac{2\pi}{\delta X N_x} b_w$$
$$\rho_y = \frac{2\pi}{\delta Y N_y} b_w \ . \tag{3.81}$$

Here, we have used b_w as an impulse-response broadening factor to account for the particular aperture weighting used in image formation. Typically, $b_w \approx 0.88$ for uniform weighting, $b_w \approx 1.46$ for Hann weighting, and $b_w \approx 1.24$ for 40 dB Taylor weighting.

Let us put this information into the context of a typical processing requirement, namely to form an image with equal spatial dimensions, scale factors, and resolution in the selected processing plane. Setting $D_x = D_y$ in Equation 3.78 implies $\delta X = \delta Y$ that, once Equation 3.77 is rearranged, yields

$$\frac{(2f_0 - B_{eff})\tan(\theta_p/2)}{N_x} = \frac{B_{eff}}{N_y} \; . \tag{3.82}$$

Equating scale factors in Equation 3.79 implies $N'_x = N'_y = N'$. Furthermore, equating resolutions in Equation 3.81 requires a square grid dimension. Specifically, $N_x = N_y = N$. For small angular apertures (typical for most SARs) $\tan(\theta_p/2) \approx \theta_p/2$. Thus, with these simplifications, Equation 3.82 reduces to

$$(\frac{f_0}{B_{eff}} - \frac{1}{2})\theta_p = 1 \; . \tag{3.83}$$

This can be further simplified to yield

$$\theta_p = \frac{1}{Q_{eff} - 0.5} \tag{3.84}$$

where, $Q_{eff} = f_0/B_{eff}$. Equation 3.84 agrees qualitatively with Equation 3.66, except for the 0.5 factor that arises from by the specific way in which we located the rectangular grid within the phase-history annulus. If the rectangular grid is located differently or covers a different amount of Fourier space from that shown in Figure 3.61, the relation between θ_p and Q_{eff} will change, as will the relations for spatial frequencies, scale factors, resolution, etc. Nevertheless, for large values of Q_{eff} (which is typical for most practical SARs, especially if a small piece of phase history is being processed), $\theta_p \approx 1/Q_{eff}$.

For image formation with equal spatial dimensions, scale factors, and resolutions, a processing methodology would be as follows. Select the number of samples per pulse to process (i.e., select N_r). In turn, this specifies the effective bandwidth, B_{eff} (and resolution, ρ) as

$$B_{eff} = \frac{\dot{f} N_r}{f_s} = \frac{c b_w}{2\rho} \tag{3.85}$$

where $\dot{f}$ is the radar chirp rate in Hz/sec. and f_s is the sample rate of the demodulated pulse in Hz. Depending on the position of the selected samples within each pulse, locate (i.e., compute) the effective center frequency f_0. Remember, f_0 is the midpoint, in frequency, of the bandwidth portion actually used. Next, knowing B_{eff} and f_0, compute the required angular span of the aperture as projected into the selected processing plane by using Equation 3.84.

Next, select any contiguous number of pulses (polar records) that span the projected angular dimension. In practice, this requires selecting any N_c pulses, that when projected into the selected processing plane, span θ_p radians. The actual pointing-vector data accompanying the SAR collection is used to establish the geometry from which the projections are computed. See Appendix C for additional information.

Finally, establish the (square) rectangular grid as shown in Figure 3.61, perform the polar-to-rectangular reformatting, zero pad and perform a square 2-D FFT of dimension N'. The resultant image will cover equal dimensions in x and y and have equal scale factors and resolutions in the desired processing plane.

Many variations of this basic processing scheme are possible, of course. All pertinent relationships are now available to establish and compute the desired image properties. These important relationships are summarized below.

Summary

- Spatial frequency span of rectangular grid (radians/m)

  1. $\Delta X = \frac{8\pi}{c}(f_0 - B_{eff}/2)P_0 \tan(\theta_p/2)$
  2. $\Delta Y = \frac{4\pi}{c}B_{eff}P_0$

- Projection scale factor

  1. P_0: Nominal length of polar radius projected into the desired processing plane relative to the unprojected radius at mid-aperture.
  2. $0 \leq P \leq 1$: The general projection scale factor depends on the instantaneous acquisition geometry.

- Effective bandwidth (Hz)

  1. B_{eff}: The effective range bandwidth actually used in the polar-to-rectangular resampling process.
  2. $0 \leq B_{eff} \leq B_T$: B_T is the transmitted bandwidth (Hz)

- Effective center frequency.
  1. f_0: Actual center frequency of offset B_{eff} selected (Hz)
- Spatial frequency spacings on rectangular grid (radians per meter per sample)
  1. $\delta X = \dfrac{\Delta X}{N_x}$
  2. $\delta Y = \dfrac{\Delta Y}{N_y}$
- Rectangular grid samples
  1. N_x: Number of x-dimension samples.
  2. N_y: Number of y-dimension samples.
- Maximum image dimensions (in meters) supported by δX and δY
  1. $D_x = \dfrac{2\pi}{\delta X}$
  2. $D_y = \dfrac{2\pi}{\delta Y}$
- Image-formation FFT lengths (samples)
  1. N'_x: FFT length in x dimension. $N'_x \geq N_x$
  2. N'_y: FFT length in y dimension. $N'_y \geq N_y$
- Image scale factors (meters per sample spacing)
  1. $S_x = \dfrac{D_x}{N'_x}$
  2. $S_y = \dfrac{D_y}{N'_y}$
- Image oversampling ratios
  1. $OSR_x = \dfrac{N'_x}{N_x}$
  2. $OSR_y = \dfrac{N'_y}{N_y}$

- Image resolutions defined by -3 dB IPR widths (meters)

  1. $\rho_x = \dfrac{2\pi}{\delta X N_x} b_w$
  2. $\rho_y = \dfrac{2\pi}{\delta Y N_y} b_w$

- Typical IPR broadening factors

  1. $b_w \approx 0.88$: Uniform-weighted aperture
  2. $b_w \approx 1.46$: Hann-weighted aperture
  3. $b_w \approx 1.24$: 40 dB Taylor-weighted aperture

- Image formation with equal spatial extent, scale factors, and resolution

  1. $N_x = N_y = N$: Square rectangular grid
  2. $N'_x = N'_y = N'$: Square 2-D FFT
  3. $Q_{eff} = \dfrac{f_0}{B_{eff}}$: Select N_r to obtain B_{eff}. Compute f_0 from location of B_{eff} in Fourier space.
  4. $\theta_p \approx \dfrac{1}{Q_{eff}}$: Select N_c to obtain required angular aperture span in the desired processing plane.
  5. $OSR = \dfrac{N'}{N}$: x and y image oversampling ratios

3.11 A TYPICAL IMAGE-FORMATION PROCEDURE

It is informative to put much of what we have discussed about image formation into perspective by describing a typical, image-formation procedure; this procedure is what a scientist, engineer, or other researcher might follow to form a high-resolution spotlight SAR image from raw data tapes. Of course, the actual sequence of events, the parameter selection and fine tuning, and the detailed processing steps depend on how the data were gathered and what the final outcome is to be. For example, forming high-resolution imagery pairs from data gathered for interferometric purposes could impose processing restrictions and fine tuning that might otherwise be unnecessary if only a single high-resolution image is needed. In the following processing sequence, we will try to define those additional steps or considerations that might be needed when there are special circumstances. We have omitted many details, in this summary, in order to emphasize major concepts.

We assume we have at our disposal the entire phase history of a spotlight collection and all the associated radar parameter information or, at least, have some subset of the entire phase history selected to match a prior collection for interferometry, or to meet specific requirements for image resolution and scale factors (see Section 3.10). The phase history is stored on a pulse-by-pulse basis with a specified number of complex samples per pulse. Each demodulated return has been properly motion compensated so that the phase of a hypothetical target at the patch center is constant over the entire synthetic aperture time. If patch-center phase stabilization has not been totally performed in the receiver, auxiliary data must be provided with the phase history to allow ground-based phase stabilization. Uncompensated platform motion will be corrected by autofocus methods at a later time in the processing sequence.

It is also assumed that a corresponding pointing vector file is provided. This file provides the coordinates of the platform on a pulse-by-pulse basis. From this file, the image-formation geometry is established. We can then determine all geometry-based processing and compensation such as: out-of-plane correction, slant or ground-plane phase-history projections, interpolation, and image formation, etc.

With these initial requirements satisfied, we can now overview a typical image-formation process. Establishing the image-formation geometry begins by specifying a ground-plane unit normal vector $\bar{z}$. The pointing vector file is read, whereby all slant-plane and ground-plane unit vectors are established by the methods discussed in Appendix C. The operator then specifies whether the processing should take place in the slant plane, the ground plane, or in some other plane by selecting a processing-plane unit-normal vector.

Once the geometry is established, the radar parameters and pointing vectors determine where the demodulated returns lie in Fourier space. A rectangular grid is then established within the projected Fourier annulus. (The Fourier annulus is projected into the selected processing plane.) The number of grid samples, the number of polar records (demodulated pulses), the number of samples per pulse, and eventually the FFT size, will determine the various properties of the subsequent image, such as resolution, area coverage, and scale factors (see Section 3.10).

Polar-to-rectangular resampling is performed according to the methods discussed earlier in this chapter. The operator selects the length of the interpolating filters and the bandwidth reduction factors so that desired transition-band sharpness and aliased energy conditions are met. After resampling, the rectangular grid is zero-padded to an appropriate size for FFT image formation. The data are Taylor weighted for sidelobe control and a 2-D FFT is performed to yield a 2-D complex SAR image. A narrow band of image samples around the border will be discarded to leave an image free from undesirable edge effects (such as, filter aliasing and amplitude roll off).

Automatic focusing (autofocus) to correct for uncompensated platform motion or propagation-induced phase errors operates on a subset of the complex image formed at this stage. (See Chapter 4 for autofocus details.) After autofocus, the entire complex SAR image is available for detection, contrast manipulation, printing, and/or interpretation. The complex image could also be used with a matching image for interferometric processing. Together these yield terrain-elevation information or coherent change-detection products (see Chapter 5).

In practice, there are many decisions that must be made during image formation. However, the above overview is common to any SAR image-formation process. We believe the overview helps put into perspective those major processing steps required to form a high-resolution image, and that it helps to clarify some of the theoretical and practical issues discussed in this book.

REFERENCES

[1] H. M. Smith, *Principles of Holography*, Wiley-Interscience, New York, 1969.

[2] R. J. Collier, C. B. Burckhardt, and L. H. Lin, *Optical Holography*, Academic Press, New York, 1971.

[3] M. Born and E. Wolf, *Principles of Optics, 4th Edition*, Pergamon Press, Oxford, 1970.

[4] J. C. Dainty, Editor, *Topics in Applied Physics: Vol. 9, Laser Speckle and Related Phenomena*, Ch. 1 and 2, Springer-Verlag, Berlin, 1984.

[5] J. W. Goodman, *Statistical Optics*, Wiley-Interscience, New York, 1985.

[6] L. J. Porcello, N. G. Massey, R. B. Innes, and M. J. Marks, "Speckle Reduction in Synthetic Aperture Radars," *J. Opt. Soc. Am.*, Vol. 66, No. 11, pp 1305-1311, November 1976.

[7] F. K. Li, C. Croft, and D. N. Held, "Comparison of Several Techniques to Obtain Multiple-look SAR Imagery," *IEEE Transactions on Geoscience and Remote Sensing*, Vol. GE-2, No. 3, pp 370-375, July 1983.

[8] J. Walker, "Range-Doppler Imaging of Rotating Objects", *IEEE Transactions on Aerospace and Electronic Systems*, Vol. AES-16, No. 1, pp. 23-51, January 1980.

[9] D. C. Munson, J. D. O'Brien, and W. K. Jenkins, "A Tomographic Formulation of Spotlight-Mode Synthetic Aperture Radar", *Proceedings of the IEEE*, Vol. 71, No. 8, pp. 917-925, August 1983.

[10] D. A. Ausherman, A. Kozma, J. L. Walker, H. M. Jones, and E. C. Poggio, "Developments in Radar Imaging," *IEEE Transactions on Aerospace and Electronic Systems*, Vol. AES-20, No. 4, pp. 363-400, July 1984.

[11] D. A. Ausherman, "SAR Digital Image Formation Processing," SPIE Vol. 528: Digital Image Processing, pp. 118-133, 1985.

[12] G. A. Mastin and D. C. Ghiglia, "A Research-oriented Spotlight Synthetic a Aperture Radar Polar Reformatter", Sandia National Laboratories Report SAND90-1793, October 1990.

[13] G. A. Mastin and D. C. Ghiglia, "An Enhanced Spotlight Synthetic Aperture Radar Polar Reformatter," Sandia National Laboratories Report SAND91-0718, March 1992.

[14] D. A. Schwartz, "Analysis and Experimental Investigation of Three Synthetic Aperture Radar Formats," Coord. Sci. Lab., Univ. Illinois, Tech. Rep. T-94, March 1980 (M.S. Thesis).

[15] A. J. Jerri, "The Shannon Sampling Theorem--Its Various Extensions and Applications: A Tutorial Review," *Proceedings of the IEEE*, Vol. 65, No. 11, pp. 1565-1596, November 1977.

[16] A. V. Oppenheim and R. W. Schafer, *Digital Signal Processing*, Prentice-Hall, New Jersey, 1975.

[17] R. W. Hamming, *Digital Filters*, Prentice-Hall, New Jersey, 1977.

[18] N. K. Bose, *Digital Filters, Theory and Applications*, Ch. 4, "Design of FIR Filters," North-Holland, 1985.

[19] J. W. Adams, R. W. Bayma, and J. E. Nelson, "Digital Filter Design for Generalized Interpolation," Proceedings of the IEEE International Conference on Circuits and Systems, pp. 1299-1302, May 1989.

[20] J. W. Adams, "A New Optimal Window," *IEEE Transactions on Signal Processing*, Vol. 39, No. 8, pp. 1753-1769, August 1991.

[21] C. E. Cooke and M. Bernfeld, *Radar Signals: An Introduction to Theory and Application*, Academic Press, New York, 1967.

4

PHASE ERRORS AND AUTOFOCUS IN SAR IMAGERY

4.1 INTRODUCTION

In previous chapters, we discovered how the steps involved in the collection and formation of SAR imagery can be explained in light of tomographic principles. We also saw that several assumptions were made in order for that paradigm to be valid. One of these assumptions limited the allowable amount of wavefront curvature; a second assumption required that the residual phase term from the deramp processing be sufficiently small. We demonstrated how the reconstructed image quality is degraded when these assumptions are not met and derived quantitative conditions on patch size, resolution, standoff range, and several other parameters (see Equations 2.74 and 2.75) that render the degradation in image quality negligible. A third assumption dealt with robust image formation. This assumption is that the amount of time required for each radar pulse to travel from the SAR platform to the patch center and back is known precisely for each transmission point along the synthetic aperture. This demodulation time was defined as $\tau_0 = 2R_0/c$, where R_0 is the distance from the SAR platform to the patch center and c is the propagation velocity of the electromagnetic wave. Up to this point, we simply ignored the consequences of inaccuracies in demodulation times that can arise from noise in real SAR systems. In this chapter we address the matter in detail, because it becomes an important issue in the design and operation of real spotlight-mode SARs.

As we will show, reconstruction of an acceptable SAR image in general requires that the *relative* uncertainties in the distance R_0 from pulse to pulse across the synthetic aperture be kept to a fraction of a wavelength of the SAR center frequency. (A constant bias in R_0 for all pulses will have no effect on reconstructing $|g(x,y)|$.) The standard approach for estimating R_0 for each pulse is to employ electronic navigation systems, which use inertial measurement units (IMUs) placed onboard the collection platform. Generally, these systems use accelerometers, the outputs of which are double-integrated to estimate platform position from pulse to pulse. Modern IMU systems may incorporate ring laser gyro technology and may also include Global Positioning System (GPS) updates to increase the system accuracy.

Even with modern IMU systems, however, determining platform position to the required tolerances over the entire synthetic aperture can prove to be a difficult task. This is especially true for SAR systems designed to produce high-resolution imagery. As the azimuthal resolution demand increases, longer aperture lengths are required. These longer lengths allow the onboard IMU system more time to *drift* during the collection period, thus producing more substantial errors than occur with shorter apertures. Consequently, methods have been developed for: 1) increasing the accuracy of the IMU systems, and 2) post-processing the reconstructed radar imagery for automated (data-driven) removal of motion-induced artifacts. The option of improving the accuracy of IMU systems does not help other situations when the demodulation errors are not directly due to platform position uncertainty. For example, propagating radar energy through atmospheric turbulence can cause random delays in the signal. These unknown delays manifest themselves in exactly the same way on a pulse-to-pulse basis as errors caused by platform position uncertainty. Image restoration techniques known as *autofocus algorithms*, on the other hand, offer an attractive alternative. Using these techniques, we can remove the effects of demodulation errors independent of the error source. In addition, autofocus techniques eliminate the significant hardware costs associated with ultra-high-accuracy navigation systems.

The purpose of this chapter is to explain the effect of demodulation errors on a SAR image and then to describe several autofocus techniques used for image restoration. In Section 4.2, a mathematical model is developed for a single received pulse of SAR data when the demodulation time τ_0 is inaccurate. Section 4.3 explains the effect on the final formed image when each pulse is corrupted by a different demodulation error. In Sections 4.4 and 4.5, we present several autofocus methods used to correct the effects of these errors in real SAR imagery. These methods include two traditional autofocus techniques as well as a more modern algorithm that overcomes the deficiencies inherent in the earlier methods.

4.2 MATHEMATICAL MODEL FOR DEMODULATION INACCURACIES

In this section we derive the relationship between a demodulation error on an individual pulse and the resulting effects that follow range compression. We discover that under most realistic SAR collection conditions, the manifestation of this error can be modeled as a constant phase on each range-compressed pulse. The phase-error model provides a basis in subsequent sections for discussing defocus effects in the reconstructed SAR image, as well as for studying the algorithms that are used for correcting these effects.

In developing the spotlight-mode SAR tomographic paradigm of Chapters 1 and 2, we showed that the result of quadrature demodulation of a returned linear FM chirp pulse is a complex signal described by:[1]

$$\bar{r}(t) = G(U) = A \int_{-u_1}^{u_1} p(u) e^{-juU} du \tag{4.1}$$

where U and t are related by

$$U = \frac{2}{c}(\omega_0 + 2\alpha(t - \tau_0)) . \tag{4.2}$$

Here, A is an attenuation factor and $p(u)$ is a projection function of the three-dimensional target reflectivity density function, $g(x, y, z)$. The projection function is associated with the particular viewing geometry for the aperture position from which the pulse was transmitted. Equations 4.1 and 4.2 state that the processed return transduces certain spatial frequencies of the Fourier transform of the projection function, $p(u)$. According to the projection-slice theorem, these are equal to certain values of the three-dimensional Fourier transform of $g(x, y, z)$ on a linear segment. The range of spatial frequencies determined is prescribed by the radar center frequency and bandwidth, and is given by

$$\frac{2}{c}(\omega_0 - \alpha\tau_c) \leq U \leq \frac{2}{c}(\omega_0 + \alpha\tau_c) . \tag{4.3}$$

(Recall that the radar bandwidth is given by $B = \alpha\tau_c/\pi$.)

Equations 4.1 through 4.3 describe the phase history of a single processed pulse when no demodulation errors are present. Next, we analyze the effect of inaccurate knowledge of the radar-to-scene distance. A measurement error of δR in this distance causes a time demodulation error of $\varepsilon = (2/c)\delta R$. Thus, instead of mixing the return

[1] This expression is obtained when the phase skew (deramp residual) term of Equation 1.37 is ignored.

signal with the expressions given by Equations 1.31 and 1.32 of Chapter 1, the return signal is mixed with

$$\cos[\omega_0(t - \tau_0 + \varepsilon) + \alpha(t - \tau_0 + \varepsilon)^2] \tag{4.4}$$

and

$$-\sin[\omega_0(t - \tau_0 + \varepsilon) + \alpha(t - \tau_0 + \varepsilon)^2]\,. \tag{4.5}$$

The output of the quadrature demodulator then becomes

$$G_\varepsilon(U) = Ae^{-j\varepsilon^2\alpha}e^{j\frac{\varepsilon c}{2}U}\int_{-u_1}^{u_1} p(u)\,e^{-juU}\,du \tag{4.6}$$

so that the corrupted and uncorrupted phase-history expressions are related as

$$G_\varepsilon(U) = e^{-j\varepsilon^2\alpha}e^{j\frac{\varepsilon c}{2}U}G(U)\,. \tag{4.7}$$

Consider first the phase term with argument $-\varepsilon^2\alpha$ that multiplies $G(U)$. As it turns out, maintaining relative-position uncertainties of the SAR platform to well less than a range-resolution cell size (e.g., 1 meter) is easily achievable by modern inertial navigation systems. This is the same as saying that the time-delay error is much less than the reciprocal of the radar bandwidth, because[2]

$$\frac{\varepsilon c}{2} \ll \frac{c}{2B} \tag{4.8}$$

implies that

$$\varepsilon \ll \frac{1}{B}\,. \tag{4.9}$$

Equation 4.9 taken with the relation $\alpha = \pi B/\tau_c$, yields the inequality

$$\varepsilon^2\alpha \ll \frac{\pi}{\tau_c B}\,. \tag{4.10}$$

The denominator on the right side of the above inequality is the *time-bandwidth* product of the transmitted waveform. In Chapter 1 we saw that the FM chirp waveforms typically employed in spotlight mode SARs have very large time-bandwidth products (nominally on the order of thousands). This indicates that $\varepsilon^2\alpha$ is much less than unity, which allows us to use the approximation $e^{-j\varepsilon^2\alpha} \approx 1$ and simplifies Equation 4.7 to

$$G_\varepsilon(U) = e^{j\frac{\varepsilon c}{2}U}G(U)\,. \tag{4.11}$$

In other words, $G_\varepsilon(U)$ is simply $G(U)$ altered by a linear phase term. Figure 4.1 illustrates this phase ramp as a function of U. This figure also re-emphasizes the fact that the data transduced are *Fourier offset* by the amount $U_0 = 2\omega_0/c$. As we

[2] Recall that the range resolution is given by $\rho = c/(2B)$.

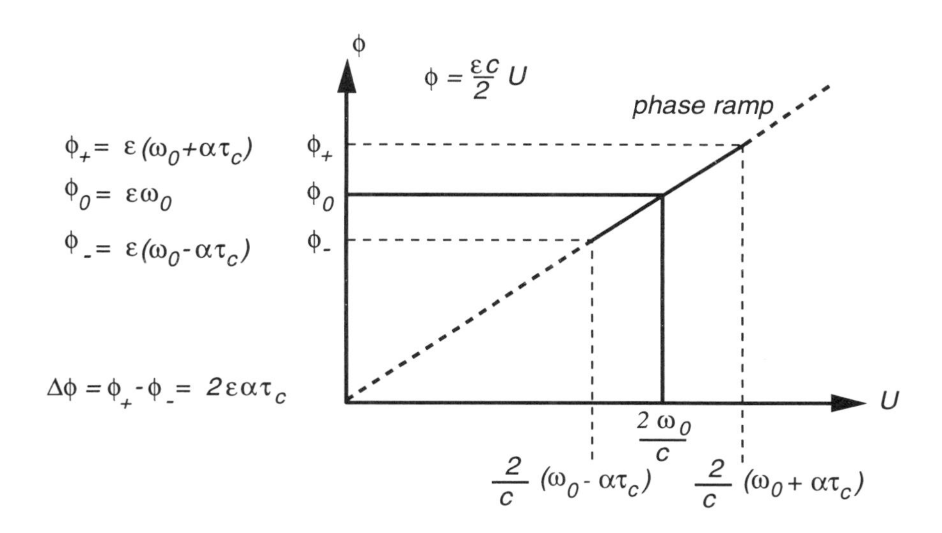

Figure 4.1 Illustration of the linear phase term applied to a single return pulse when $\varepsilon \neq 0$. The figure illustrates that the data of interest are bandlimited and centered at $\frac{2\omega_0}{c}$.

saw in Chapters 2 and 3, the received data are actually treated as baseband versions of Equations 4.1 and 4.2 in the image-formation process, wherein the offset spatial frequency U_0 is removed. That is, the origin of the spatial-frequency domain is taken to be the center of the support of the phase-history data. The basebanded data for the case of no demodulation errors can therefore be represented by

$$G_b(U) = G(U + U_0)\, W(U) \tag{4.12}$$

where $W(U)$ is a windowing function that imposes the restricted range of U given by

$$-\frac{2}{c}\alpha\tau_c < U < \frac{2}{c}\alpha\tau_c \; .$$

The basebanded version of $G_\varepsilon(U)$ is represented by the equation

$$\begin{aligned} G_{b\varepsilon}(U) &= G_\varepsilon(U + U_0)W(U) \\ &= e^{j\varepsilon\omega_0} e^{j\frac{\varepsilon c}{2}U} G(U + U_0)W(U) \\ &= e^{j\varepsilon\omega_0} e^{j\frac{\varepsilon c}{2}U} G_b(U) \; . \end{aligned} \tag{4.13}$$

This equation indicates that the linear phase term on the corrupted basebanded signal

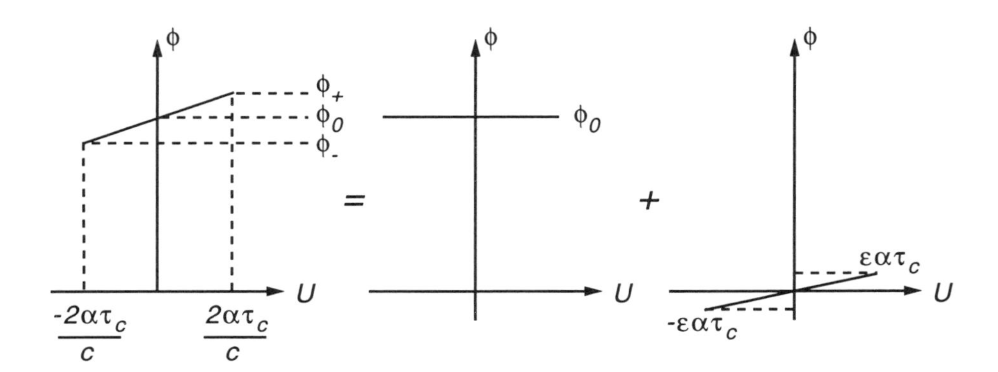

Figure 4.2 Degrading phase function for a single pulse following removal of offset frequency (basebanding). The phase can be decomposed into a constant part, $\varepsilon\omega_0$, and a ramp with slope $\varepsilon c/2$.

can be expressed as the sum of a constant phase $\varepsilon\omega_0$ and a phase ramp with slope $\varepsilon c/2$ as shown in Figure 4.2.

We are ultimately concerned with how these phase terms effect the *range-compressed* data in the image-formation process. The range-compressed pulse is calculated by taking the one-dimensional inverse Fourier transform[3] of $G_{b\varepsilon}(U)$. Denoting the corrupted and uncorrupted range compressed pulses as $g_\varepsilon(u)$ and $g(u)$, and using the relationship of Equation 4.13 we have

$$\begin{aligned} g_\varepsilon(u) &= \mathcal{F}^{-1}\{G_{b\varepsilon}(U)\} \\ &= e^{j\varepsilon\omega_0}\mathcal{F}^{-1}\{e^{j\frac{\varepsilon c}{2}U}G_b(U)\} \end{aligned} \tag{4.14}$$

so that the compressed pulses are related as

$$g_\varepsilon(u) = e^{j\varepsilon\omega_0}g(u+\frac{\varepsilon c}{2})\,. \tag{4.15}$$

Thus, the corrupted range-compressed pulse has been altered by a constant phase and also shifted by an amount equal to the platform position uncertainty, $\delta R = \varepsilon c/2$. Figure 4.3 illustrates this effect by showing two targets in both a corrupted and uncorrupted pulse. Note that a target with complex reflectivity a_1, positioned at u_1 in the uncorrupted range-compressed pulse, is moved to position $u_1 - c\varepsilon/2$ and is modified by a phase term of $\varepsilon\omega_0$ in the corrupted pulse.

[3] We acknowledge that the polar reformatting step is performed before either range or azimuth compression is executed. In many cases, however, the effects of the reformatting on the phase-error analysis are negligible.

Equation 4.8 indicates that the shift in the range-compressed pulse due to a demodulation error is small enough to be ignored. It then remains for us to analyze the effects of the phase-error term $\varepsilon\omega_0$ on the reconstructed image. Accordingly, we ignore the shift and model the corrupted compressed pulse simply as

$$g_\varepsilon(u) = e^{j\varepsilon\omega_0} g(u) . \tag{4.16}$$

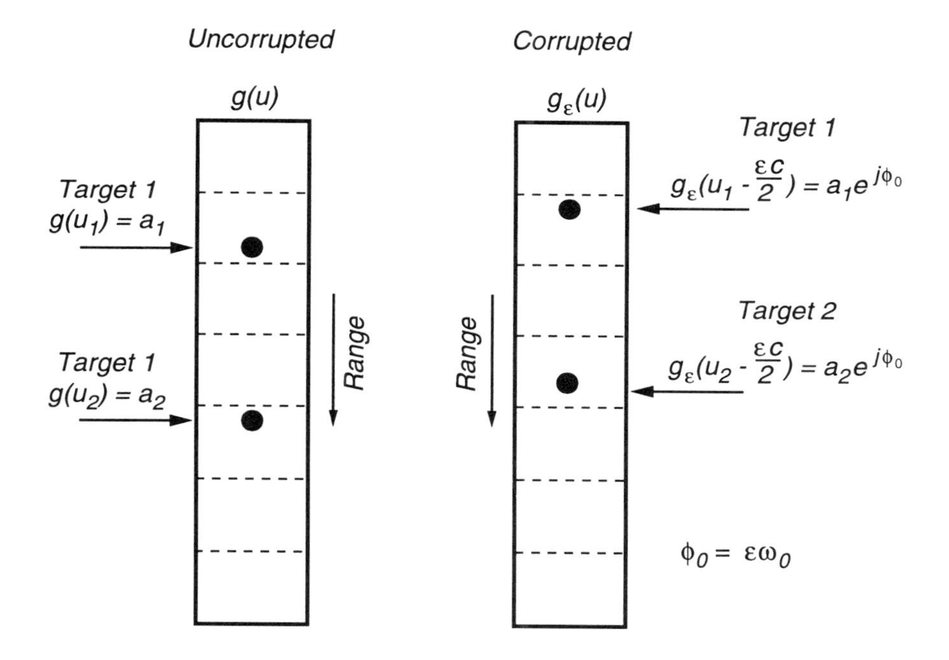

Figure 4.3 Illustration comparing a range-compressed pulse with and without the effects of demodulation-time error. In this hypothetical case, there exist two targets with complex responses a_1 and a_2 located at positions u_1 and u_2, respectively. Comparing the two cases shows that the corrupted pulse is related to the uncorrupted one by a shift and a constant phase-error term. In practice, the size of the shift, $\varepsilon c/2$, is a small fraction of the resolution range cell size; thus, its effects are negligible. As a result, the net effect of the demodulation-time error is to impose a constant phase, $\phi_0 = \varepsilon\omega_0$, across the range-compressed pulse.

As we continue our analysis of $g_\varepsilon(u)$ in the following section, we will find it beneficial to change our notation to reflect that the demodulated range-compressed data are sampled instead of continuous. To this end, we will denote the sampled uncorrupted and corrupted range-compressed data for range line k and aperture position m as $\bar{g}(k,m)$ and $\bar{g}_\varepsilon(k,m)$, respectively. (The reader should carefully note that the notation, $g(k,n)$, denotes image-domain data, i.e., data that are compressed in both dimensions, with n as the cross-range index. The overbar notation, $\bar{g}(k, m)$, denotes data that are range-compressed but not azimuth-compressed, with m as the aperture-position index.) Rewriting Equation 4.16 using this new notation we have

$$\bar{g}_{\varepsilon}(k, m) = e^{j\phi(m)} \bar{g}(k, m) \tag{4.17}$$

where $\phi(m) = \varepsilon(m)\, \omega_0$.

Summary of Critical Ideas from Section 4.2

- **An incorrect demodulation time on a given received pulse applies a phase term to the corresponding demodulated data.**
- **Because the demodulated data are Fourier offset, the imposed phase term can be modeled as a constant phase plus a phase ramp on the bandlimited data of interest.**
- **With the offset ($U_0 = 2\omega_0/c$) removed, the phase ramp portion of the error results in a *shift* in the range-compressed data, while the constant phase portion of the error remains as a constant phase in the range-compressed data.**
- **The shift in the range-compressed data due to the phase ramp is ignored, because the shift amount is typically a small fraction of a range resolution cell.**

4.3 PHASE ERRORS INDUCED ACROSS THE APERTURE

The analysis of the previous section describes how an error in the demodulation time τ_0 results in a range-compressed pulse with a constant phase error imposed. Because a spotlight-mode SAR collection comprises many pulses and because each pulse is generally subjected to a different error in demodulation time, a phase function in the form of a two-dimensional *ribbon* is induced on the range-compressed phase-history data. The function is constant in the range dimension, but can have an arbitrary variation in the aperture-position dimension. This concept is illustrated in Figure

4.4. Each dot in the figure represents the phase-error value for given aperture and range positions. The continuous curves illustrate the same sequence, or slice, of phase errors taken along the aperture dimension at every range value. We refer to this one-dimensional variation of phase vs. aperture position as the *aperture phase-error function* , or simply as the phase-error function.

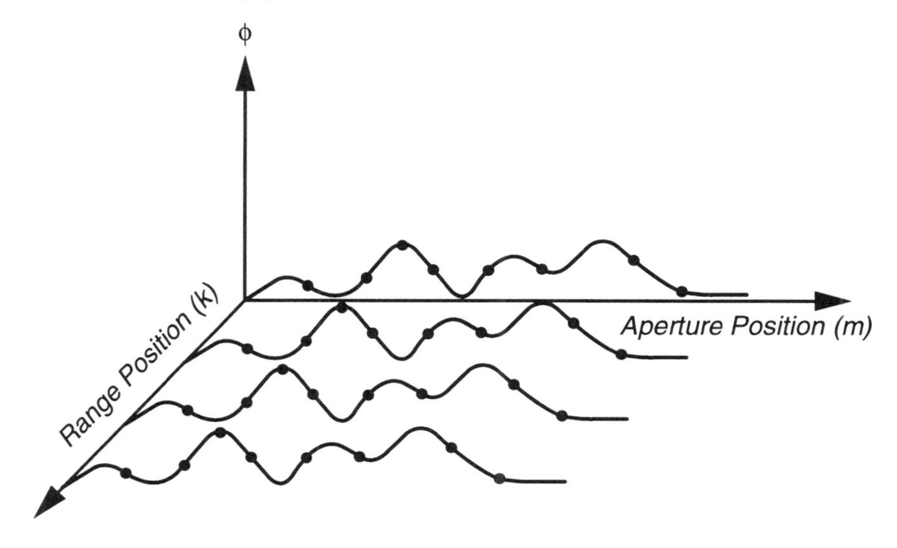

Figure 4.4 Illustration of phase error as a function of aperture position and range in the range-compressed data space. Because the phase function is constant in range, a *slice* in the aperture direction produces the phase-error function.

It remains to be seen how the presence of an aperture phase-error function affects the reconstructed SAR image. The final step in forming the image from range-compressed data is an inverse Fourier transform in the cross-range dimension. (Recall that the two-dimensional Fourier transformation from phase-history domain data to the image domain is separable and computed as two one-dimensional transforms, range compression and cross-range compression.) Using a well-known convolution theorem, we can compute the inverse discrete Fourier transform of Equation 4.17 as

$$\begin{aligned} g_\varepsilon(k,n) &= IFFT_m\{\bar{g}_\varepsilon(k,m)\} \\ &= IFFT_m\{e^{j\phi(m)}\bar{g}(k,m)\} \\ &= IFFT_m\{e^{j\phi(m)}\} \otimes g(k,n) \end{aligned} \tag{4.18}$$

where $\otimes$ indicates the discrete convolution operation and $g_\varepsilon(k,n)$ and $g(k,n)$ denote the corrupted and uncorrupted image-domain data. (The notation $IFFT_m\{\cdot\}$ denotes the one-dimensional inverse Fast Fourier Transform across the aperture position dimension m. Later in this chapter, we use the notation FFT_n to denote the forward Fast Fourier Transform on the (image-domain) cross-range dimension n.) The result is that the uncorrupted image-domain data are convolved in the cross-range dimension with the inverse Fourier transform of a phase-only function, the argument of which is the aperture phase-error function, $\phi(m)$. The output of this convolution is a *defocused* version of the uncorrupted image data, where the defocus occurs in the cross-range direction. Both the shape and the amplitude of the phase-error function dictate the characteristics of the target defocus.

As suggested in the previous section, phase errors in SAR imagery occur as the result of inaccurate knowledge of the SAR platform position as each transmitted pulse in the synthetic aperture is processed by quadrature demodulation. One mode in which such position uncertainties are commonly generated arises from errors in measuring the aircraft velocity. For the analysis of these velocity-induced phase errors, we consider the simplified two-dimensional collection geometry shown in Figure 4.5. The aircraft is d meters from the center of the aperture, which is R_0 meters from the scene patch center. The demodulation time τ_0 at any point in the aperture is easily shown to be

$$\begin{aligned} \tau_0 &= \frac{2}{c}\sqrt{R_0^2 + d^2} \\ \tau_0 &\approx \frac{1}{c}\left(2R_0 + \frac{d^2}{R_0}\right) \\ &= \frac{1}{c}\left(2R_0 + \frac{(Vt)^2}{R_0}\right) \end{aligned} \tag{4.19}$$

where V is the measured aircraft velocity.

If the actual aircraft velocity is V_a, then the correct demodulation time should be

$$\tau_0' \approx \frac{1}{c}\left(2R_0 + \frac{(V_a t)^2}{R_0}\right) \tag{4.20}$$

This leads to a demodulation timing error of

$$\varepsilon(t) = \tau_0 - \tau_0' = \frac{1}{cR_0}(V_a^2 - V^2)\, t^2 \tag{4.21}$$

resulting in a *quadratic function* of aperture time t (also called *slow-time*), or equivalently, a quadratic function of aperture position, $V_a t$ (denoted as m in the sampled data). Converting this timing error to the equivalent phase error using the relationship $\phi(t) = \varepsilon(t)\,\omega_0$, we obtain

$$\phi(t) = \varepsilon(t)\,\omega_0 = \frac{\omega_0}{cR_0}(V_a^2 - V^2)\, t^2 \tag{4.22}$$

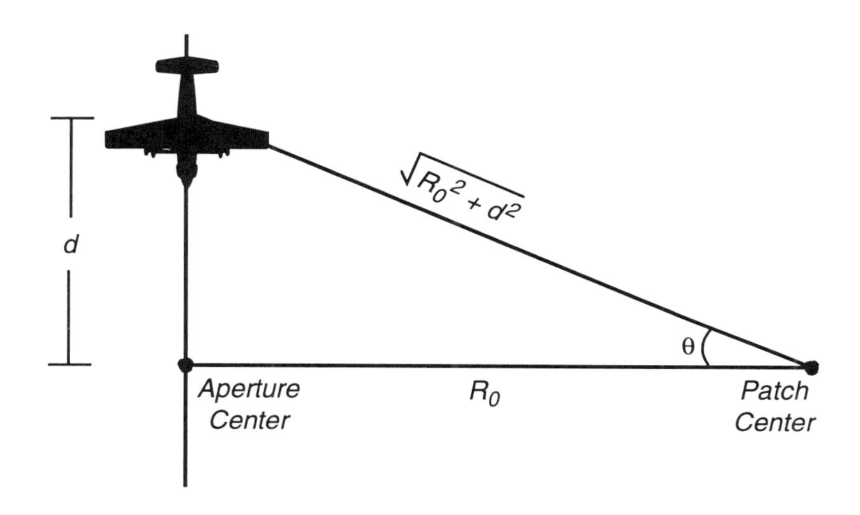

Figure 4.5 Illustration of a simplified two-dimensional collection geometry. The aircraft is d meters from the aperture center and the aperture center is R_0 meters from the patch center.

We now examine the magnitude of velocity error that would result in significant image defocus. Recall that earlier we analyzed the effects of a quadratic phase term on image quality when we discussed the limitations of the tomographic paradigm in Section 2.6. In Figure 2.29 we demonstrated that for a peak quadratic phase error of $\pi/4$ radians or less, the amount of degradation was negligible. Applying this same criterion to the expression of Equation 4.22 yields the maximum velocity error that results in no significant degradation of image quality. We start by rewriting Equation 4.22 as

$$\begin{aligned} \phi(t) &= \frac{\omega_0}{cR_0}(V_a + V)(V_a - V)\, t^2 \\ &\approx \frac{\omega_0}{cR_0}(2V_a)(\delta V)\, t^2 \end{aligned} \tag{4.23}$$

where δV is the velocity error . If the total aperture time is T, the synthetic aperture length is $L_a = V_a T$ and the angular diversity of the aperture is $\Delta\theta = L_a/R_0$. Combining these expressions with the fact that the azimuthal resolution is related to

$\Delta\theta$ as $\rho_x = \lambda/(2\Delta\theta)$, we obtain

$$\phi_{max} = \frac{\pi}{4} = \frac{\omega_0}{cR_0}(2V_a)(\delta V_{max})\frac{T^2}{4} \tag{4.24}$$

so that

$$\delta V_{max} = \frac{\lambda R_0 V_a}{4(V_a T)^2} = \frac{\lambda R_0^2 V_a}{4 R_0 L_a^2} = \frac{V_a}{\lambda R_0}\left(\frac{\lambda}{2\Delta\theta}\right)^2 = \frac{V_a}{\lambda R_0}\rho_x^2 \tag{4.25}$$

The above expression for δV_{max} indicates that it has a *quadratic dependence on the azimuthal resolution*. Therefore, it is clear why high-resolution SAR systems are particularly susceptible to demodulation errors. The effects of velocity errors are exacerbated as the required aperture length increases.

An interesting additional insight into this quadratic defocus due to a velocity error is obtained when the phase error is expressed as a function of the along-the-line-of-sight (range) position errors, $\delta R(t)$. That is, Equation 4.22 is equivalent to

$$\phi(t) = \varepsilon(t)\,\omega_0 = \frac{2\,\delta R(t)}{c}\omega_0 = \frac{4\pi}{\lambda}\,\delta R(t)\;. \tag{4.26}$$

The corresponding maximum position error along the aperture that results in negligible defocus effects (equivalent to less than $\pi/4$ radians of peak quadratic phase error) is then calculated as

$$\phi_{max} = \frac{\pi}{4} = \frac{4\pi}{\lambda}\delta R_{max} \tag{4.27}$$

so that

$$\delta R_{max} = \frac{\lambda}{16}\;. \tag{4.28}$$

Equation 4.28 states that no more than one sixteenth of a wavelength of *relative* range position error along the synthetic aperture can be tolerated if significant defocus is to be avoided. Again, a "dc" bias in the range position measurement across the aperture is unimportant. It is the "ac" error component that counts here. Interestingly, this result is independent of the range, resolution, or actual aircraft velocity.

Figure 4.6(a) shows a well-focused SAR image of an urban scene. Figure 4.6(b) shows the defocused image that results from applying the quadratic phase error of Figure 4.7, according to the mathematical model for aperture phase errors developed in the previous section (Equation 4.17). The phase-error function is shown in units of radians, as well as in equivalent relative position uncertainty (expressed as wavelengths) required to produce this error (see Equation 4.26). Figure 4.8 shows the

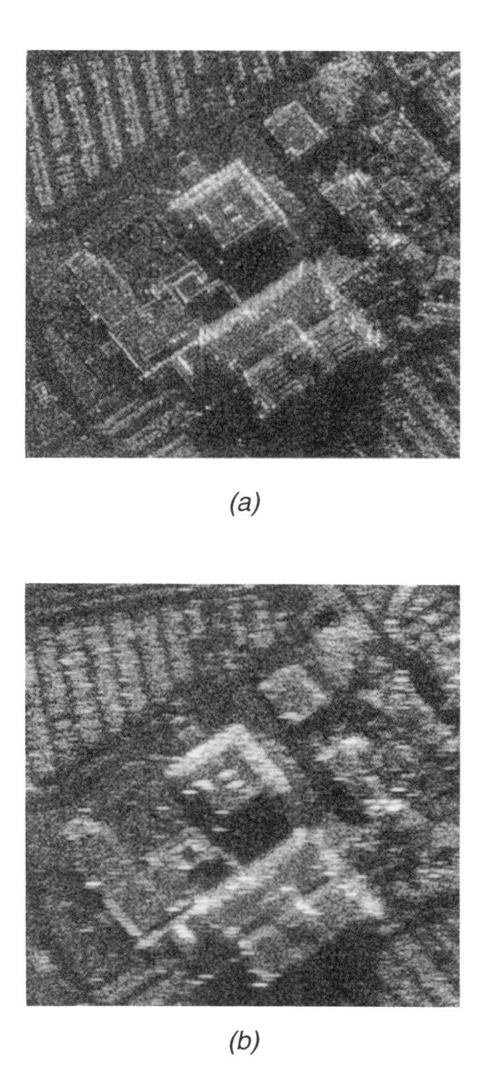

(a)

(b)

Figure 4.6 SAR image of an urban scene. (a) Well focused image. (b) Degraded image caused by a quadratic phase error.

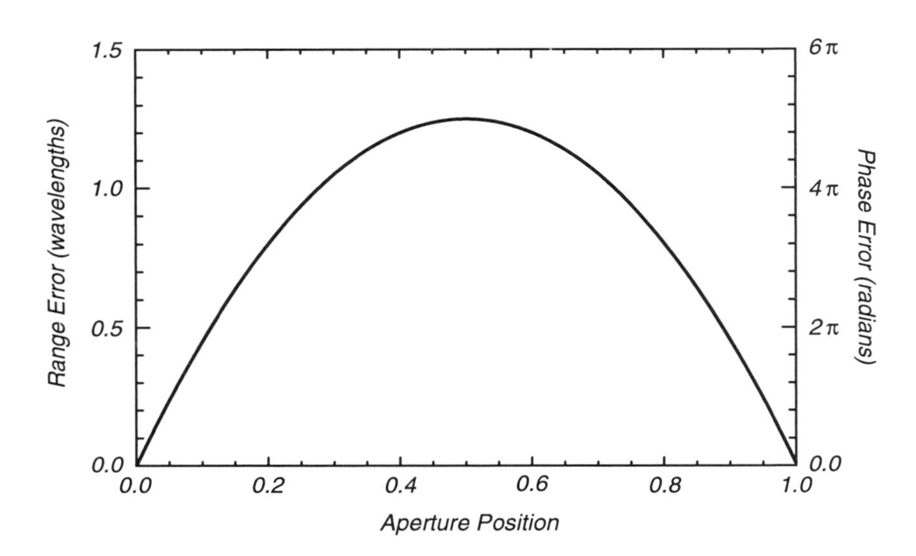

Figure 4.7 Quadratic phase-error function used to degrade image. Note that an error of one wavelength corresponds to 4π radians of error, because of the two-way propagation (see Equation 4.26).

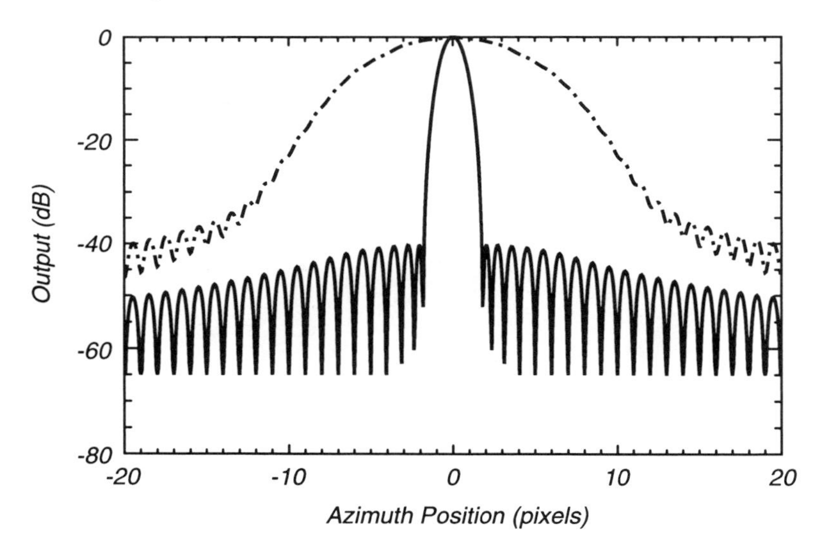

Figure 4.8 Impulse response (IPR) from focused image (solid line) and impulse response from the same image degraded by a quadratic phase error (dashed line). The responses have been produced by weighting the aperture domain data with a 40 dB Taylor window function.

effects of defocus as broadening of the IPR of a point reflector in the scene. In this case, the mainlobe is broadened by a factor of approximately five. This example clearly demonstrates our prediction of how severely a SAR image can be defocused by a quadratic aperture phase function corresponding to only slightly more than one wavelength (peak) of relative range uncertainty. Of course, a constant aircraft velocity error as described in the above example is not the only mode of navigational uncertainty that arises in inertial measurement systems. More complicated uncompensated motion errors can induce phase functions that are not quadratic, but that are instead more rapidly varying functions of aperture position. In addition, propagation of microwave energy through the troposphere or ionosphere can result in phase errors due to random transmission path delays (see references [1] through [4]). These errors are typically modeled as a *power-law* random process, which has a spectral density of the form $S_{\Phi\Phi}(\omega) = c_1\omega^{-p}$, where c_1 is a constant, p is the spectral index, and ω represents spatial frequency. In other words, the spectral density of the phase-error function decreases linearly with slope $-p$ as a function of spatial frequency when plotted on a log-log scale.

Propagation-induced phase errors generally have a higher frequency content, i.e., more rapid variation across the synthetic aperture, than do those generated by aircraft position uncertainties. Figure 4.9 illustrates an image formed by corrupting the urban scene of Figure 4.6(a) with the simulated propagation-induced phase error illustrated in Figure 4.10. The degradation of an ideal point-target response from this phase error is shown in Figure 4.11. Clearly, these types of errors raise the sidelobes of a target rather than simply spreading the mainlobe, as does a quadratic phase function (see Figure 4.8). In this case, the peak sidelobe level is raised by 30 dB. Visually, this results in a loss of contrast in the image.

The two examples illustrate that the possible types of degrading phase errors in spotlight-mode SAR data may vary from low-frequency (platform-motion induced) to high-frequency (propagation-induced) functions. Increasing the accuracy of the onboard IMU system will certainly decrease any low-frequency demodulation problems produced by platform position uncertainty. On the other hand, a *data-driven* restoration algorithm has the potential for extracting phase errors independent of their origin. This is the subject of *autofocus*, or automatic phase-error correction. We examine autofocus extensively in the remaining sections of this chapter.

Figure 4.9 Urban SAR scene degraded by a rapidly varying (high frequency) phase error. The peak-to-peak phase variation corresponds to only 1/4 wavelength of relative range error.

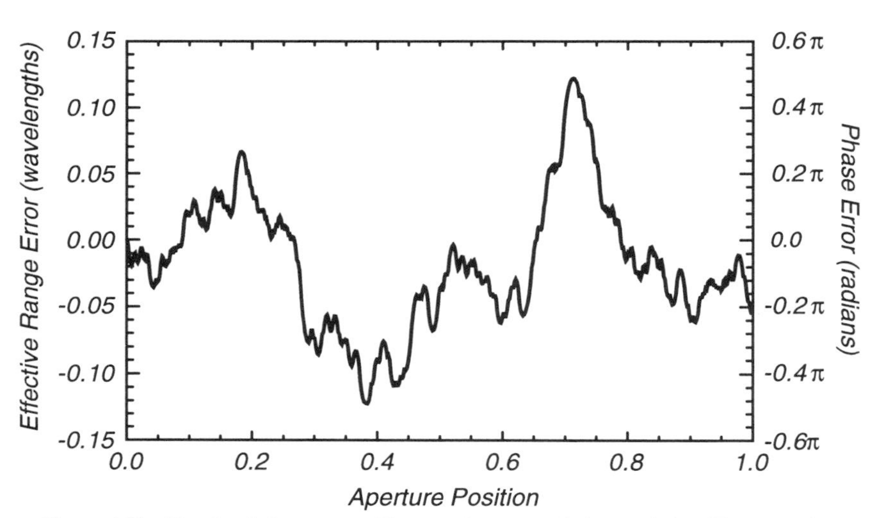

Figure 4.10 Simulated phase error with power-law spectral characteristics. The spectral index used here is $p = 2.5$.

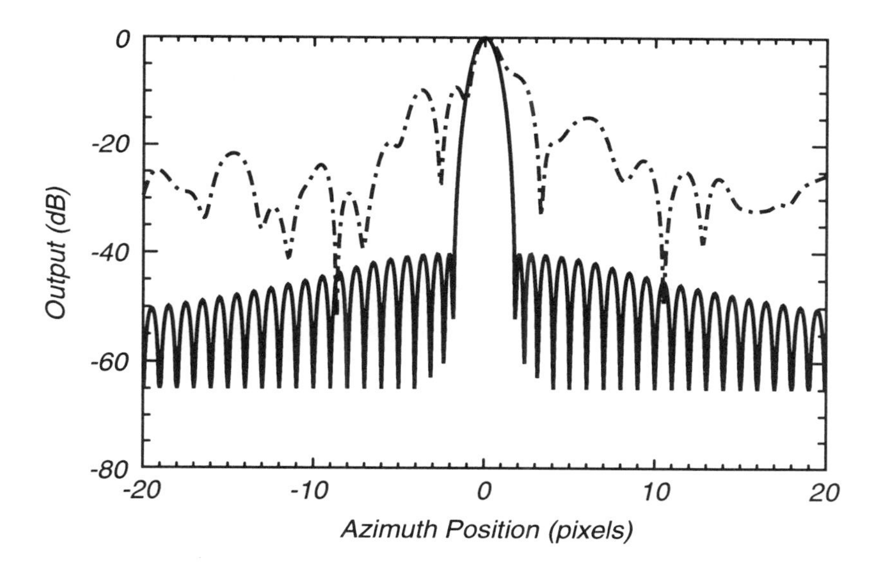

Figure 4.11 Impulse response (dashed line) for the image degraded by the high-frequency (power-law) phase-error function of Figure 4.10. IPR of the original well-focused image is shown by the solid line. Note that the defocus raises the peak sidelobe level by 30 dB in this case.

Summary of Critical Ideas from Section 4.3

- **Demodulation timing errors that vary from pulse to pulse across a synthetic aperture collection induce a one-dimensional phase-error function across the range-compressed data domain. The phase errors are a function of aperture position but are not a function of range.**

- **The effect of an aperture phase-error function on the reconstructed (azimuthally-compressed) image is to cause defocus in the cross-range dimension.**

- **The defocus effect can be viewed as the convolution of the uncorrupted image-domain data (scene reflectivity function) with the inverse Fourier transform of $e^{j\phi(m)}$.**

Summary of Critical Ideas from Section 4.3 (cont'd)

- **Different types of phase-error functions can result in rather different defocusing effects.**
- **An often-encountered phase-error function in SAR systems is a quadratic phase error. One source of this error is an incorrect platform velocity measurement. This phase error tends to spread the mainlobe of the IPR, resulting in a blurring of the image.**
- **"High-frequency" phase errors can also be induced across the aperture due to propagation effects or due to certain forms of uncompensated platform-motion errors. These phase errors do not spread the mainlobe of the IPR as much as they raise the sidelobe level, resulting in a loss of contrast in the image.**

4.4 PHASE CORRECTION USING CONVENTIONAL APPROACHES

This section outlines two traditional methods used for correcting (autofocusing) SAR imagery that has been defocused by phase errors. These methods are the *inverse filtering* and the *map-drift* techniques. The fundamental concepts of both algorithms are presented along with a discussion of the limitations that each present. In Section 4.5, we describe a modern robust methodology for phase-error correction that overcomes the deficiencies of both of the traditional approaches.

4.4.1 Phase Estimation Using Inverse Filtering

The inverse filtering technique is a simple and straightforward approach to phase-error correction. This method is based on information contained in the defocus effects of a single point target. Recall from Equation 4.18 that the corrupted image data may be modeled as the original image data convolved with the inverse Fourier transform of $e^{j\phi(m)}$. Mathematically, our model is

$$g_\varepsilon(k,n) = h(n) \otimes g(k,n) \tag{4.29}$$

where

$$h(n) = IFFT_m\{e^{j\phi(m)}\} \tag{4.30}$$

is the blur function, $\otimes$ indicates the discrete convolution operation, m indexes aperture position, and n is the cross-range image-domain index.

The requirement of any autofocus algorithm is to produce an estimate of $g(k,n)$, given only the corrupted image-domain observation $g_\varepsilon(k,n)$. Because we assume that no information about the corrupting phase error $\phi(m)$ is available *a priori*, it appears at the outset that this problem has no general solution. The inverse filtering technique circumvents this problem by assuming that the effects of the phase-error function on a single point target can be isolated in the defocused image. Consider an idealized situation where a single point target exists on range line k at the center cross-range position. The image-domain data for that range line is given by

$$g(k,n) = a\,\delta(n) \tag{4.31}$$

where

$$\delta(n) = \begin{cases} 1 & \text{if } n = 0 \\ 0 & \text{if } n \neq 0 \end{cases}$$

and a is a complex constant representing the point target reflectivity. The corresponding corrupted data for that range line is

$$\begin{aligned} g_\varepsilon(k,n) &= a\,IFFT_m\{e^{j\phi(m)}\} \otimes \delta(n) \\ &= a\,IFFT_m\{e^{j\phi(m)}\}\ . \end{aligned} \tag{4.32}$$

The phase error $\phi(m)$ can now be determined to within a constant by simply measuring the phase of the Fourier transform of $g_\varepsilon(k,n)$, i.e.

$$\begin{aligned} \hat{\phi}(m) &= \angle\{FFT_n\{g_\varepsilon(k,n)\}\} \\ &= \angle\{a\,e^{j\phi(m)}\} \\ &= \angle a + \phi(m) \end{aligned} \tag{4.33}$$

where $\angle$ indicates the argument or phase of the complex number.

In reality of course, one can seldom rely on a SAR image having a single isolated point target positioned at the center of the scene on any range line. However, the ideal inverse filtering concept can be approximated if a strong point-like target that is reasonably well-isolated from other surrounding targets and clutter can be located. To this end, a window is applied so that all image data outside of the window are set to zero. The width of the window is based on an estimate of the nominal support of the blurring function. These modified data are then placed in an array with the center of the window at the array center, and are azimuthally decompressed (i.e., transformed to the range-compressed domain) via one-dimensional Fourier transformation. The phase error is then measured directly as prescribed by Equation 4.33.

Finally, the entire image is corrected by multiplying all range lines of the range-compressed data by the complex conjugate of the measured phase function, followed by azimuth compression. (Recall that our model for the phase-error function assumes that it does not vary with range position.)

The phase correction process outlined above is described mathematically as

$$\begin{aligned}\bar{g}_c(k,m) &= \bar{g}_\varepsilon(k,m)e^{-j\hat{\phi}(m)} \\ &= \bar{g}_\varepsilon(k,m)e^{-j(\phi(m)+\phi_r(m)+\angle a)}\end{aligned} \tag{4.34}$$

where $\bar{g}_c(k,m)$ is the corrected range-compressed data, $\bar{g}_\varepsilon(k,m)$ is the uncorrected (corrupted) range-compressed data, and $\hat{\phi}(m)$ is the estimated phase-error function. In general, the estimated phase will be equal to the actual phase-error function, plus the unknown target phase $\angle a$, plus a residual phase function $\phi_r(m)$. This residual phase is the result of the selected windowed data never being totally free of competing clutter or interference from neighboring point targets.

The final estimate of the corrected image is then given by

$$g_c(k,n) = IFFT_m\{\bar{g}_c(k,m)\} = e^{-j\angle a}g(k,n) \otimes b_r(n) \tag{4.35}$$

where

$$b_r(n) = IFFT_m\{e^{-j\phi_r(m)}\} \ . \tag{4.36}$$

The corrected image is equal to a convolution of the original, uncorrupted image with a residual blurring function, $b_r(n)$. (The unknown constant phase $\angle a$ clearly has no effect on the *magnitude* of the reconstructed image.) If the point target selected for the synthesis of the inverse filter is well isolated and large compared to the surrounding clutter level, the effects of the residual blurring function will be negligible and the image will be restored completely.

An example illustrating such a case where the inverse filtering technique works relatively well is presented in Figure 4.12. Figure 4.12(a) shows a well-focused urban scene. Figure 4.12(b) shows the scene that has been artificially degraded by the phase-error function indicated by the dotted line in Figure 4.13. This particular scene contains many bright reflectors, one of which was selected for the inverse filtering operation. Its location is depicted by the arrow in the figure. The target defocus (blur) width was estimated to be 64 pixels. The 64 pixels of data (on the same range line) surrounding this target were extracted and placed in an array. The data were Fourier transformed in the cross-range direction and the phase-error function was estimated by measuring the phase of the range-compressed data at each aperture position. The measured phase error is compared to the actual applied phase in Figure 4.13. Clearly, there is overall good agreement between the applied and the estimated phase error. The result of using the estimated phase to correct the degraded image is shown in Figure 4.12(c). As would be expected, the selected target is well-focused. The remaining image features are also generally well-focused, but it is clear that some residual streaking remains. This indicates that the phase-error estimate contains some residual error, due to the fact that the selected defocused target data were not completely free of interfering clutter. The insert of Figure 4.13 confirms this.

Although the relatively straightforward theoretical concepts of inverse filtering make it appealing, the results of applying this method in practice to a large class of images and phase-error functions fall far short of expectations. There are at least two reasons for this unspectacular performance. First, it is often difficult to find a *strong* point target in real defocused SAR imagery. This is especially true if the image is that of a rural scene or if it is corrupted by a relatively large phase-error function. Second, because targets are rarely *totally isolated*, there almost always exist components of surrounding clutter in the neighborhood of the selected target, which result in residual defocus effects.

Figures 4.14 and 4.15 illustrate the results of the inverse filtering technique for a typical rural SAR scene. The original scene shown in Figure 4.15(a) was artificially corrupted by the low-order phase error used in the previous example. The resulting corrupted image is shown in Figure 4.15(b). The candidate target selected for use in the inverse filtering technique is identified by the arrow placed in the images of the figure. Although this target is relatively bright, it is not isolated to the extent that we can assume that the surrounding clutter level is negligible.

The data supporting the blur of this particular target were extracted from the image (approximately 60 pixels), centered in a separate one-dimensional array, zero-padded to size 64 pixels and Fourier transformed to the range-compressed domain. Figure 4.14 compares the actual corrupting phase to that which is measured by the inverse filtering approach. Using this estimated corrupting phase, shown by the solid line of Figure 4.14, we correct the degraded image by multiplying the range-compressed data by the complex conjugate of the estimated phase. After transforming the corrected range-compressed data to the image domain, we obtain the image shown in Figure 4.15(c). It is clear that the refocused image is not fully restored by this technique. Again, this is primarily due to the fact that the point selected for phase estimation is not free of competing clutter. This example also illustrates that defocus effects in the imagery can make locating suitable point targets quite difficult.

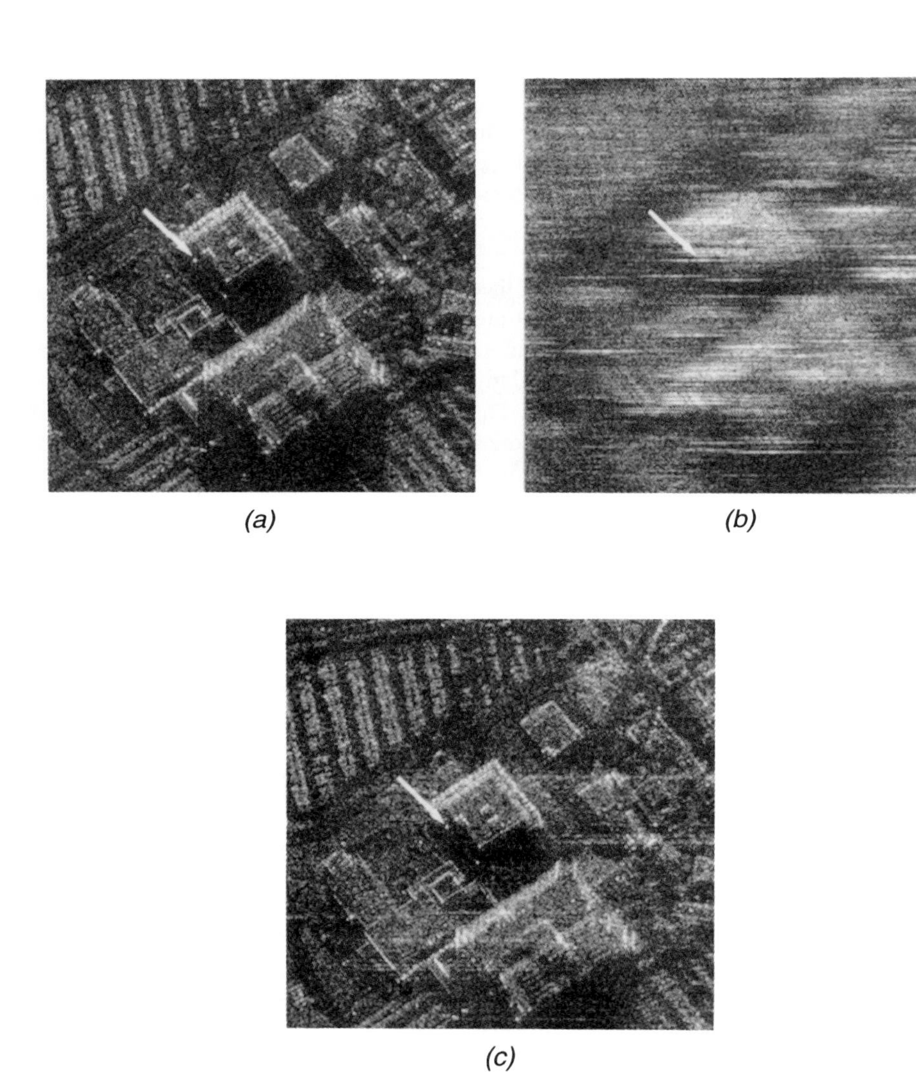

Figure 4.12 Performance of inverse filtering technique for autofocus. SAR images of an urban scene. (a) Original well-focused SAR image. (b) Defocused image. The arrow indicates the target selected to estimate the degrading phase using inverse filtering. (c) Refocused scene using phase estimated from inverse filtering technique. Residual defocus effects (streaking) are clearly evident.

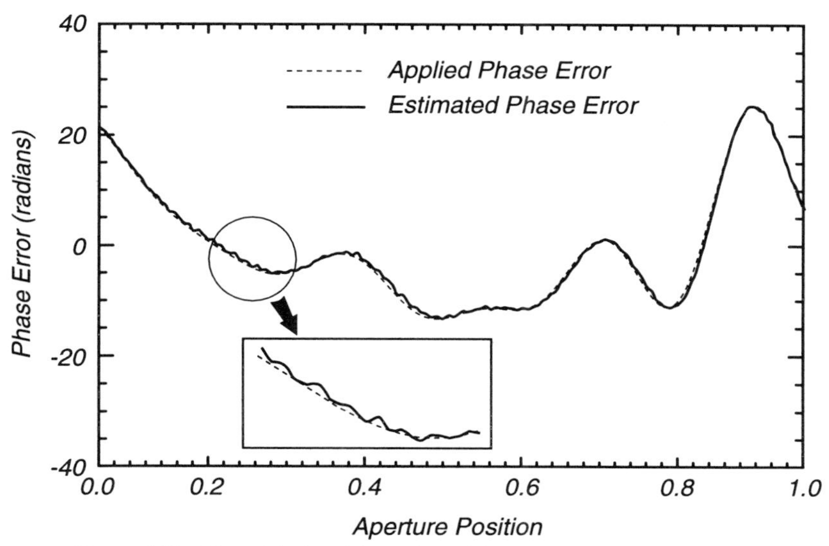

Figure 4.13 The original and the estimated low-order phase-error function. The estimated phase error was calculated from the urban scene using the inverse filtering approach. Insert shows residual high-frequency estimation errors that result in streaking effects in refocused image.

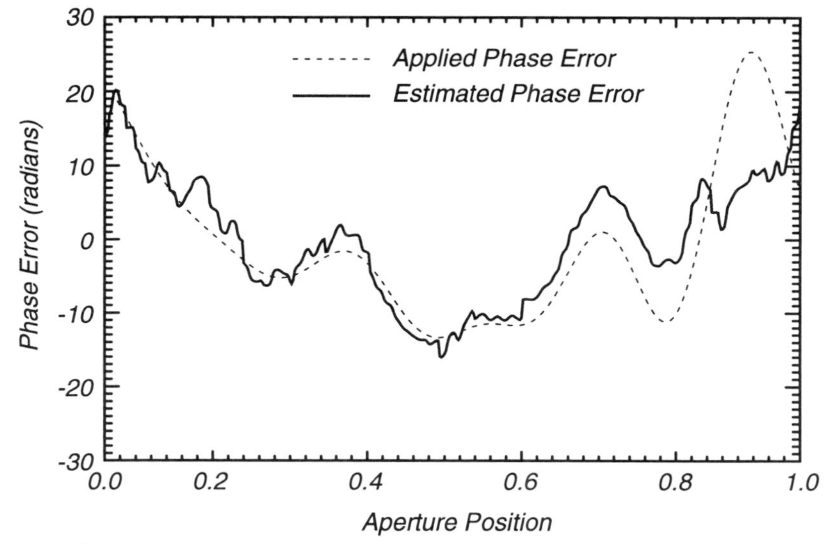

Figure 4.14 The original and estimated phase-error function used on the rural scene. The estimated phase error was calculated using the inverse filtering approach.

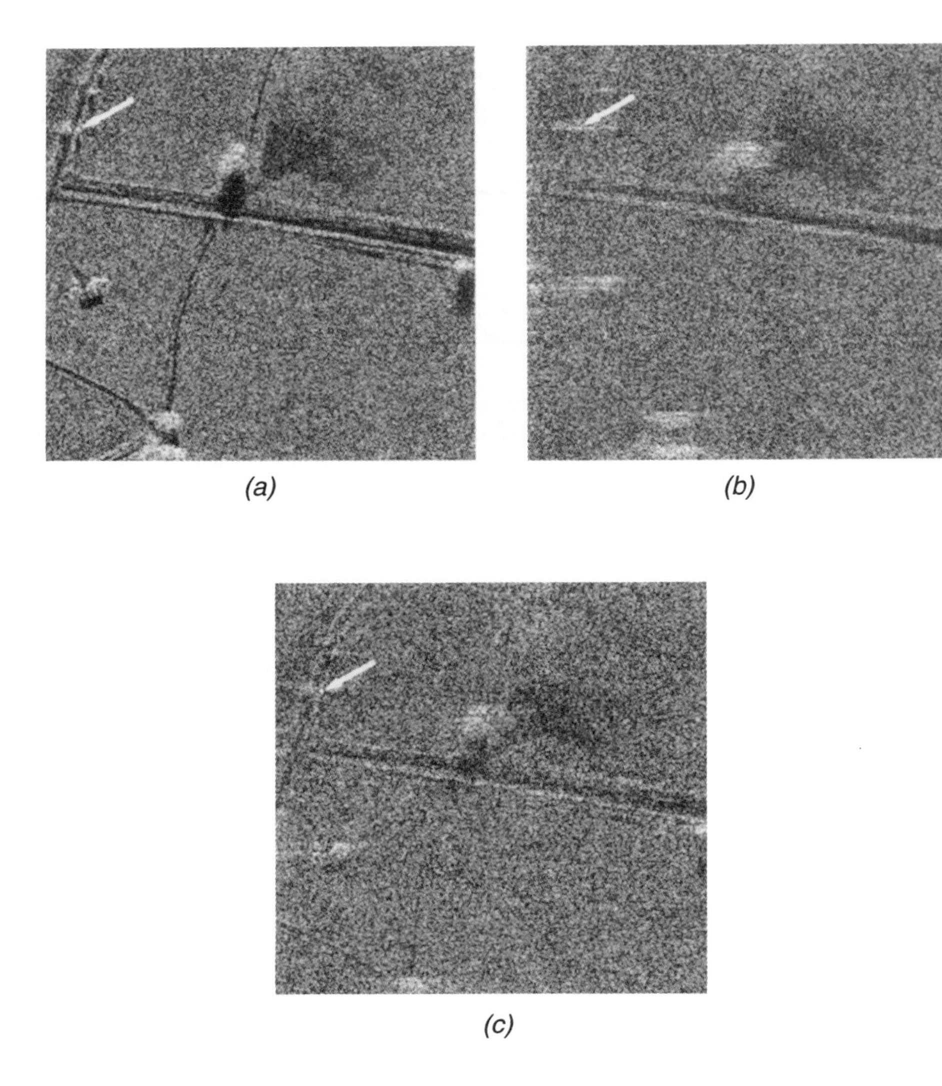

Figure 4.15 SAR images of rural scene. (a) Original well-focused SAR image. (b) Defocused image. The arrow indicates the target selected to estimate the degrading phase using inverse filtering. (c) Focused scene using phase estimated from inverse filtering technique. Full restoration of the image is clearly not achieved in this case.

4.4.2 Subaperture-Based Techniques

One of the earliest established approaches to data-driven SAR phase correction was based on sub-aperture processing. This class of algorithms is commonly referred to as *map drift* (see references [5] and [6]). The primary assumption of this approach is that the aperture phase-error function can be completely described by a finite polynomial expansion. This parametric assumption simplifies the problem because estimates of the polynomial coefficients are sufficient to determine the phase-error function. Unlike the inverse filtering technique that uses data contained in a single range line, map-drift methods use all the SAR image data to estimate the phase error. This would logically lead to a more robust estimate of the degrading phase error.

The nominal map-drift implementation assumes that the degrading phase error is purely quadratic.[4] This assumption enables us to model the phase error as $\phi(m) = bm^2$, where m indexes the aperture position and b is the unknown quadratic parameter. Given this model, we need only estimate the single parameter b to obtain the estimated phase-error function.

The map-drift approach relies on two fundamental properties of SAR images. First, the magnitude of a SAR image formed from a subset of aperture domain data is macroscopically similar to the magnitude of the original image, except that it is of lower resolution. (The reader will recognize this as a direct result of the holographic property of SAR imagery discussed in Section 3.3). Second, a linear aperture phase function causes a spatial shift, or *drift*, in the image domain, as a consequence of the shift property of Fourier transforms.

The basic tenets of the map-drift procedure can be explained as follows. Assume that the SAR image is corrupted with the quadratic phase error shown at the top of Figure 4.16. If the aperture domain is divided into two equal portions, the resulting subapertures are each corrupted by half of the original quadratic phase error. Because half of a quadratic phase error is composed primarily of a linear component (see dashed plot of Figure 4.16), the corresponding lower-resolution images formed by compressing the half-aperture data will be shifted in cross-range relative to the original. The amount of shift is proportional to the slope of the linear term, so that the parameter b can be estimated by determining how much one sub-image has shifted in cross-range relative to the other sub-image. This is done by cross-correlating the two images and by finding the location of the peak in the cross-correlation function to sub-pixel accuracy. For the quadratic phase-error case, images will be shifted by one pixel for every π radians of peak quadratic phase error. The cross-correlation

[4] Actually, this could include a constant and/or a linear term as well. Neither of these degrade the image quality. The constant term has no effect on the image magnitude, while the linear term simply shifts the image data.

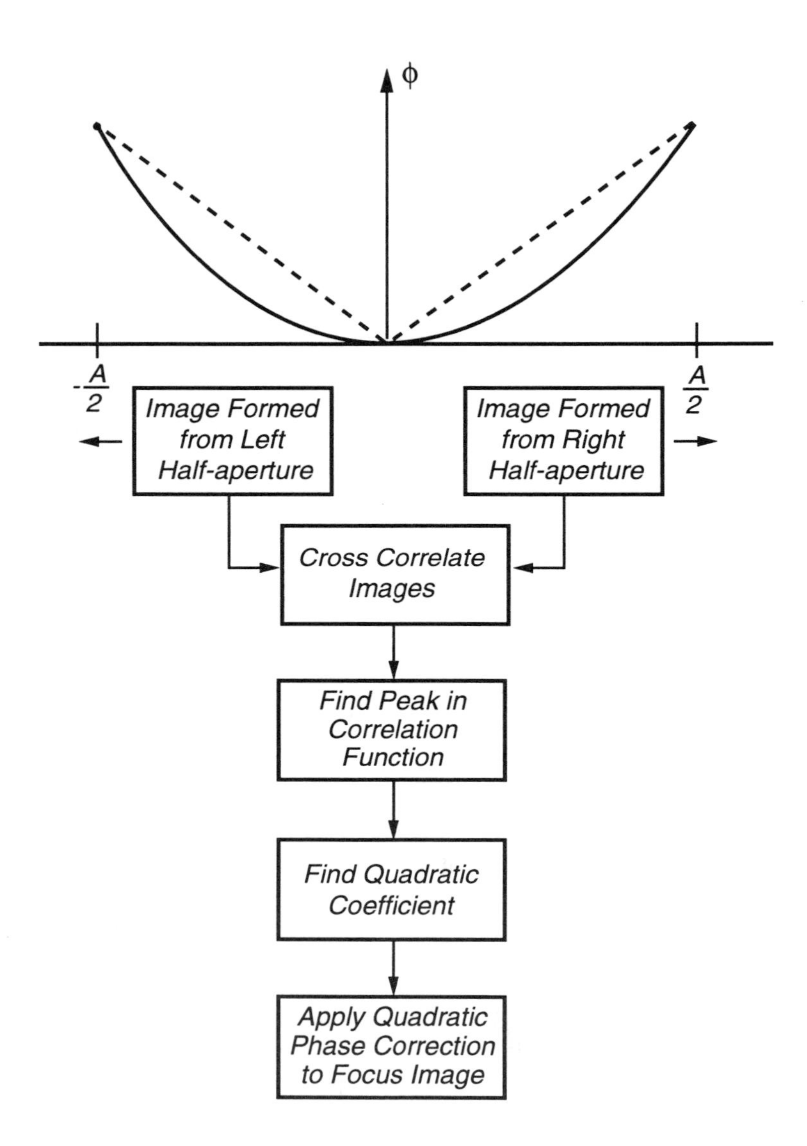

Figure 4.16 Diagram of steps used to implement the map-drift autofocus technique. This particular implementation assumes the phase error is quadratic.

procedure yields an estimate, $\hat{b}$, for the quadratic coefficient. In turn, this provides an estimate of the entire phase-error function as $\hat{\phi}(m) = \hat{b}\, m^2$. We are then able to correct the degraded image by multiplying the corrupted range-compressed data by the complex conjugate of the estimated phase function. Finally, we should note that the entire procedure outlined in the flow chart of Figure 4.16 can be performed iteratively. As the image becomes better focused, the cross-correlation procedure can be done more accurately, producing a better estimate of the phase error, and so on.

Figure 4.17 shows results from the map-drift technique for quadratic phase errors. Figure 4.17(a) shows an image degraded by the quadratic phase-error function represented by the graph of Figure 4.17(c). Figure 4.17(b) shows the image that results from applying map-drift autofocus. In this case, the estimated quadratic function is very nearly the actual one imposed on the image, as indicated in the graph of Figure 4.17(c). Accordingly, the refocused image represents an essentially perfect restoration.

As would be expected, the use of the quadratic map-drift algorithm becomes problematic if the phase error is assumed to be quadratic, but in fact contains components of higher order. An example of this situation is shown in Figure 4.18. Figure 4.18(a) is the same scene of Figure 4.17, but with the SAR data now corrupted by the low-order phase function shown by the dotted line in Figure 4.18(c). This phase error is composed mostly of a quadratic term along with some lower-amplitude terms of higher order. The quadratic map-drift algorithm estimates the quadratic phase error shown by the solid line in Figure 4.18(c). Correcting the image using this estimated quadratic results in the image of Figure 4.18(b), which clearly is not well-focused, i.e., the higher-order terms are not estimated by the algorithm and cause substantial residual defocus effects.

There is a way in which the map-drift concept can be extended to estimate phase errors composed of higher-order terms. This is accomplished by subdividing the aperture domain into many small subapertures instead of two half-apertures. In general, if we desire n coefficients (beginning with the quadratic coefficient) in the phase-error model, then $n + 1$ subapertures are required to estimate these coefficients. The drifts between the subapertures are again computed by correlating the corresponding low-resolution images, and are then used to estimate the n coefficient values. As n becomes larger, the subapertures become proportionately narrower. In turn, the cross-correlations of the resulting low-resolution images yield noisier estimates of the polynomial coefficients. In this way, the order of the model polynomial for which reliable results can be obtained eventually becomes limited [6].

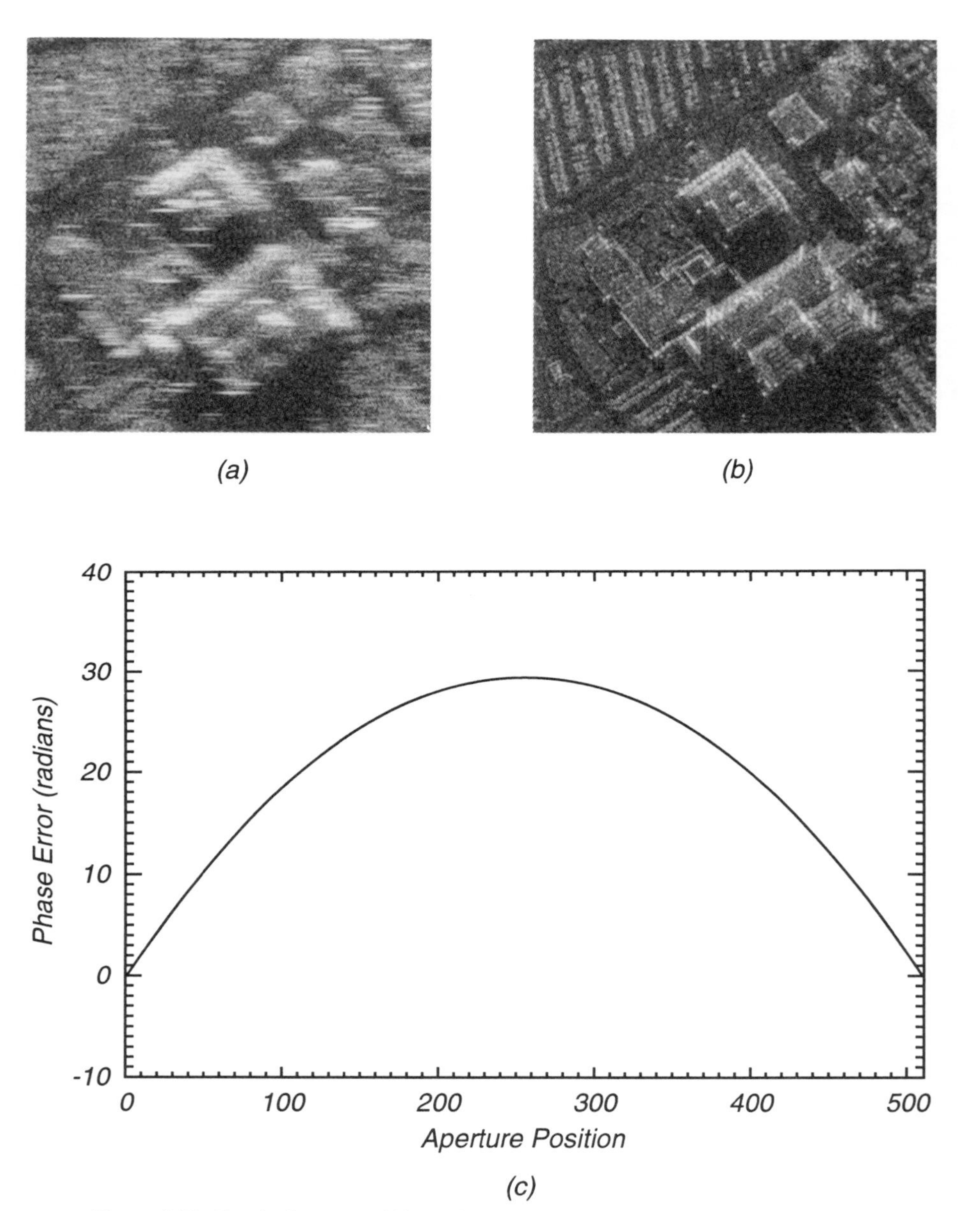

Figure 4.17 Results from map-drift autofocus technique for the case of quadratic phase-error function. (a) Degraded image from quadratic phase error. (b) Corrected image using quadratic phase from the map-drift technique. (c) Applied and estimated phase-error functions are indistinguishable.

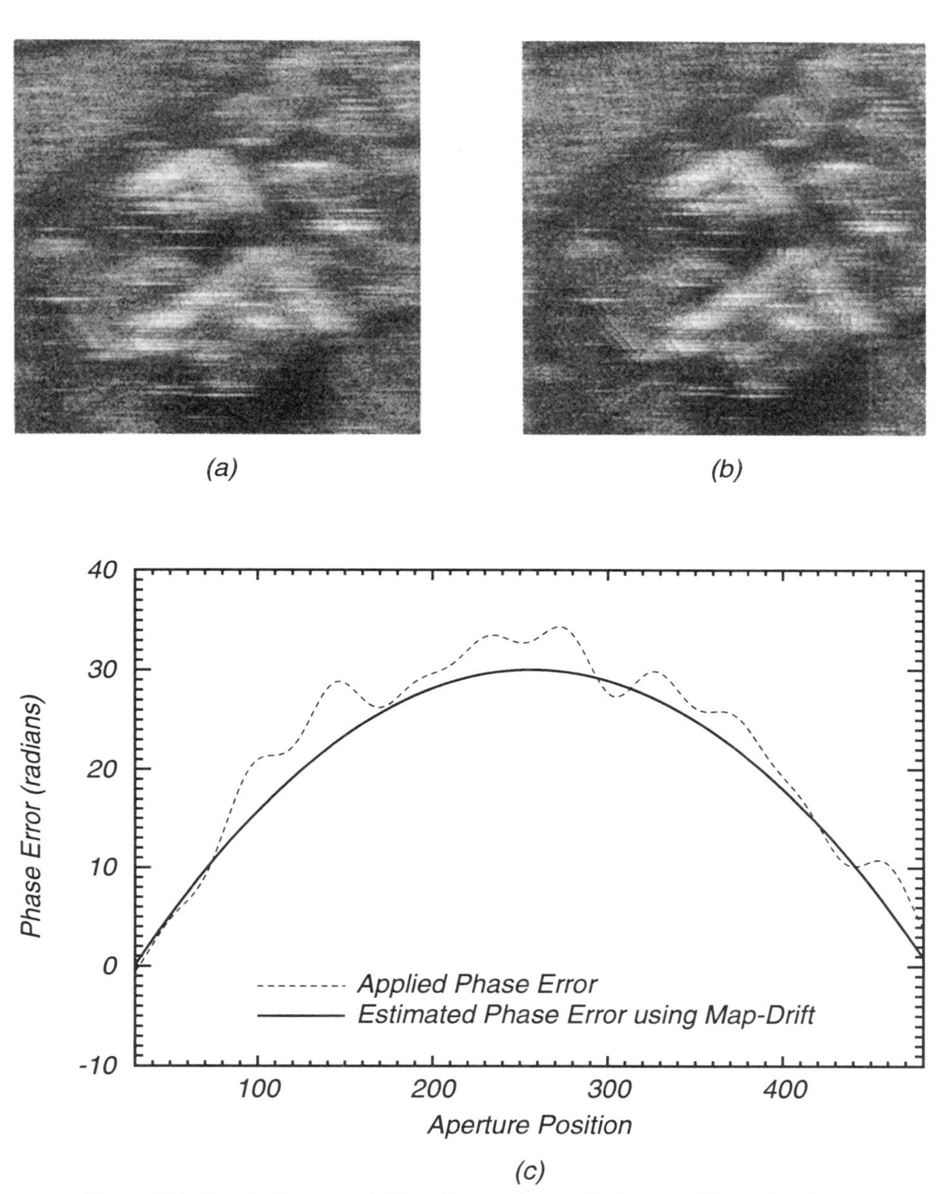

Figure 4.18 Results from map-drift autofocus technique for the case of low-order phase-error function. (a) Degraded image from low-order phase-error function. (b) Corrected image using quadratic phase from the map-drift technique. (c) Applied and estimated phase-error functions. Note substantial residual defocus due to higher-order terms not estimated by the quadratic map-drift algorithm.

In summary, the map-drift method can perform well in situations where the aperture phase-error function can be adequately described as either purely quadratic or at most as a polynomial of low order (say up to order 5). For situations in which high-order phase functions are involved, map drift (or any other parametric technique) is not well-suited. Instead, a robust, non-parametric technique known as *phase gradient autofocus (PGA)* becomes the algorithm of choice. This method is the subject of Section 4.5.

Summary of Critical Ideas from Section 4.4

- **The two general classes of conventional methods used for phase correction are the inverse filtering method and a subaperture-based technique known as the map-drift method.**
- **Inverse Filtering**
  - **The inverse filtering method works well if an isolated point reflector exists in the scene and if this target and its associated point spread function can be adequately isolated. Unfortunately, these conditions exist in only a small percentage of SAR collections.**
  - **Because the phase-error estimate from the inverse filtering method is estimated from a single target, the method is highly susceptible to noise.**
  - **The method has the ability to estimate low-frequency as well as high-frequency phase errors.**
- **Map Drift**
  - **The map-drift method is relatively robust because it is able to use all the data in the SAR scene to estimate the phase error and because it does not require target selection by an operator.**
  - **The map-drift method is parametric, meaning that it assumes that the phase error is described by a polynomial. As a consequence, this method is limited to estimating low order phase-error functions, because the subapertures required to estimate high-order polynomial coefficients become so small that the correlation measures are inaccurate.**

4.5 PHASE CORRECTION USING PGA

This section describes the phase gradient autofocus (PGA) algorithm for automatic estimation and correction of phase errors in spotlight-mode SAR imagery. First published in 1989 (see reference [7]), PGA represents a significant departure from previously published autofocus techniques in several ways. It is non-parametric (unlike map-drift); it takes advantage of the redundancy of the phase-error function by averaging across many range cells (unlike inverse filtering); and it is derived using one of the methods of formal optimal estimation theory known as *maximum-likelihood estimation* (unlike any previous methods).

The motivation for PGA begins by considering the inadequacies of the inverse filtering technique. Taken by itself, any single target in a defocused image may not be strong enough or sufficiently isolated from the other targets to provide a reasonable estimate of the phase-error function. This was illustrated by one of the examples of the previous section. One critical observation, however, strongly suggests that there must be some way of combining the information from all these targets to obtain a far superior estimate of the phase error compared to that achievable from a single target. This is the realization that *every* target in the image is effectively corrupted by the *same* blur function. Because the defocus of image-domain targets on all range lines is the result of the convolution of the image data with the same kernel (namely, the inverse Fourier transform of $e^{j\phi(m)}$), every point target is turned into the same blur pattern.

The question that naturally arises is: "How can information from a multiplicity of targets be processed simultaneously to provide an optimal estimate of the common blurring kernel, or equivalently, the aperture phase-error function?" A straightforward ad hoc procedure would be to use the inverse filtering approach on several separate targets followed by a simple averaging of the resultant phase-error estimates. However, this averaging may not yield the desired result. This follows because for very low signal-to-clutter ratios, the phase estimate from any individual target effectively becomes a random variable uniformly distributed between $-\pi$ and π. An average of such phases will therefore not be a useful estimate of the true common phase term. Only if one or more of the targets in the average has substantial signal strength relative to the surrounding clutter will a meaningful phase estimate be produced by this procedure.

The PGA algorithm uses a different averaging strategy than that of the simple ad hoc method suggested above for obtaining a phase-error estimate. It does so by combining the defocus information from a multiplicity of image targets with an optimality that is rooted in the tenets of statistical estimation theory. We will show later that the estimator used in PGA performs as well as any estimation scheme possibly can, given the mathematical (statistical) model that we employ for the targets and the clutter of the SAR image. This is accomplished by comparing the performance of PGA to the so-called *Cramer-Rao lower bound* on estimator variance. As will be demonstrated, the estimator used in PGA works very well even if *all* the targets used in the estimation have signal-to-clutter ratios that are less than unity.

The PGA algorithm consists of several critical steps that include: center (circular) shifting, windowing, phase difference estimation, and iterative correction. These algorithmic steps are shown in the flow diagram of Figure 4.19. In the following subsections, we demonstrate how each processing step contributes to the robustness of the PGA autofocus methodology.

4.5.1 Center Shifting and Windowing

The fundamental concept of the PGA algorithm is to average information across many targets for determining the phase-error estimate. To this end, the algorithm attempts to isolate a number of single targets in the image for use in the estimation process. A nominal procedure (but by no means the only one possible) for selecting targets is to choose one on each range line. This isolation procedure is accomplished by center shifting and windowing (see steps 2 and 3 in Figure 4.19).

Strong targets relative to the surrounding clutter naturally provide better potential for phase estimation then do targets with strength only on the order of the clutter. Therefore, PGA selects the strongest target on each range line and circularly shifts it to the scene center. This circular-shifting operation preserves the effects of the phase error on the selected target while simultaneously removing any linear phase component (in the azimuth dimension of the range-compressed domain) associated with that target. The removal of the linear phase component is desirable when the common aperture phase-error function is estimated using integration across the range dimension. An example of a defocused image and its corresponding circularly-shifted image is shown in Figure 4.20. The shifting operation in essence creates a new image wherein all the targets to be used in the estimation process are aligned and stacked in the center of the scene.

The next step in PGA is a windowing operation. The intent of windowing is to preserve the information contained in the center-shifted targets that describes the blurring kernel, while simultaneously rejecting information from all other surrounding clutter and targets. Those phase-error functions that produce more severe defocusing effects have wider cross-range blur footprints. (As we saw in Section 4.3,

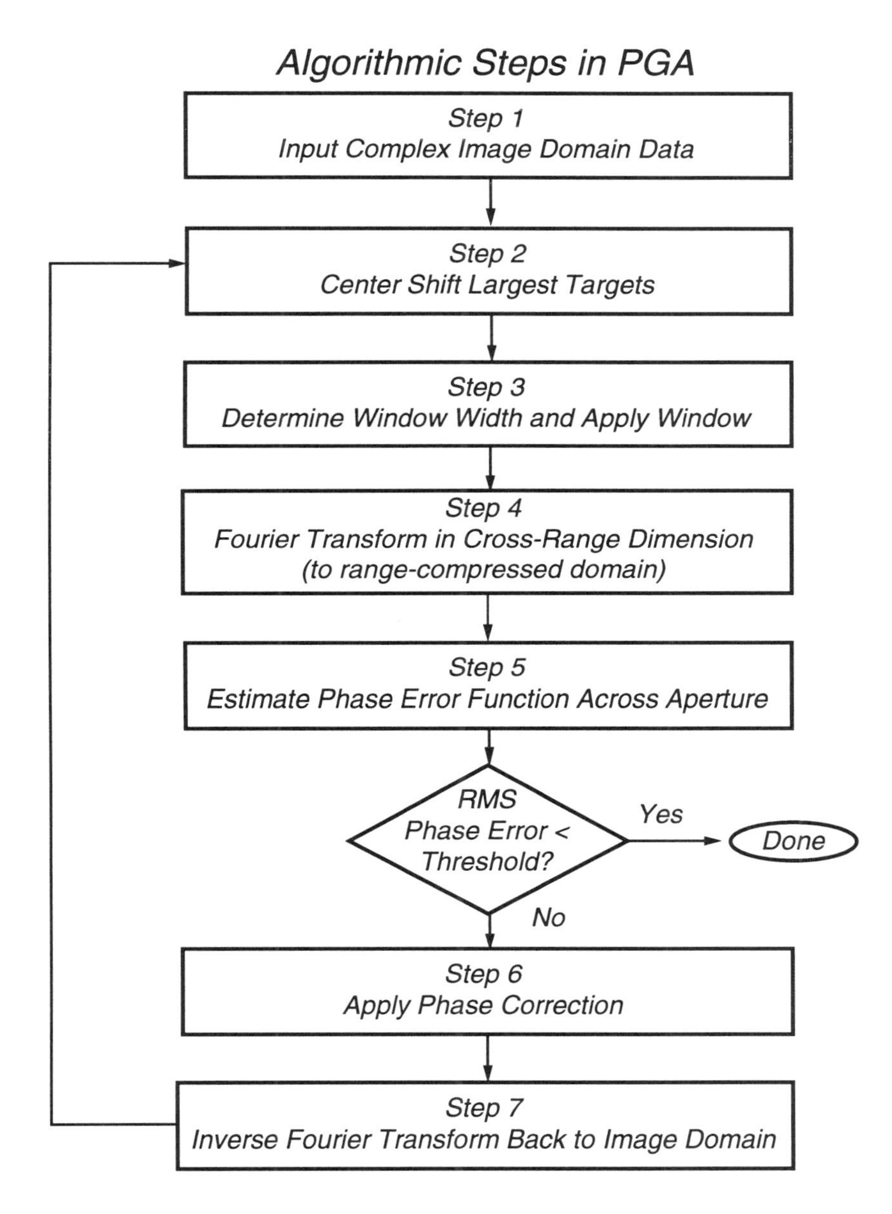

Figure 4.19 Flow diagram of the phase gradient autofocus (PGA) algorithm.

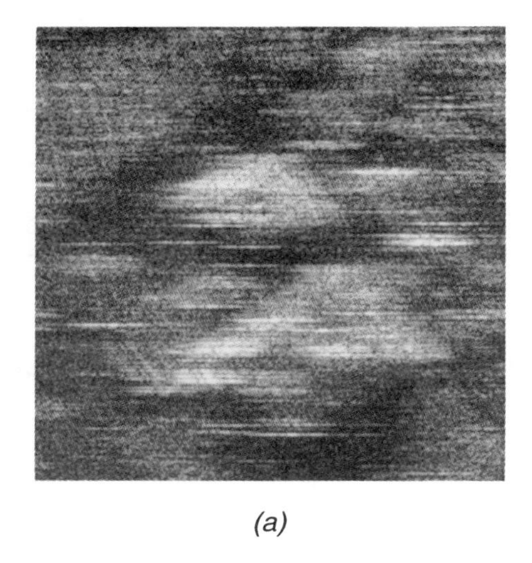

(a)

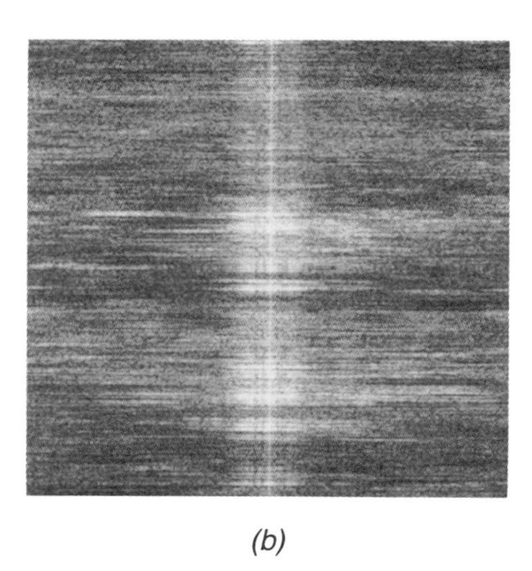

(b)

Figure 4.20 SAR images of an urban scene. (a) Degraded image. (b) Corresponding circularly shifted image.

both the amplitude and the frequency of the aperture phase-error function determine the severity of its defocus effects.) The PGA algorithm simply applies a rectangular window to the center-shifted data, which has the effect of rejecting data outside the window and uniformly weighting all data inside the window. Because the data outside the support of the blur footprint represent noise to any estimation scheme, the effective signal-to-noise ratio (i.e., target-to-clutter ratio) of the image data employed for phase estimation is thereby increased. The difficult part of this process is to determine the window width. If the window is too wide, an unnecessarily large amount of noise is admitted. On the other hand, it is important to capture all the center-shifted target energy so that none of the defocus effect on the selected target is lost.

One method for automatic determination of the window width uses *non-coherent averaging*. This concept is illustrated in Figure 4.21. The upper portion of the figure shows a defocused image following center shifting. It is clear from the picture that there is some nominal support of the blurring effect that is the same for targets on all range lines. Again, this is true because each center-shifted target has been defocused by the same phase-error function. The window width is estimated by simply summing the magnitude of the image data in the range direction for every cross-range position. This is described by

$$s(n) = \sum_{k=1}^{N} |g(k,n)|^2, \tag{4.37}$$

where $g(k,n)$ is the circularly-shifted complex image data, N represents the total number of range lines, and n represents the image-domain cross-range position index.

Because of the center-shifting process, $s(n)$ will tend to achieve its maximum value at the center of the image and will typically exhibit a plateau having approximately the same width as the blur footprint, as shown in the plot of Figure 4.21. Thus, the blur width can be estimated by thresholding $s(n)$ at some chosen level below its peak. Typically, the level selected is -10 dB (see reference [13] for details).

The non-coherent averaging method works well for images that have been degraded by relatively low-order phase errors (such as a quadratic). This is because these types of errors cause a distinct broadening of the mainlobe of the point target response, which is readily detected using the averaging process. High-order phase errors, however, do not significantly broaden the mainlobe of the IPR; instead, they raise its sidelobes. This generally results in a loss of image contrast. As a result, the target energy can be spread through the entire image, making the 10 dB drop-off point in $s(n)$ rather useless for determining the window width. As we will see later, an alternate scheme is employed in such cases.

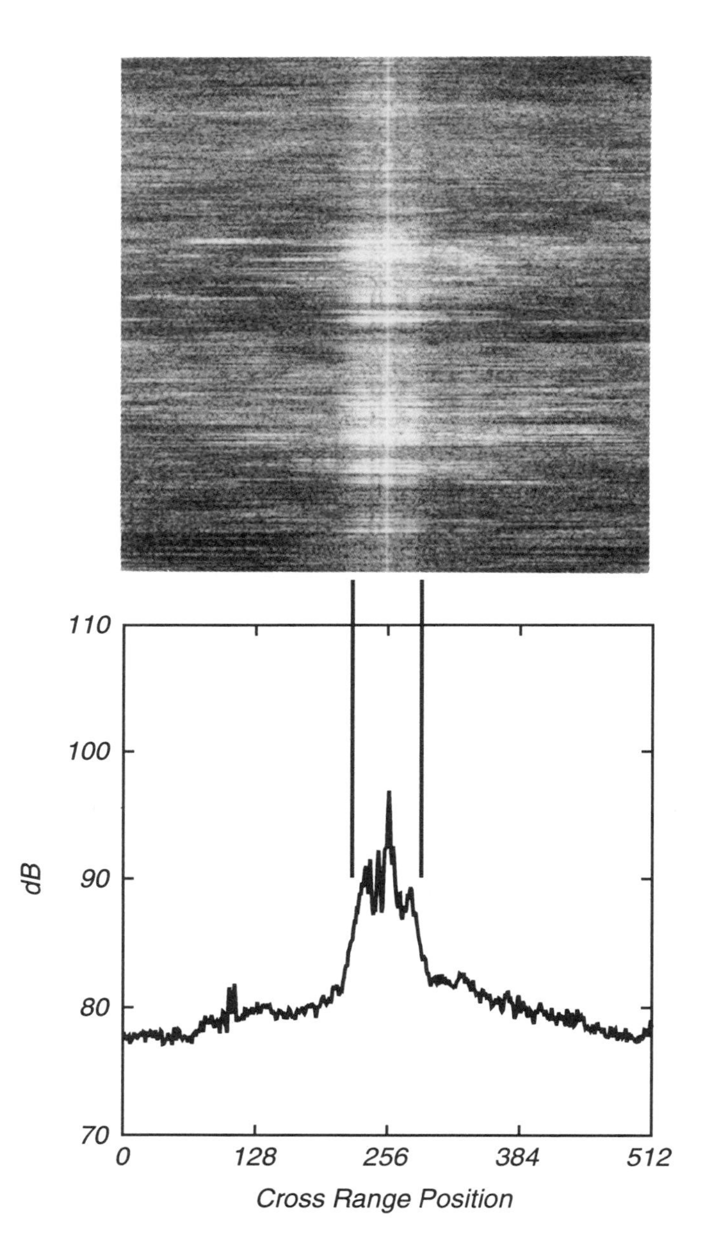

Figure 4.21 Window-width determination using non-coherent averaging. The upper portion of the figure shows the magnitude of the center-shifted image. The resulting one-dimensional range average is plotted in the lower portion of the figure.

4.5.2 Phase Estimation

The center-shifting and windowing operations isolate single defocused targets and position them at the center of each range line. Then, these targets are decompressed using a one-dimensional discrete Fourier transform (in the cross-range dimension) on each range line. This operation (indicated as Step 4 of the PGA flow chart) is generally implemented by finding the next power of two greater than the selected window size and using the fast Fourier transform (FFT) method. The model for the center-shifted, windowed and transformed data is then taken to be

$$\bar{g}(k,m) = a(k)e^{j\phi(m)} + \eta(k,m) \tag{4.38}$$

where k represents the range line, and m is the aperture position index. Note that the target complex reflectivity $a(k)$ is different for each range line. The noise term $\eta(k,m)$ models the interfering clutter that surrounds a selected point target. The data set is composed of N range lines and M cross-range columns.

Given the observations described by the model in Equation 4.38 along with certain statistical assumptions regarding the complex random variables $a(k)$ and $\eta(k,m)$, we can derive an optimal scheme for estimating the degrading phase-error function $\phi(m)$. In particular, the noise $\eta(k,m)$ is assumed to be zero mean, white, Gaussian, and independent of the signal term $a(k)$. In reference [8], a rigorous justification for this noise model is derived. In short, it is assumed that within the image-domain window, the target is a point reflector surrounded by clutter targets, which are modeled as independent identically distributed (iid) white Gaussian noise. Once Fourier transformation (azimuthal decompression) occurs, the target signal becomes constant (equal to $a(k)$) while the noise remains iid white Gaussian, due to the fact that the Fourier transform operation is unitary.[5]

A direct approach to the phase estimation problem is to use all N x M data points to derive a maximum-likelihood (ML) estimate of $\phi(m)$. A simpler approach, and the one that has become the most widely used in real SAR systems, is to use data on two adjacent pulses at a time to estimate the *phase difference* between them. These differences may then be integrated to obtain an estimate for the entire $\phi(m)$. This approach can be shown to be a special case of the more general procedure that uses all the data simultaneously. We will discuss the details of the simpler algorithm here. The reader should consult reference [8] for details on the more general algorithm.

In the phase difference algorithm, $\Delta\phi(m) = \phi(m) - \phi(m-1)$ is estimated for each value of m. The expression for the ML estimator of this adjacent-pulse phase

[5] This same argument will again be used to construct a noise model for SAR interferometry in Chapter 5.

difference (derived in Appendix F) is found to be[6]

$$\widehat{\Delta\phi}(m) = \angle \sum_{k=1}^{N} \{\bar{g}^*(k, m-1)\, \bar{g}(k, m)\} \tag{4.39}$$

where the superscript $*$ indicates the complex conjugate operation, and $\widehat{\Delta\phi}(m)$ denotes the ML estimate of $\Delta\phi(m)$. Appendix F shows that this estimator has several desirable properties. Specifically, it is unbiased and is efficient over a broad range of signal-to-clutter ratios.

Once the estimate for $\Delta\phi(m)$ is obtained for all m, the entire aperture phase error is estimated by integrating the $\widehat{\Delta\phi}(m)$ values. Mathematically, this is given by the expression

$$\hat{\phi}(m) = \sum_{l=2}^{m} \widehat{\Delta\phi}(l) \quad ; \quad \hat{\phi}(1) \doteq 0\, . \tag{4.40}$$

Thus, the phase estimation portion of the algorithm (Step 5 in the flow diagram) consists of estimating the phase differences for all adjacent pulses using M-1 applications of Equation 4.39, followed by integrating (summing) the result using Equation 4.40.

The expression for the estimator in Equation 4.39 is quite simple and has a straightforward explanation. The angle between two complex numbers, $\bar{g}(k, m-1)$ and $\bar{g}(k, m)$, can be obtained as $\Delta\phi(m) = \angle\{\bar{g}^*(k, m-1)\bar{g}(k, m)\}$. For the case at hand where we are presented with N such observations, the optimal strategy (in the ML sense) for obtaining an estimate of the common phase difference is to sum over all the data first and then compute the angle.

The vector interpretation of the phase difference measurement for a given aperture position m is shown in Figure 4.22. This diagram shows that measuring $\widehat{\Delta\phi}(m)$ is equivalent to measuring the angle of vector C, which is composed of the sum of signal vector A and noise vector B. The signal vector is obtained as a sum of N smaller vectors, each of which correspond to the information contributed by the center-shifted target on a given range line. The individual vectors *all* have an angle of $\Delta\phi(m)$ and have individual magnitudes determined by the magnitude of the selected target on a given range line. As a result, they will add *coherently* to produce the signal vector. It can be shown [9] that for the case of large values of N,

[6] The original PGA algorithm [7] used a different optimal estimator based on the phase derivative. In [8], the ML phase difference estimator was introduced.

the noise becomes a small probability *cloud* at the tip of the summed signal vector. The radius of the noise cloud is much smaller than the length of the signal vector, enabling a useful estimate of $\Delta\phi$ to be made.

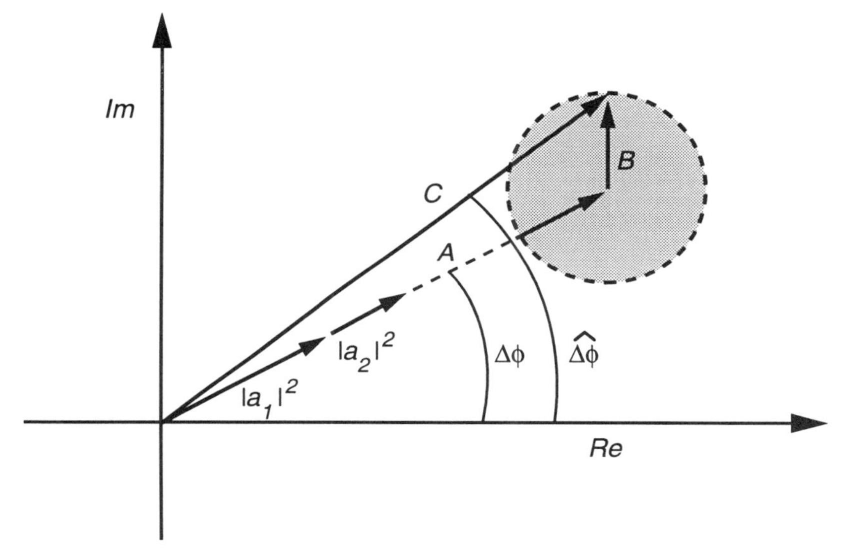

Figure 4.22 Vector interpretation of the phase difference measurement. The vertical direction is the imaginary part and the horizontal direction is the real part.

The ML estimation scheme outlined above and interpreted via Figure 4.22 should be carefully contrasted to the ad hoc scheme suggested in the introduction to this section. Simply *averaging the phases* of a set of targets all having poor signal-to-clutter ratio does not lead to a useful estimate of the common phase. By contrast, the ML estimator works extremely well even in situations when all the individual target-to-clutter ratios are well below 0 dB, which is the condition when the noise vector and the signal vector for an individual target are of similar size. We will demonstrate this performance using SAR imagery in Section 4.5.4 and also via simulations in Appendix F.

4.5.3 Correcting and Iterating

The phase-error estimates produced by Equations 4.39 and 4.40 may now be used to correct the degraded image. This process is identical for all phase correction algorithms and simply involves multiplying the degraded range-compressed data by the complex conjugate of our phase-error estimate (see Step 6 in the PGA flow diagram). Mathematically, this is given by the operation

$$\bar{g}_c(k,m) = \bar{g}_\varepsilon(k,m)e^{-j\hat{\phi}(m)} \tag{4.41}$$

where $\bar{g}_c(k,m)$ denotes the *corrected* range compressed data.

Substituting the corrupted range-compressed model found in Equation 4.17 into Equation 4.41 we have

$$\bar{g}_c(k,m) = \bar{g}(k,m)e^{-j(\hat{\phi}(m)-\phi(m))} \; . \tag{4.42}$$

If the phase-error estimate $\hat{\phi}(m)$ is equal to the true phase error $\phi(m)$, the corrected range-compressed data are equal to the uncorrupted range-compressed data and the image will have been perfectly restored. All that remains is to transform the corrected range-compressed data to the image domain using one-dimensional inverse Fourier transformation, as shown in Step 7 of the PGA flow diagram.

Of course, the phase-error estimate is only as good as the assumptions that are used in its derivation. In practice, it is very difficult to precisely center shift the target when it has been degraded by a phase-error function. This is because a phase-degraded target is "smeared" so that its true peak cannot be located without some error. In addition, it is not unusual in real SAR imagery to have two strong targets on the same range line spaced sufficiently close together in cross range that the windowing procedure captures a portion of the neighboring target. For these reasons, it is often necessary to execute the PGA algorithm in an iterative manner, in an attempt to converge on the correct center shifting and target isolation. Therefore, after the image is corrected using the initial ML estimate of the phase-error function, the entire process is repeated on this refocused image. A sequence of *incremental* phase-error functions is thus computed. On each iteration, $\hat{\phi}(m)$ is updated by addition of the current incremental estimate. As $\hat{\phi}(m)$ becomes more accurate, the targets become more compact (less smeared out), allowing more accurate center shifting and smaller windowing to be used with each iteration. The decreasing window width also decreases the probability of capturing undesired competing targets within the window. An estimated incremental phase-error function with small total energy implies that the marginal benefit from the phase error produced at this iteration is negligible. The algorithm is terminated when the energy (or rms value) of the estimated incremental phase-error function on any iteration falls below a prescribed threshold.

4.5.4 PGA Examples

This section presents two examples of the PGA algorithm. The first example illustrates the critical steps of PGA as it operates on a cultural scene degraded by the low-order phase-error function shown by the solid line in Figure 4.25(c). Figures 4.23(a)-(d) illustrate the results of these steps for the first iteration only. The original degraded scene is shown in Figure 4.23(a). The result of circular shifting the largest targets on each range line to the center of the scene (Step 2 in PGA flow diagram) is shown in Figure 4.23(b). These circularly-shifted data are then non-coherently averaged to produce the one-dimensional signal shown in 4.23(d), from which the window width is estimated. The data are then windowed, Fourier transformed to the aperture domain, and used to estimate the degrading phase-error function (steps 3 through 5 of the PGA flow diagram). This phase-error function is used to correct the original degraded range-compressed data, which are then inverse Fourier transformed to the image domain. The result of this correction is shown in Figure 4.23(c), which is a clear improvement over the original image in Figure 4.23(a).

The energy of the estimated phase-error function on the first iteration was large enough to warrant another iteration. The results of the steps of the second iteration are shown in Figure 4.24(a)-(d). The degraded image used as input to the second iteration is simply the corrected image of the first iteration shown in Figure 4.24(a). Note that the non-coherent average on the second iteration yields a much sharper peak than does the initial iteration. This is because the targets are now better focused. The sharp peak indicates a smaller window can be used on this iteration. In general, as the iterations progress, the non-coherent average peak becomes much more defined, resulting in a decreasing window width. In order to completely focus this image, a third iteration was necessary to extract a slight amount of residual phase error. The steps of that iteration are not shown. The original degraded image along with the final corrected image is shown in Figure 4.25(a) and (b), and the resulting estimated phase-error function is shown by the dashed line in Figure 4.25(c).

The second example presents a phase correction problem that is in general very difficult to solve. This involves a high-order phase-error function imposed on a scene with virtually no distinguishable point targets. The simulated phase-error function has power-law frequency properties that can be produced by propagation effects and is shown by the solid line in Figure 4.28(c).

Figure 4.26(a)-(d) shows the results of the critical steps of the algorithm for the first iteration. Note that the result of the non-coherent average shown in 4.26(d) does not exhibit the "plateau" characteristic that was observed in the previous example. This is due to the fact that high-frequency phase errors (such as those imposed on this

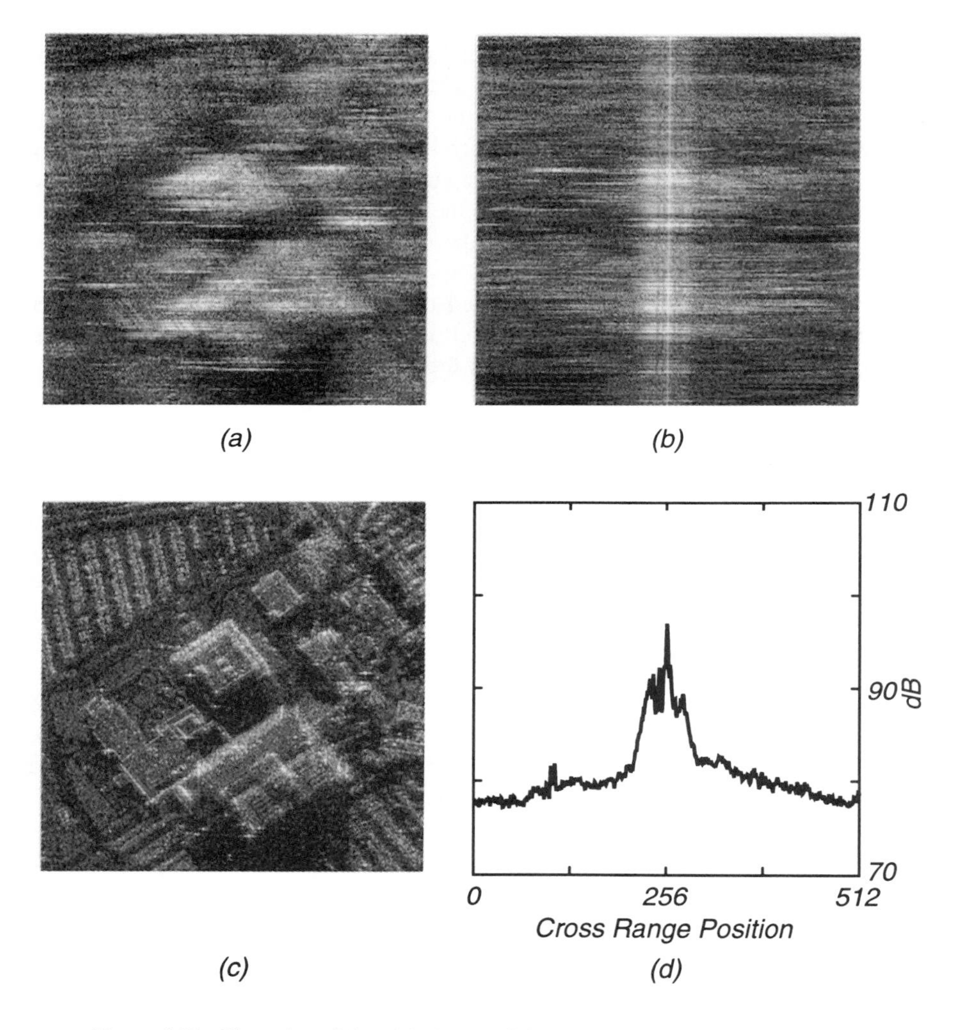

Figure 4.23 Illustration of the critical steps of the PGA algorithm on the first iteration for an urban scene. (a) Initial degraded image. (b) Center-shifted image. (c) Resulting corrected image after the first iteration. (d) Incoherent average of center-shifted image.

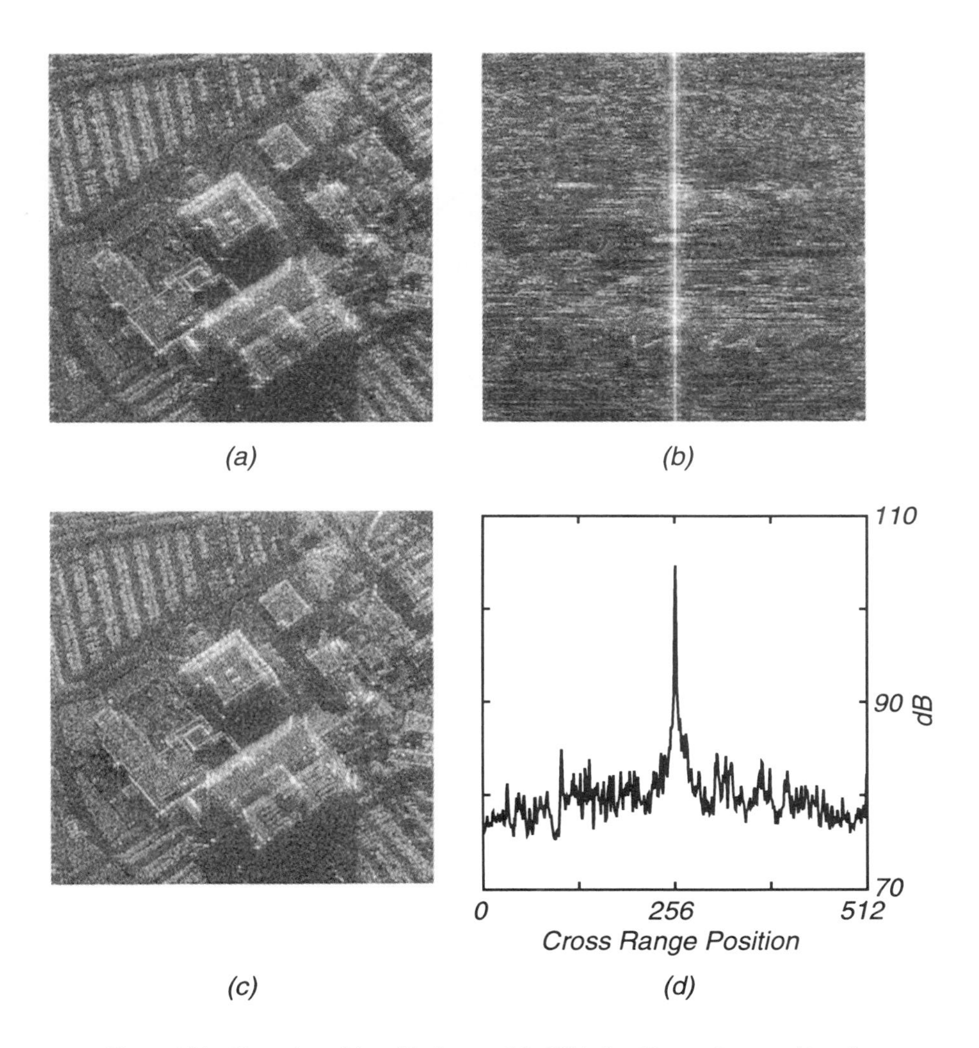

Figure 4.24 Illustration of the critical steps of the PGA algorithm on the second iteration for an urban scene. (a) Initial degraded image (result of the first iteration). (b) Center-shifted image. (c) Resulting corrected image after the second iteration. (d) Incoherent average of center-shifted image.

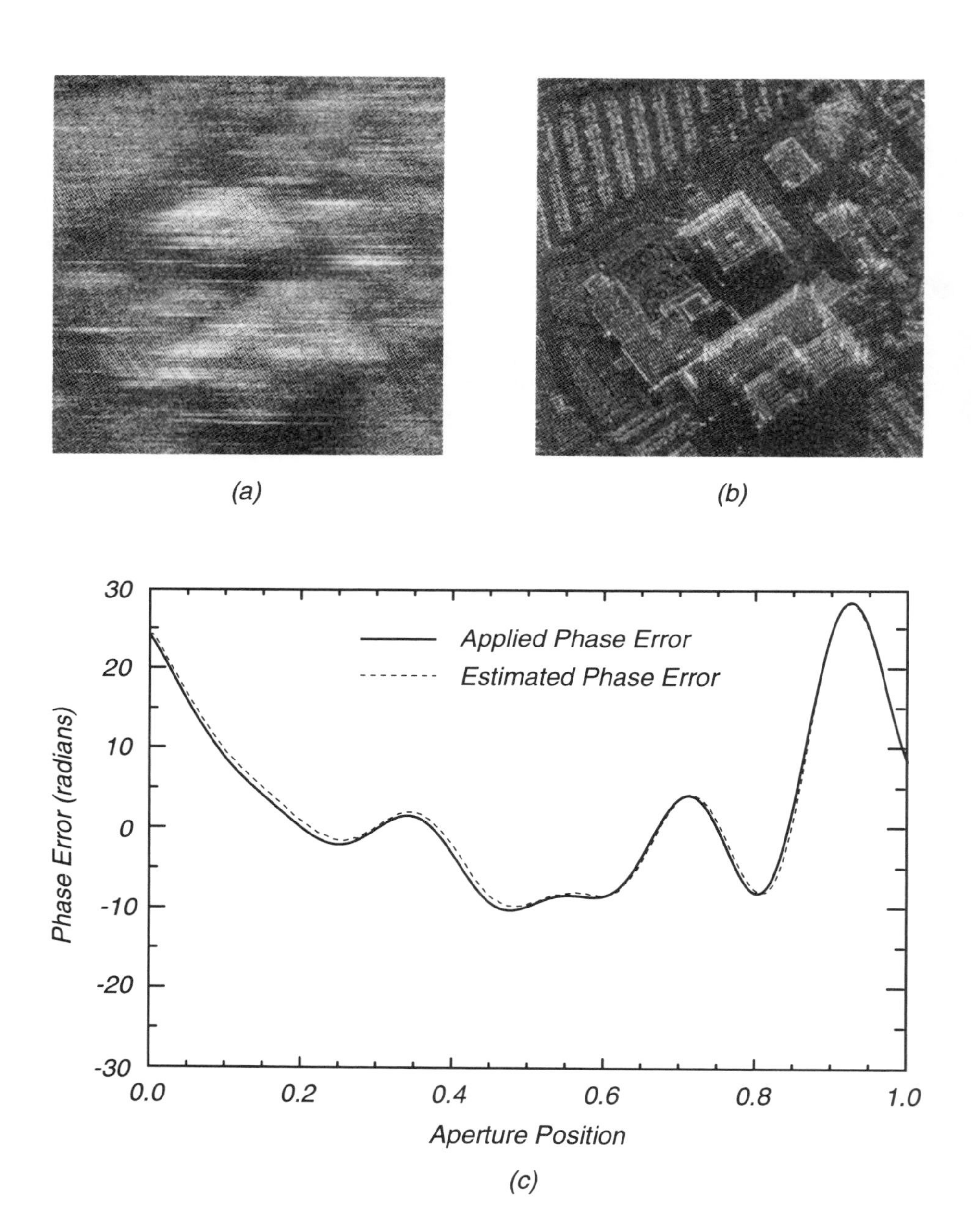

Figure 4.25 PGA results after three iterations. (a) Defocused image. (b) Result of refocus with PGA after three iterations. Note that the scene is essentially restored to full focus. (c) The actual and estimated aperture phase-error functions. This phase-error function is one typical of low-order.

scene) tend to raise the sidelobes of the point target rather than simply to broaden the mainlobe. Therefore, in this case, the non-coherent averaging method was not used to determine the window width. A *progressive windowing* scheme where the window size was simply halved at each iteration was used instead.

The results of the critical steps of the second iteration are shown in Figure 4.27. Again, the non-coherent averaging method gives little or no information about the appropriate window size due to the residual high-frequency phase error. Three iterations were required to focus the image completely. The resulting focused image produced by the PGA algorithm is compared to the original image in Figure 4.28(a) and (b), while the estimated phase-error function is shown by the dotted line in Figure 4.28(c). Clearly, the PGA algorithm was able to combine the information in all the targets to estimate the degrading phase error quite accurately.

In summary, the PGA algorithm has been shown to be an effective and robust tool for correcting phase errors in spotlight-mode SAR imagery. It offers several distinct advantages over simple inverse filtering: 1) it does not require an initial isolation of any single, large point targets in the defocused image; 2) it uses a large portion of the image data (i.e., whatever the windowing process keeps); 3) it estimates the phase in an optimal and robust manner that is grounded in estimation theory; and 4) it exploits the power of iteration to refine the phase estimate. In addition, it is superior to map-drift methods because it is not parametric (order based). Thus, it can accommodate phase-error functions of arbitrary complexity (spatial-frequency content). Currently, PGA is being adopted into a significant number of operational SAR systems.

Finally, we comment on a number of other non-parametric autofocus algorithms that have some similarity to PGA. These include "shear averaging" [10], "spatial correlation" [11], and "ROPE" [12]. The first two employ the same form of the ML estimator for phase differences used in PGA, although the papers describing them do not demonstrate this optimality. In addition, these algorithms do not include the critical PGA steps of center shifting, windowing, or iterating. As a result, their performance is nearly always inferior to that of PGA. The ROPE algorithm suffers for similar reasons. The reader is referred to the paper by Wahl, Eichel, Ghiglia, and Jakowatz [13], where it is clearly demonstrated that leaving out any of the critical processing steps of PGA generally leads to a serious lack of performance.

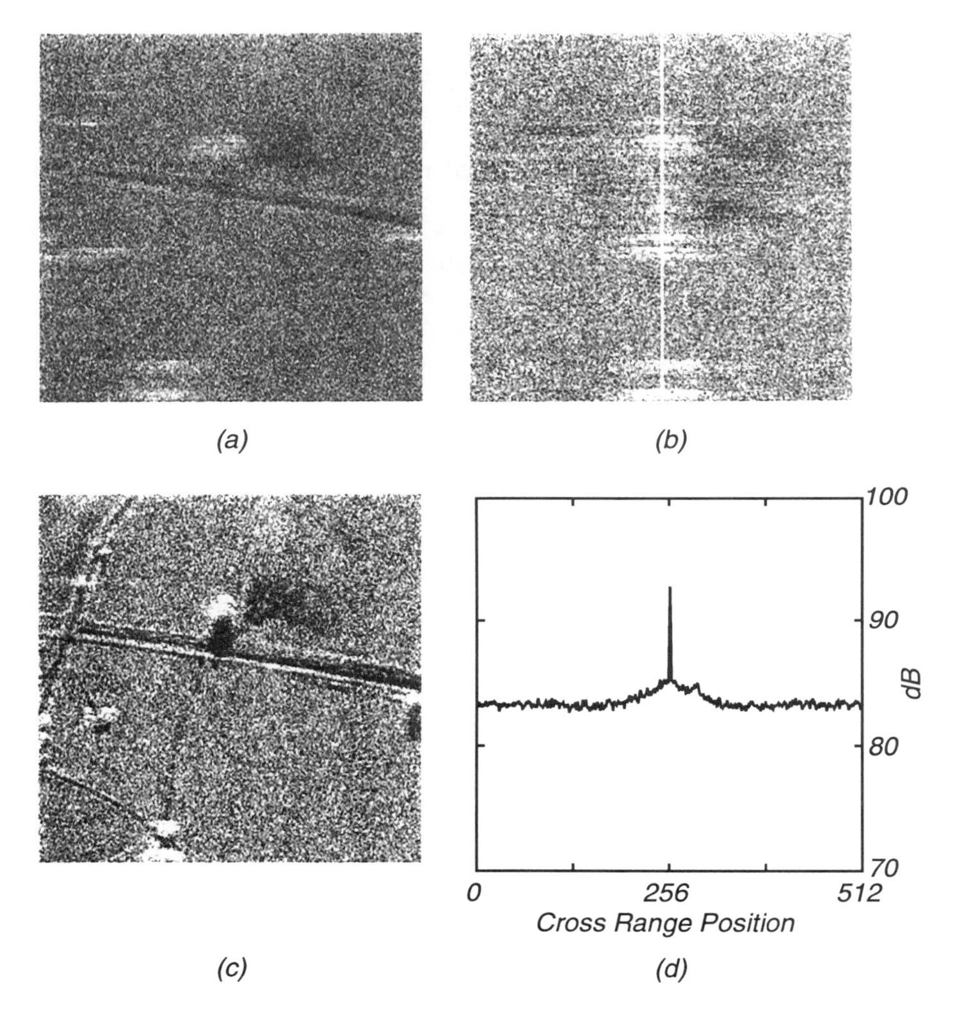

Figure 4.26 Illustration of the critical steps of PGA algorithm on the first iteration for a rural scene. (a) Initial degraded image. (b) Center-shifted image. (c) Resulting corrected image after the first iteration. (d) Incoherent average of center-shifted image.

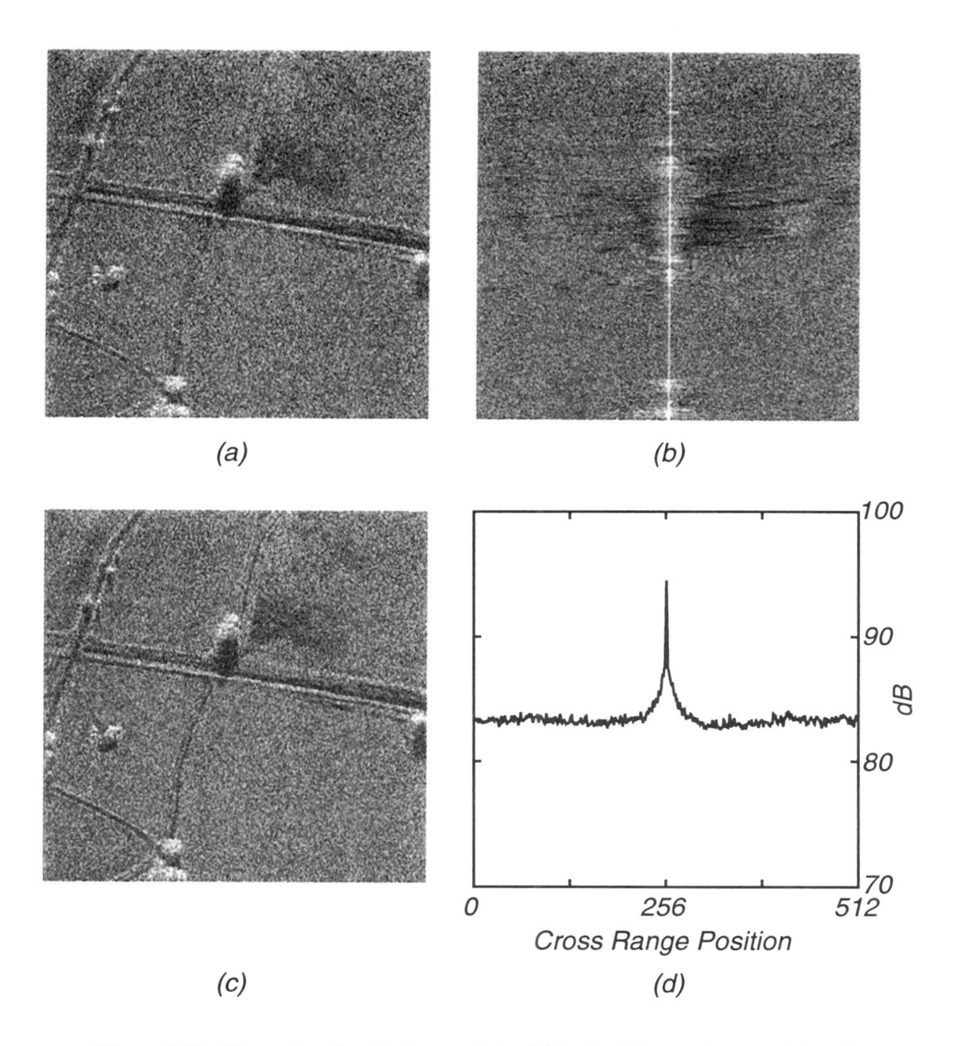

Figure 4.27 Illustration of critical steps of the PGA algorithm on the second iteration for a rural scene. (a) Initial degraded image (result of first iteration). (b) Center-shifted image. (c) Resulting corrected image after the second iteration. (d) Incoherent average of center-shifted image.

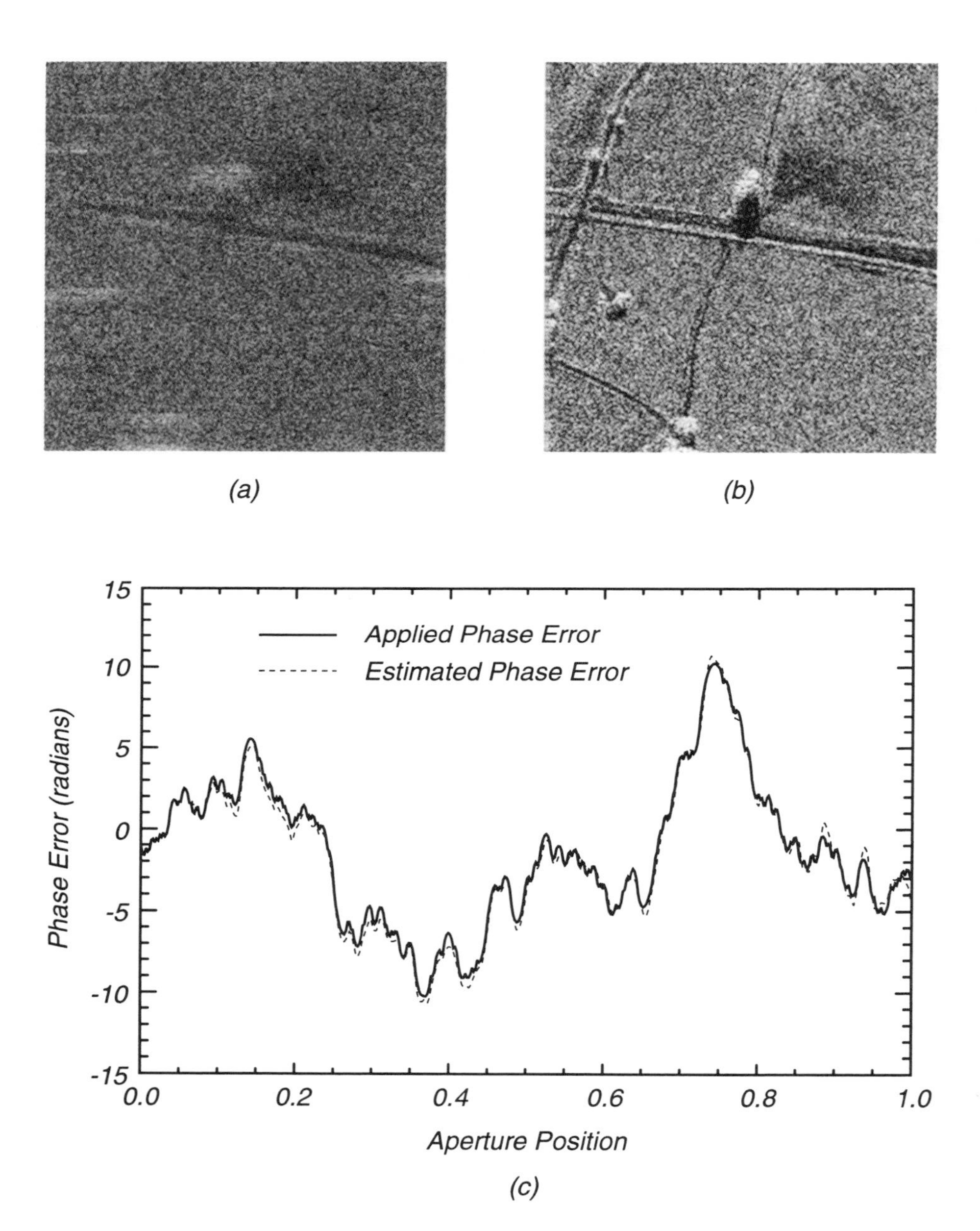

Figure 4.28 Results of PGA after third iteration. (a) Defocused image. (b) Result of refocusing using PGA after three iterations. In this case, the applied phase error was a high-order function. (c) The actual and estimated phase-error functions.

Summary of Critical Ideas from Section 4.5

- **The PGA autofocus algorithm represents a significant departure from previous techniques. It is a non-parametric estimator that exploits redundancy of the aperture phase-error function across multiple range lines.**
- **The PGA algorithm consists of four critical steps: center (circular) shifting, windowing, phase estimation, and iterative correction.**
- **The phase estimation portion of the algorithm has been shown to be optimal in a maximum-likelihood sense.**
- **The maximum-likelihood estimator used in PGA meets the Cramer-Rao lower bound over a wide range of signal-to-clutter ratios. The Cramer-Rao lower bound provides a theoretical limit on the performance of any estimation scheme and is discussed in detail in Appendix F.**

REFERENCES

[1] L. J. Porcello, "Turbulence-Induced Phase Errors in Synthetic-Aperture Radars," *IEEE Transactions on Aerospace and Electronic Systems*, Vol. AES-6, pp. 636-644, 1970.

[2] C. V. Jakowatz, Jr., P. H. Eichel, and D. C. Ghiglia, *Proceedings of SPIE, Millimeter Wave and Synthetic Aperture Radar*, Vol. 1101, 1989.

[3] C. L. Rino and J. F. Vickrey, "Recent Results in Auroral-Zone Scintillation Studies," *J. Atmos. Terr. Phys.*, 1982.

[4] J. Aarons, "Global Morphology of Ionospheric Scintillations," *Proceedings of IEEE*, Vol. 70, pp. 360-378, 1982.

[5] G. A. Bendor, T. W. Gedra, "Single-Pass Fine-Resolution SAR Autofocus," *Proceedings of IEEE National Aerospace and Electronics Conference NAECON*, Dayton, Ohio, pp. 482-488, May 1983.

[6] T. C. Calloway and G. Donohoe, "Subaperture Autofocus for Synthetic Aperture Radar," *IEEE Transactions on Aerospace and Electronic Systems*, Vol. 30, No. 2, pp. 617-621, April 1994.

[7] P. H. Eichel, D. C. Ghiglia, and C. V. Jakowatz, Jr., "Speckle Processing Method for Synthetic Aperture Radar Phase Correction," *Optics Letters*, Vol. 14, pp. 1101-1103, January 1989.

[8] C. V. Jakowatz, Jr., and D. E. Wahl, "An Eigenvector Method for Maximum Likelihood Estimation of Phase Errors in SAR Imagery," *J. Opt. Soc. Am. A*, Vol. 10, No. 12, pp. 2539-2546, December 1993.

[9] J. W. Goodman, *Statistical Optics*, John Wiley and Sons, New York, pp. 54-55, 1985.

[10] J. Fienup, "Phase Error Correction by Shear Averaging," *Technical Digest Series, Signal Recovery and Synthesis I*, Vol. 15, pp. 134-137, June 1987.

[11] E. H. Attia and B. D. Steinberg, "Self-Cohering Large Antenna Arrays Using the Spatial Correlation Properties of Radar Clutter," *IEEE Transactions on Antenna Propagation*, AP-37, pp. 30-38, January 1989.

[12] W. Press, "Recovery of SAR Images with One Dimension of Unknown Phases," JASON Program Office Report JSR-91-175 DTIC AD-B164984, March 1992.

[13] D. E. Wahl, P. H. Eichel, D. C. Ghiglia, and C. V. Jakowatz, Jr., "Phase Gradient Autofocus - A Robust Tool for High Resolution SAR Phase Correction," *IEEE Transactions on Aerospace and Electronic Systems*, Vol. 30, No. 3, pp. 827-834, July 1994.

[14] N. R. Goodman, "Statistical Analysis Based on a Certain Multivariate Complex Gaussian Distribution (An Introduction)," *Annals of Statistical Analysis*, Vol. 34, pp. 152-177, 1963.

[15] H. Van Trees, *Detection, Estimation, and Modulation Theory, Part I*, John Wiley and Sons, New York, 1968.

5

INTERFEROMETRIC PROCESSING OF SAR DATA

5.1 INTRODUCTION

We saw in Chapter 2 that, because it is a *coherent* imaging system, a synthetic aperture radar transduces the *complex* reflectivity of the illuminated scene. This reflectivity function, however, is modulated by phase terms that capture the imaging geometry. As a result, an interesting possibility arises when two SAR images are made of the same scene using very close geometries. It becomes possible in this case to *interfere* the two images in such a way as to cancel the scene reflectivity which is common to both and to recover the geometric information that contains the scene topography transduced by the image-domain phase data. We will use the acronym *IFSAR* to refer to this type of processing.

As a terrain-mapping technique, the interferometric approach is attractive for several reasons. First, because this form of elevation estimation involves wavelength properties of the imaging system and because the wavelengths typically employed by SAR sensors are relatively short (on the order of centimeters), the accuracy afforded can be quite exceptional. Second, the spatial sampling rate of such a survey can approach that of the SAR itself. Finally, and perhaps most importantly, the requisite data processing can be completely automated, avoiding the labor-intensive interaction common to optical, stereoscopic-mapping techniques.

The same properties of complex SAR images that make interferometric terrain mapping possible can also be used in two-pass collections to perform a very sensitive form of change detection. The extent to which the scene reflectivity function is common between the two images of an interferometric pair collected at different times becomes a measure of change over that time interval. Such change maps, owing their sensitivity to the phase as well as to the magnitude of the returned energy, can reveal very subtle physical changes on the earth's surface. We will present an example of this phenomenon later in the chapter. As in the case of terrain mapping, the required data processing can be completely automated.

The use of a synthetic aperture radar in a single-pass interferometric mode for topographic mapping was first proposed by L. Graham in 1974 [1]. Since the mid-1980s, this area of research and development has enjoyed robust growth. However, as with much of the SAR literature, most of the applications deal with low-resolution strip-map systems. In this chapter, we examine the theory, signal processing, and applications of interferometry to spotlight-mode SARs. Building on the tomographic formulations of Chapter 2, we present a unified view of this newest of SAR utilities. Our approach differs substantially from that found in the archival literature, not only because we deal with spotlight-mode as opposed to strip-map SAR here, but also because, in our view, nearly all the critical ideas of IFSAR can be explored in the context of transform theory without considering a radar at all. As we have seen before, the SAR is merely a device for collecting spatial-frequency-domain data of a particular scene. The information one can extract from these data, as well as the methodology used to perform this extraction, is governed by the properties of Fourier transforms.

The development of this chapter will be as follows. We begin by reviewing the basic spotlight-mode SAR image reconstruction equations of Chapter 2 together with some aspects of image formation from Chapter 3, in order to introduce the concept of a *pair* of *interfering* complex images. All aspects of the reconstructed image function are examined in detail with particular emphasis placed on the phase terms. We then take up the issues related to processing interferometric pairs, including registration, parameter estimation, 2-D phase unwrapping, topographic scaling, and corrections for both wavefront curvature and layover effects. Following this, we examine the important topic of interferometric change detection. Finally, we discuss stereoscopic mapping using SAR imagery as a related but fundamentally different technique for terrain height estimation.

5.2 REVIEW OF THEORETICAL FUNDAMENTALS

As this chapter unfolds, we will see that Equation 2.72, repeated below as Equation 5.1, contains essentially everything we need to know from the radar to perform IFSAR processing. We begin this chapter by examining its terms from this viewpoint:

$$g(x_1, y_1) = s_A(x, y) \otimes \left[r(x, y) e^{-j\beta Y_0 h(x,y)} e^{-jyY_0} \right] \tag{5.1}$$

where the height-dependent translations (layover effects) in cross-range and range between the scene space (x, y) and the reconstructed image space (x_1, y_1) are given by $x_1 = x + \alpha h(x, y)$ and $y_1 = y + \beta h(x, y)$, with $\alpha = \tan \eta$ and $\beta = \tan \psi$. Recall that η and ψ are the angles that describe the slant-plane orientation (see Figure 2.28), and that ψ is also the depression angle at the aperture center. In addition, $Y_0 = (4\pi/\lambda)\cos\psi$ is the spatial-frequency offset in the ground-plane-projected phase-history space, $h(x, y)$ is the terrain height function, and $s_A(x, y)$ is the *sinc* function obtained as the inverse Fourier transform of the aperture region A, the region of Fourier space over which the phase-history data are collected.[1]

5.2.1 Image Equations for Coherent Pairs

In interferometric applications, we will always be interested in the interaction of two or more images, i.e. we will be *interfering* one image with one or more others. Therefore, it is instructive to highlight those terms in Equation 5.1 that vary from one image to another. In the next equation, we have subscripted with index i the terms we wish to study:

$$g_i(x_i, y_i) = s_{A_i}(x, y) \otimes \left[r(x, y) e^{-j\beta_i Y_{0_i} h(x,y)} e^{-jyY_{0_i}} \right] . \tag{5.2}$$

As we saw in Chapter 2, the translation equations

$$\begin{aligned} x_i &= x + \alpha_i h(x, y) \\ y_i &= y + \beta_i h(x, y) \end{aligned} \tag{5.3}$$

indicate that the reflectivities at the true location (x, y) *layover* onto the image location (x_i, y_i). Given two image functions g_1 and g_2, the effect of this translation is to cause a *misregistration* of the reflectivity function as transduced by the two images. Unless the slant-plane orientation angles η_i and ψ_i are identical for the two collections, the reflectivity for a given true location (x, y) appears in different image locations: $(x_1, y_1) \neq (x_2, y_2)$. Making matters more complicated is the fact that this misregistration depends on the local height $h(x, y)$. For a one-pass, two-antenna interferometer, this is of no consequence, because the collection angles are very nearly identical. As we shall see for a two-pass single antenna collection, however, these angles can differ substantially. In this case, we are presented with a formidable registration problem because the displacements are spatially variant, depending on the unknown height function.

In a practical situation, there are, of course, other sources of misregistration for a two-pass collection. Slight uncertainties in platform position and velocity estimates, as

[1] As in Chapter 2, we note that Equation 5.1 is simplified in that it assumes multiple targets in the physical scene do not layover onto the same point in the image space by superposition.

well as antenna aiming errors, generally lead to a systematic translation and rotation between the reconstructed images. Fortunately, an affine transformation (first order 2-D polynomial) adequately models most such situations. We return to this question as well as the more complicated problem of spatially-variant misregistration later in this chapter.

Next, consider the reconstructed complex reflectivity function represented by Equation 5.2. Recall from Chapter 2 that the reconstructed complex reflectivity function is an apertured, or narrowband, version of the actual function $r(x, y)$. That is, the reconstruction can be viewed as the result of: (1) performing a Fourier transformation of $r(x, y)\ exp[-j\beta Y_o h(x, y)]$, (2) retaining the Fourier-plane data only in the region defined by the polar collection annulus, denoted by A, and (3) taking the inverse Fourier transformation, with the origin positioned at the center of this apertured data. Given *two* SAR phase-history collections, then, what can be said of their corresponding reconstructed, apertured functions? Two possibilities suggest themselves. First, the physical terrain reflectivity function can be different in the two data sets. This would not be expected for a one-pass, two-antenna collection, because the collection of g_1 and g_2 occur simultaneously or very nearly so. For a two-pass or multi-pass collection, however, the terrain reflectivity can undergo a temporal change. This is an important phenomenon and will be studied in some detail later in the chapter when we discuss the technique known as *coherent change detection*. Second, another possibility is that the underlying function $r(x, y)$ does not change, but that the aperture regions of support in the spatial-frequency domain A_1 and A_2 do. This occurs in two-pass collections because the positions of A_1 and A_2 in frequency space depend on the collection geometry angles. For example, Figure 5.1 shows the effect of different depression angles on the Y position of these regions.

The collection surface in spatial-frequency three-space is determined by the center frequency, the bandwidth, the angular extent of the aperture, the depression angle, and the squint angle/velocity vector (see Figure 2.18 in Chapter 2). The platform velocity vector enters here because it determines the orientation of the slant plane. These parameters completely specify the region of support of spatial-frequency-domain data transduced by the radar. If any of these parameters vary from one collection to the next, the region of support moves in frequency space. Some change in collection angles is usually intentional. For example, we will see that terrain-mapping sensitivity depends on a small but non-zero difference in depression angles. In addition, lack of precise platform position data contributes to unintended differences in A_1 and A_2. We examine the ramifications of the position differences of A_1 and A_2 in Section 5.2.2.

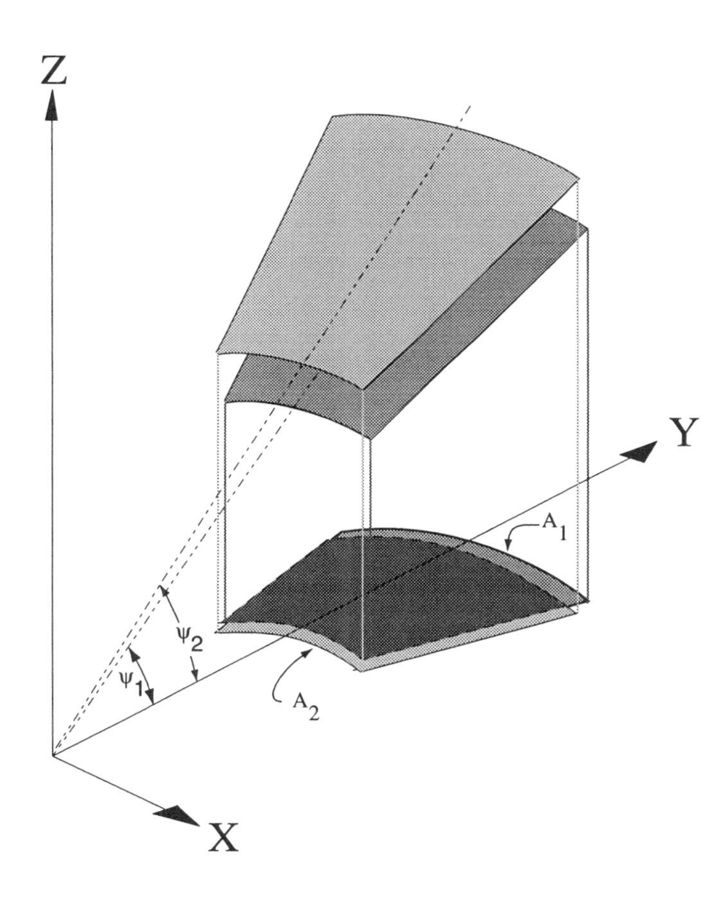

Figure 5.1 SAR collection surface in spatial-frequency three-space.

We are now in a position to consider how the imaging equations corresponding to an pair of SAR images (Equations 5.2 and 5.3) can be utilized to extract information about the terrain. We provide the motivation for three key techniques here, and expand upon each of these in the remaining sections of the chapter. First, suppose that the images are collected with a small (but non-zero) difference in depression angle, and that the two data sets are demodulated using a common offset Y_0. We may write the imaging equations as

$$\begin{aligned} g_1(x_1, y_1) &= s_{A_1}(x, y) \otimes \left[r(x, y)e^{-j\beta_1 Y_0 h(x,y)}e^{-jyY_0}\right] \\ g_2(x_2, y_2) &= s_{A_2}(x, y) \otimes \left[r(x, y)e^{-j\beta_2 Y_0 h(x,y)}e^{-jyY_0}\right] . \end{aligned} \tag{5.4}$$

Multiplying and dividing $r(x, y)$ in the second equation by $e^{-j\beta_1 Y_0 h(x,y)}$, we have the equivalent equations

$$\begin{aligned} g_1(x_1, y_1) &= s_{A_1}(x, y) \otimes \left[r(x, y)e^{-j\beta_1 Y_0 h(x,y)}e^{-jyY_0}\right] \\ g_2(x_2, y_2) &= s_{A_2}(x, y) \otimes \left[r(x, y)e^{-j\beta_1 Y_0 h(x,y)}e^{j(\beta_1-\beta_2)Y_0 h(x,y)}e^{-jyY_0}\right] . \end{aligned} \tag{5.5}$$

Now, if we postulate a slowly varying function $h(x, y)$, and because the depression angle difference $(\beta_1 - \beta_2)$ is also small, the term $e^{j(\beta_1-\beta_2)Y_0 h(x,y)}$ may be considered constant over the width of the impulse response function (i.e., the support of $s_A(x, y)$). When this constant term is brought out of the convolution integral we can rewrite these equations as

$$\begin{aligned} g_1(x_1, y_1) &= r_{A_1}(x, y) \\ g_2(x_2, y_2) &= r_{A_2}(x, y)\, e^{j(\beta_1-\beta_2)Y_0 h(x,y)} \end{aligned} \tag{5.6}$$

where $r_{A_1}(x, y)$ and $r_{A_2}(x, y)$ are defined as:

$$\begin{aligned} r_{A_1}(x, y) &\triangleq s_{A_1}(x, y) \otimes \left[r(x, y)e^{-j\beta_1 Y_0 h(x,y)}e^{-jyY_0}\right] \\ r_{A_2}(x, y) &\triangleq s_{A_2}(x, y) \otimes \left[r(x, y)e^{-j\beta_1 Y_0 h(x,y)}e^{-jyY_0}\right] . \end{aligned} \tag{5.7}$$

If we use an image-domain registration procedure to accommodate the height-dependent translations between the coordinates (x_1, y_1) and (x_2, y_2), then the image Equations 5.6 differ by the *sinc* functions $s_{A_1}(x, y)$ and $s_{A_2}(x, y)$, and by a phase shift of

$$\Delta\phi = (\beta_1 - \beta_2)Y_0 h(x, y) . \tag{5.8}$$

We will investigate the ramifications of the *sinc* function apertures in Section 5.2.2. Equation 5.8, however, indicates that the phase shift between the two images is a linear function of the terrain height, so that $h(x, y)$ can be recovered by computing image-domain phase differences (i.e. *interfering* the images). This is the essence of *terrain-height mapping* using interferometric SAR imagery. It is this mode of IFSAR that has received the greatest amount of attention in the literature during the past decade. This is due perhaps to demands for higher-accuracy digital terrain elevation maps of the earth surface and the all-weather capability of SAR.

A second mode of interferometric processing arises when the physical terrain reflectivity function $r(x, y)$ actually changes between the collection of the two SAR images. Such changes could arise from human activity (e.g., driving vehicles over open terrain) or from environmental factors. As we will see later, even fairly subtle physical effects can be detected by measurement of the *spatial correlation* of the two complex images. Where correlations are relatively small, the indication is that some change to $r(x, y)$ must have occurred. This mode of IFSAR processing is called *coherent* or *interferometric change detection.* Yet a third mode of IFSAR involves situations wherein the height function $h(x, y)$ changes slightly between the two SAR collections. For example, a section of earth can undergo a vertical displacement by a small amount due to effects of an earthquake. In this case, a phase difference term similar to the one derived in Equation 5.6 emerges. (Once again, the image phases may only be differenced following an image registration procedure.) However, now the resulting phase term is a linear function of the height *change* at each position in the scene. This form of IFSAR processing will be referred to as *terrain-motion mapping*. Finally, we note that in all of the above IFSAR modes, the pre-processing step of image registration is crucial. For a certain class of IFSAR collections, the result of the registration procedure is itself a computed map of the terrain height that may be shown to be a SAR stereoscopic product.

In summary and based on Equations 5.2 and 5.3, two SAR images of the same nominal terrain can differ in four ways. First, a difference in depression or squint angles can lead to a misregistration of the images that is a local function of the terrain height. We will later interpret the by-product of the data-driven image registration procedure required for IFSAR as classical SAR stereoscopy. Second, a difference in depression angles results in image-domain phase shifts that are directly proportional to terrain height, which gives rise to interferometric terrain mapping. Third, the terrain reflectivity function itself can vary from one collection to the next when the two collections are separated in time. This variation leads to the possibility of interferometric change detection. Finally, if the depression angle and radar wavelength are kept constant between two temporally separated collections, we can transduce temporal changes in the terrain-height function, which becomes the basis of interferometric terrain-motion mapping. The remainder of this chapter is devoted to an exposition of these phenomenologies.

5.2.2 Collection Constraints

The fact that two SAR image collections of a given terrain segment obtained for interferometric processing often possess somewhat different spatial-frequency regions of support A_1 and A_2 is generally regarded as a nuisance. This is because we can infer no useful information about the scene from this difference. Indeed, unless the effects of the difference are removed from the standard IFSAR products, they introduce a bias into the quantities we desire to measure.

A closer scrutiny of this question is warranted, however, because this analysis leads directly to an understanding of imaging geometry constraints for IFSAR applications. The fundamental question here is: How must two-pass or multi-pass collection geometries be constrained, if at all, in order to render interferometric processing possible? At the outset, it will be appreciated that the entire question of collection constraints applies only to two-pass or multi-pass experiments. That is, single pass (two-antenna) collections generally meet these constraints by construction.

One-Pass vs. Two-Pass Interferometry

We have already used the terminology *one-pass* and *two-pass* IFSAR in several contexts. We now make the definitions of these terms more precise. To perform any type of interferometric processing, we require two or more *independent* images as input. Interferometric processing draws its utility from the fact that interference of two complex functions causes common elements of the functions to be cancelled, magnifying the differences, or vice versa. Interferometric processing is meaningful only as applied to images formed from distinct phase histories, wherein the inter-image differences are functions of quantities we wish to measure. It is not meaningful for the case of a single SAR image collected with a single antenna.

Two distinct SAR images can be produced in a number of ways. They can be separated temporally, spatially, or a combination temporal/spatial separation can be used. This separation of interferometric collections is largely a matter of experimental design. Certain terrain features are best measured with zero spatial separation. Others require specific, e.g., relatively large, separations. Separation in time allows still other aspects of the imaged area to be transduced. Small separations of either type are accomplished with the two-receiver, or single-pass, interferometer, wherein either one or both antennas are used as a transmitter. If the two-antenna phase centers are separated in the along-track direction, a (small) temporal spacing between the resultant images is achieved. Such an along-track interferometer can be very effective as a *moving target indicator* (MTI), although we will not pursue the details of such a device here. On the other hand, if the antennas are separated in a direction orthogonal to both the flight path and line-of-sight vector, a *change in depression angle* between the images results. It is this mode of IFSAR with which we concern ourselves here.

Large separations cannot be accommodated with a two-antenna system, due to obvious limitations of physical size of the SAR platform. To generate a large spatial separation, we could collect images simultaneously using two platforms, or could collect the images at different times using the same platform. The latter obviously also accomplishes a large temporal separation between images. Either scenario will be referred to as two-pass. Thus, the essential defining characteristic of a one-pass collection is that it is performed by one imaging pass of a single vehicle. This collection mode is shown in Figure 5.2. Two-pass collections are performed by si-

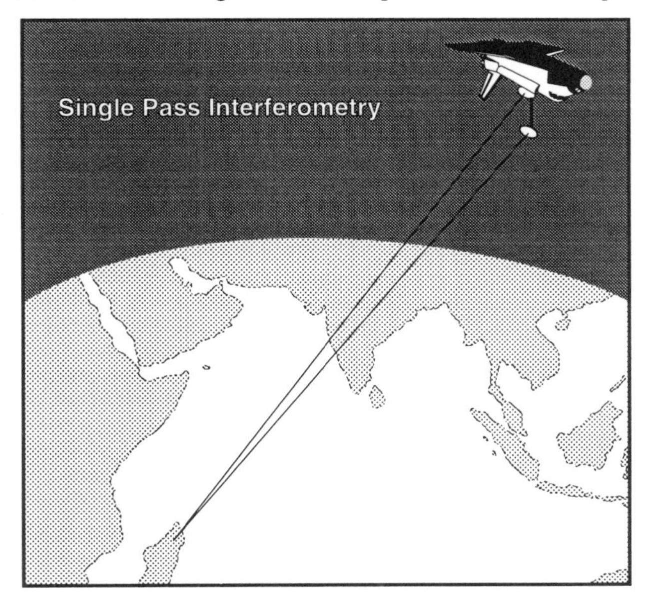

Figure 5.2 Basic collection mode for one-pass SAR interferometry using two antennas on the same platform to obtain both images of the pair simultaneously.

multaneous collections using two vehicles, or two imaging passes using one vehicle (see Figure 5.3). Generally speaking, two-pass data processing is more complicated than one-pass data processing, owing to the effects of larger separations in space and time between the image collections and to the fact that the precise orientation of the antenna phase centers is not rigidly fixed. Table 5.1 summarizes the possible collection scenarios and their applications. As this chapter unfolds, it will become

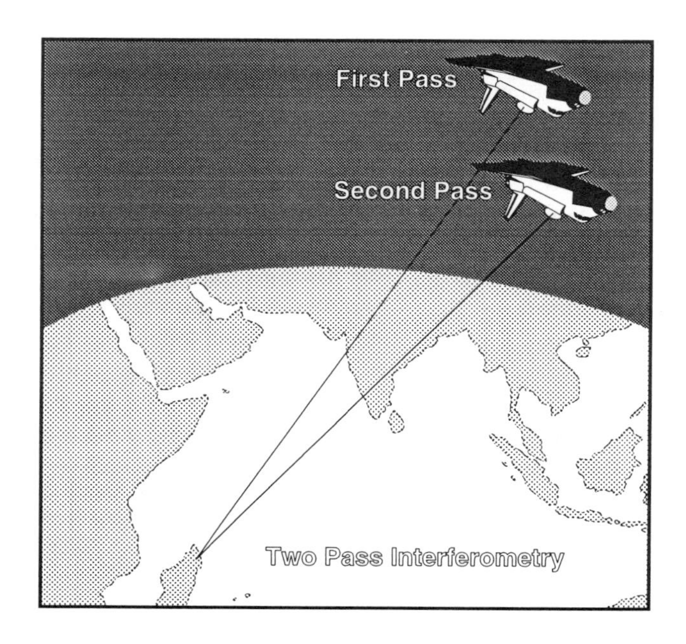

Figure 5.3 Basic collection mode for two-pass SAR interferometry using conventional single-antenna system.

Table 5.1 Interferometric Collection Scenarios and Their Applications

Collection Geometry	Temporal Separation	Spatial Separation
One-Pass	Moving Targets	Topographic Mapping
Two-Pass	Change Detection Terrain Motion	High-Precision Topographic Mapping

clear why these applications are associated with the various collection types in this way.

Collection Geometry and A_i

Returning to the question of collection constraints, we have already seen in Figure 5.1 and in related discussions the relationship between collection parameters and the frequency-domain region of support A. Given that two-pass collections can involve substantially different regions A_i, the central issue here involves the possible associated effects on the reconstructed reflectivity functions $r_{A_i}(x, y)$. To examine this matter, we must consider the properties of radar reflectivity from complex, distributed targets. In this discussion, we assume that the illuminated terrain is rough relative to the radar wavelength. For specular or discrete point targets, the analysis and conclusions would be quite different. However, the typical situation of natural terrain illuminated at microwave frequencies adequately fits the former situation (see Reference [2]).

The classical radar model of rough terrain, in which each resolution cell contains many independently positioned scatterers, results in a stochastic description of the returned energy. This model [3] yields an exponential probability density function for the backscattered power from each cell, and a uniform density function for the phase. More central to this discussion is the fact that resolution cells are *statistically independent* [4]. This is nothing more than the classical statistical description of speckle noise (see Reference [5]). Because natural terrain exhibits many scattering centers, this description is physically plausible. Both the physical dimensions of these centers as well as the radar wavelength are much smaller than the dimensions of a resolution cell. Furthermore, the precise locations of the scattering centers in one resolution cell are completely random with respect to those of adjacent cells.

Given that the backscattered energy of natural terrain can be reasonably modeled as above, what can we say about its spatial transform domain? We present a qualitative argument that addresses this question, although the concepts can be made mathematically rigorous in a straightforward manner. If the image-domain reflectivity can be modeled as a wide sense stationary (WSS) random process with statistically independent samples, then its Fourier transform is also independent and wide sense stationary. (This follows because the Fourier transform is a *unitary* operator.) Therefore, samples of the phase-history data for the SAR collection may be considered independent and WSS, at least for natural, rough terrain.

Now, given two frequency-domain regions of support A_1 and A_2, we can partition them into three component regions: $A_1 \cap A_2$ (their intersection), $A_1 \setminus A_2$ (A_1 setminus A_2), and $A_2 \setminus A_1$ (A_2 setminus A_1). Figure 5.4 shows the X-Y extent of the spatial-frequency data, wherein these three regions are clear. A mismatch of the two collections is illustrated, with differences in both the depression angle, as represented by the Y-displacement, and in the azimuthal viewing angle, as represented by the rotation effect. Note that a small rotation of this form is essentially the same as a displacement in the X direction. From the foregoing discussion, these non-overlapping regions of the spatial-frequency domain are statistically independent. Thus, the reconstructed reflectivity functions $r_{A_1}(x, y)$ and $r_{A_2}(x, y)$ *having been apertured in the frequency domain by apertures* A_1 *and* A_2, both differ from the reflectivity function $r(x, y)\ exp[-j\beta Y_0 h(x, y)]$ that results from transforming data over the entire frequency plane. However, these two apertured reflectivity functions have a compo-

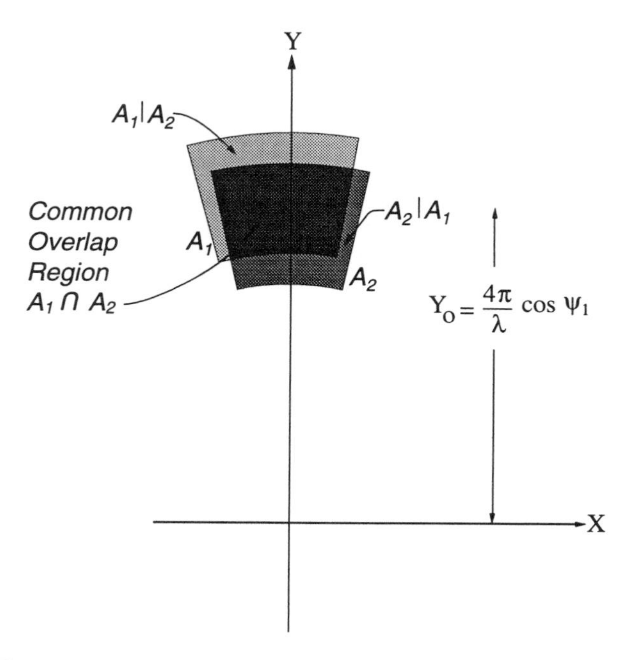

Figure 5.4 X-Y extent of spatial-frequency data for an interferometric collection.

nent in common (arising from the transform of the overlapped region $A_1 \cap A_2$), and an uncorrelated component (arising from $A_1 \setminus A_2$ or $A_2 \setminus A_1$). The net result is that if we are to have any hope of interferometrically processing a pair of SAR images with the intention of cancelling a common reflectivity function, then the geometry of the collections must be designed such that the aperture regions of support A_1 and A_2 *are largely overlapped in their* X, Y *coordinates.* Those portions of the aperture regions that are not overlapped ($A_1 \setminus A_2$ and $A_2 \setminus A_1$) contribute uncorrelated components to the reconstructed reflectivity functions, because the phase-history samples are independent. If the regions A_1 and A_2 are totally disjoint, the reconstructed reflectivity functions will be completely decorrelated, and interferometric processing is impossible. We stress that this argument applies to scenes composed of natural terrain. Discrete specular targets (such as corner reflectors) will correlate over much larger angles than these considerations dictate.

How can we insure that the aperture support functions overlap? There are several considerations. Recall that the aperture region in spatial-frequency three-space is determined by the center frequency, as well as by the depression angle and the squint angle/velocity vector. The center frequency and depression angle determine the distance from the origin to the center of the region A. The depression angle enters here because it is the position of A in the X-Y plane that is of concern (see

Figures 2.19 and 5.4). To provide substantial overlap, two collections must have close (but not necessarily identical) center frequencies. Indeed, if one collection is made at a somewhat higher depression angle than the other, a slightly higher center frequency for that collection will result in *more* overlap of the aperture regions. By making the Y coordinates of the center frequencies for the two collections equal, we can solve for the precise center frequency:

$$\frac{4\pi}{\lambda_1} \cos \psi_1 = \frac{4\pi}{\lambda_2} \cos \psi_2 \tag{5.9}$$

$$\lambda_2 = \frac{\cos \psi_2}{\cos \psi_1} \lambda_1 \ . \tag{5.10}$$

This observation, that a change in radar center frequency can partially compensate for the so-called baseline decorrelation from different depression angles, was simultaneously advanced by Prati, et al. [7] and Eichel, et al. [8]. The latter describes a real-time hardware implementation of this concept wherein the radar center frequency is slaved to the instantaneous depression angle by the motion-compensation hardware in real time. This scheme cannot totally compensate for baseline decorrelation because it ignores the effect of terrain variation in the vertical direction, but is very effective for reasonably flat terrain. A useful physical interpretation of Equation 5.9 is that it equates the effective radar wavelength *over the ground*. For flat terrain, the compensation is "perfect." However, if the ground is sloping toward or away from the SAR platform, the actual grazing angle of the energy with the ground is not equal to ψ, so this compensation is not quite what is needed. In an average sense, though, we can achieve a greater percentage of overlap of the aperture regions for collections with substantial interferometric baselines by offsetting the center frequencies in this manner.

So much for the Y direction, but what about the X direction? Here, the squint angle and/or velocity vector for the two collections must be chosen so that the azimuth angles from the sensor to the reference point on the ground are nearly equal (see Figure 2.21). We can accomplish this by making the velocity vectors equal (parallel ground tracks) and using the same squint angle for the two collections, or by using different velocity vectors (intersecting ground tracks) and compensating with different squint angles. Both collection geometries work (see Reference [8]), but the latter introduces other effects that must be accommodated in the processing, as we shall see.

All these considerations may be stated succinctly by insisting that the *boresight* of the two collections be relatively close. Exactly how close they must be is related to the system resolution in an interesting, if perhaps counter-intuitive way. The extents of the regions A_i are a function of the bandwidth and the angular span of the synthetic aperture. The former is related to the range resolution and the latter to the cross-range resolution. The finer the system resolution the larger the aperture region. Thus, fine-resolution systems can accommodate *larger* mismatches in boresight angles for a given percentage of overlap of A_1 and A_2. Stated another way, low-resolution SARs necessitate *finer* control over the collection geometry to achieve substantial overlap. In *all* cases, the angular differences must be quite small. For a one-meter resolution X-band system, for example, azimuth and depression angle differences must be held to within a few tenths of a degree to insure sufficient overlap. Furthermore, from Figure 5.4 we can conclude that the difference in depression angle $\Delta\psi$ becomes more tightly constrained as ψ increases, because the fractional mismatch in the Y dimension is

$$\frac{\delta Y}{\Delta Y} = \frac{\cos(\psi) - \cos(\psi + \Delta\psi)}{cos\psi} \approx \tan\psi \sin\Delta\psi \,. \tag{5.11}$$

Summary of Critical Ideas from Section 5.2

- **The radar reconstruction equations**

$$\begin{aligned} g(x_1, y_1) &= s_A(x,y) \otimes \left[r(x,y) e^{-j\beta Y_0 h(x,y)} e^{-jyY_0} \right] \\ x_1 &= x + \tan\eta \; h(x,y) \\ y_1 &= y + \tan\beta \; h(x,y) \end{aligned}$$

derived by means of the tomographic formulation in Chapter 2, not only provide an accurate description of a spotlight-mode SAR image, but also incorporate a virtually complete interferometric model as well.

Summary of Critical Ideas from Section 5.2 (cont'd)

- **Two SAR images of the same nominal terrain can differ in four ways based on this equation:**

  1. **Depression or squint angle differences lead to height-dependent image misregistration (making stereoscopy possible).**
  2. **For a non-changing height function, a difference in depression angles results in image-domain phase shifts that are proportional to terrain height. This phenomenon is exploited in interferometric terrain-mapping.**
  3. **Reflectivity functions can vary temporally from one collection to the next in two-pass IFSAR. This is the foundation for interferometric change detection.**
  4. **The terrain height function $h(x, y)$ may vary temporally from one collection to the next in two-pass IFSAR. This property leads to interferometric terrain-motion mapping.**

- **Two or more *independent* SAR images are required for interferometry. The collections of these images can be separated in space or in time or in both. The collections may be accomplished by using two antenna apertures and two receiver channels on a platform (single-pass collection), or by a single-channel system that overflies a region several times (two- or multi-pass collection).**

- **Interferometric processing is critically dependent on collection geometries that result in highly overlapped spatial-frequency regions of support A_i. This condition gives rise to the very important phenomenon of *coherent speckle*, which becomes relevant to the subjects of image correlation and registration.**

5.3 INTERFEROMETRIC PAIR PROCESSING: REGISTRATION

In Section 5.2, we learned from the reconstructed image equation (Equation 5.2) the kinds of information that can be extracted by interferometrically processing a pair of SAR images. We also learned how those images must be collected to support these applications. We now turn our attention to the details of processing the data. We will examine both the theory and the practice of interferometric data processing for terrain mapping and change detection. In addition, we will briefly touch on stereoscopic terrain mapping, because that technique involves processing steps that are common to those used for crossed-ground-track IFSAR.

After image formation, the principal steps required for interferometric data processing are: (1) image registration; (2) parameter estimation; and (3) 2-D phase unwrapping. (Phase unwrapping applies to terrain-elevation extraction, but not to coherent change detection.) In the case of one-pass collections, the first of these tasks is obviated by the fact that the formed images should already agree spatially. For two-pass collections, the registration step can be quite formidable, depending on the particular geometries used. However, it is still possible to achieve sub-pixel registration accuracy over large areas with automated processing because of the high correlation between the reconstructed reflectivity functions of interferometric pairs. Using terrain reflectivity models we have seen previously, we will derive maximum-likelihood estimators for the two most important interferometric parameters: the image-to-image phase difference and the temporal-change parameter. Finally, we will discuss 2-D phase unwrapping. Most of the literature on this latter subject stems from the classical approach of *path following*. Unfortunately, this avenue has led to the invention of highly complicated *ad hoc* decision rules intended to combat the difficulties imposed by inconsistencies in noisy data. Instead, we will concentrate on a newer methodology that is based on a least-squares formulation of the problem. The elegant method that arises from this approach shows much promise for fast, accurate, and robust phase unwrapping.[2]

5.3.1 Image Formation Issues

We have already considered many aspects of image formation for spotlight-mode SAR in Chapters 2 and 3. We will briefly take up this topic again to consider particular image-formation issues relevant to IFSAR pair processing. These include

[2] Phase unwrapping is still an area of active research. There remain a host of theoretical as well as practical issues that are not totally resolved by any present phase unwrapping methodology.

phase preservation, frequency-domain aperture trimming, common baseband translation, and attention to the image reconstruction plane, resolution, and spatial sampling rates.

The image-formation product of any SAR system is a complex reflectivity map. Traditionally, SAR system designers have emphasized quality of the reconstructed reflectivity function only so far as its magnitude is concerned. Interferometric applications, by contrast, necessitate accurate phase values as well. Image-formation processors that do not maintain phase integrity throughout cannot be used for subsequent interferometric processing. This state of affairs was acutely recognized by the strip-mapping SAR community in the late 1980s. This recognition spawned a number of papers in the literature on so-called *phase-preserving* image-formation processors [9][10].

Fortunately, for spotlight-mode applications, phase preservation has not been as big an issue. The derivation leading to Equation 2.72 of Chapter 2 demonstrates via the projection-slice theorem that the reconstructed image $g(x_1, y_1)$ produced by polar formatting and Fourier transformation is, in fact, an apertured version of the terrain reflectivity function. This function is modulated by a phase term that involves only terrain height and imaging geometry. As long as high quality, linear-phase interpolators and filters are used in the polar reformatter as described in Chapter 3, the image-formation process *per se* introduces no phase distortion in the output product. (Of course, the tomographic formulation itself ignores wavefront curvature. We will deal with that error source in Section 5.4.4. as a post-processing step.)

We have already seen in the previous section how collection constraints arise for IFSAR image pairs. For one-pass collections, these constraints are automatically met by the precise geometry imposed by the collection platform. Two-pass collections, however, place a significant burden on the navigation subsystem of the SAR platform. Not only must the platform repeat its position on the second pass to match the first accurately, but the relative separation of the collections must be known precisely enough to enable accurate scaling of the interferometric results. We further explore this issue in Section 5.4.3.

In practice, it is difficult to match the imaging geometries precisely for a two-pass IFSAR pair. We argued in Section 5.2.2 that the non-overlapping portions of the frequency domain apertures A_1 and A_2 yield uncorrelated image-domain information. This is the genesis of what has been termed *baseline decorrelation* [11],[12],[13]. However, this decorrelation can be reduced by appropriate steps in the image-formation processor. One effective method of dealing with any unavoidable mismatch is to retain for image formation only those raw data samples lying in the overlapped region: $A_1 \cap A_2$ (see Figure 5.4). Of course, because the resulting *trimmed* apertures span a smaller region of frequency space, the images formed will exhibit a somewhat reduced resolution. In most cases, this tradeoff between resolution and noise reduction due to baseline decorrelation is most advantageous. Furthermore, the parameter estimation techniques to be discussed later involve spatial averaging anyway, so that a slight loss in resolution is typically of no consequence.

A second more sophisticated means of dealing with baseline decorrelation, at least in the range dimension, is to perform the two image collections with different center frequencies, so as to satisfy Equation 5.10. As discussed in Section 5.2.2, this requires hardware capable of slaving the transmitted chirp center frequency to the instantaneous depression angle on a pulse-by-pulse basis.

A third consideration for image formation of coherent pairs is to use a common baseband translation Y_0 in performing the 2-D Fourier transform. When the polar reformatted data are loaded into a rectangular array for transformation, care must be exercised to place samples at the same spatial frequencies into common array indices for the two collections. Preventing *offsets* in the spatial-frequency-domain in this way prevents the subsequent generation of a *plane wave* in phase, i.e., a linear-phase term in the image domain. Of course, the accuracy to which we can position the two data sets in the frequency domain depends entirely on the accuracy of the pointing vector information we obtain from the platform navigation system.

Because subsequent processing steps will involve registering the two images, it seems reasonable to form the images in a common plane. This simplifies the registration process by eliminating large differences in scale factors and skew between the images. For two-pass collections that possess a common ground-track angle, one of the two slant planes may suffice for this purpose. For crossed ground tracks, it generally makes more sense to use the ground plane for the common image plane. By employing the technique of frequency-domain projection discussed in Sections 2.4.4, 3.8, and 3.9, ground-plane imagery is produced with virtually no impact on processing time.

A final consideration for image formation is the matching (to the accuracy supported by the pointing vector information) of the image-domain sampling rates, impulse-response functions, and over-sample ratios for the pair of images. From Section 3.10, it should be clear that this can be accomplished by equating the frequency-domain aperture areas A_i, the number of rectangular output samples from the polar formatter, the array dimensions used for the 2-D FFT, and using a common image formation plane normal vector. Having two images with closely matched sampling rates eases the processing burden in the registration operation that follows image formation.

5.3.2 Registration

SAR interferometry is fundamentally concerned with obtaining information from the difference in phase of the complex reflectivity of two images. To accomplish this, the complex reflectivity functions of the two images must be *registered*, i.e. one discrete image must be interpolated so that samples corresponding to a given scatterer on the ground have the same indices in both images. Because resolution cells are uncorrelated for natural terrain (see Section 5.2.2), this registration must be accurate to less than an impulse-response width everywhere. For single-pass, displaced phase center systems, this is not a burdensome requirement, particularly at low spatial resolutions. For large images and for two-pass collections, especially those with significantly different ground-track crossing angles, however, this registration requirement can become very demanding. Platform navigation and imaging geometry errors introduce misregistrations that are large compared to the IPR width, especially with high-resolution spotlight systems. Thus, interferometric processing depends on *data-driven, automatic* registration techniques. In this section, we explore some effective methods for accomplishing this task.

Introduction to Registration

Data-driven image registration techniques abound in the literature [4]. Generally, image-pair registration is accomplished in three steps. First, control points are generated. Measurements are made of the local displacement vector from one image to the other in many places throughout the scene. Second, the control-point displacements are then used to calculate a *warping function*, which maps a location in one image to the corresponding location in the other. Finally, the first image is resampled so that it overlays the second precisely. We will denote the image to be warped as the source image and that to which it is warped as the target image.

The control-point measurements can be made using projection properties [14], feature identification [15], or correlation. Variations include multi-resolution, recursive, and iterative methods. Projection-based techniques separately compute x and y displacements by computing 1-D projections and then by correlating the projection functions. Computationally, this is faster than performing 2-D correlations, but the inherent integration that occurs in the projection process reduces the accuracy of the resulting registration. It is the high spatial-frequency content of the data that allows sub-pixel accuracy to be achieved. Methods that rely on feature extraction and matching also seem ill-suited to this purpose. Except for areas with extensive man-made structures, high-resolution SAR images typically contain relatively few distinctive features. Therefore, a dense control-point set may be difficult to generate.

This state of affairs leads us to 2-D correlation as the method of choice. For the application of SAR interferometry, spatial correlations are particularly well suited. The reason for this has to do with the concept of *coherent speckle*. In Section 5.2.2, we saw that if two images have the same aperture region of support, then barring temporal changes between the collections, they will have the same apertured complex reflectivity function $r_A(x, y)$. But as we know, all SAR images exhibit the phenomenon of speckle for resolution cells composed of random collections of scatterers. Thus, we conclude that two such images have *the same spatial speckle pattern*. Furthermore, speckle possesses a great deal of energy at high spatial frequencies. These are the perfect conditions for achieving high performance from a correlator, i.e., the functions to be correlated are identical[3] and possess a great deal of high-frequency content. Such correlations can be performed virtually anywhere in the image, and thus satisfy the control-point density requirement. Finally, the spatial cross-correlation function can be computed using transform methods for high computational efficiency.

For two-pass SAR imagery, there is a strong case to be made for multi-resolution registration processing. Because of navigation errors and antenna-pointing errors, two images taken of the same nominal terrain can have a relative translation and/or rotation that is significant. Because the 2-D correlation method for obtaining local displacement estimates for control points requires that the image data substantially overlap within the correlation window, a large initial translation or rotation can necessitate large correlation windows when full-resolution imagery is employed. Not only is this inefficient, but the final displacement vector computed may represent an average value over the window and so does not constitute a local measurement. In multi-resolution processing, control points are initially generated using downsampled imagery, thus they accommodate large translations and rotations. From these control points, an affine transformation is calculated. This guides the selection of correlation windows for the next-higher-resolution control point generation stage.

The phenomenon of layover is an important consideration for registering two-pass interferometric pairs. In Reference [16], an azimuthal shift in crossed-ground-track image pairs is derived and shown to be linearly related to altitude. It may be seen that this effect can be understood simply as differential layover between the images. From Equation 5.3 of Section 5.2.1, we know that two images collected with different slant-plane angles η_i or ψ_i will have different layover transformations. Consider a scatterer at true location $(x, y, h(x, y))$. After image formation, it will be positioned

[3] In point of fact, the complex image functions are not really identical. We note from Equation 5.2 that image-domain phase terms are also present in the reconstructed image. In addition, some receiver noise will always be present. Shortly, we will show how these terms can be dealt with.

in the target image at coordinates (x_1, y_1), where

$$\begin{aligned} x_1 &= x + \alpha_1 h(x,y) \\ y_1 &= y + \beta_1 h(x,y) \ . \end{aligned} \tag{5.12}$$

Thus, the amounts of layover in the cross-range and range directions are

$$\begin{aligned} \Delta x &= h(x,y) \tan \eta_1 \\ \Delta y &= h(x,y) \tan \psi_1 \ . \end{aligned} \tag{5.13}$$

Similar equations would describe the layover observed in the source image. Therefore, the *differential* layover (misregistration) between the two images is

$$\begin{aligned} y_2 - y_1 &= h(x,y)(\tan \psi_2 - \tan \psi_1) \\ x_2 - x_1 &= h(x,y)(\tan \psi_2 \tan \theta_{g_2} - \tan \psi_1 \tan \theta_{g_1}) \ . \end{aligned} \tag{5.14}$$

In the above equations, we use the relationship $\tan \eta = \tan \psi \tan \theta_g$, where θ_g is the ground-plane squint angle as shown in Figure 2.28. Because of the collection constraints (Section 5.2.2), IFSAR pairs are always imaged with approximately the same mid-aperture boresight. Therefore, the grazing angles are nominally equal. This implies that the range component of differential layover $y_2 - y_1$ is nearly zero. However, the cross-range component $x_2 - x_1$ depends on the collection ground tracks. As a result we have

$$\begin{aligned} y_2 - y_1 &\cong 0 \\ x_2 - x_1 &\cong h(x,y) \tan \psi (\tan \theta_{g_2} - \tan \theta_{g_1}) \ . \end{aligned} \tag{5.15}$$

If the collection ground tracks are parallel for both images (have the same squint angle), then there would be no differential shift of the cross-range position of the target. If the ground tracks are crossed, then the squint angles for the two images differ, and a differential shift of the cross-range position of any target with non-zero altitude would be observed. It can be seen, in fact, that this phenomenon is identical to that exploited in classical, stereo-pair processing of optical imagery to derive elevation (see Reference [17]). Note that the squint angles do not have to be equal and opposite with respect to broadside. As long as the boresights of the two collections are the same, any difference in ground tracks will result in a difference in squint, which will lead to a differential azimuthal shift as a function of elevation.

Control-Point Generation

The automatic generation of accurate control points throughout the scene of interest is crucial to two-pass IFSAR processing. In some cases, thousands of control points must be computed per image pair. This is especially important if we wish to use a local warp (discussed below), but is useful even when a global warp is employed, because the determination of warp coefficients is more robust given an over-determined input set. The image patches used for each control-point calculation must be necessarily small to insure it is a local computation and not averaged over a large area. Small image patches can lead to difficulties when the potential displacements between the source and the target images are large. As previously mentioned, this situation can be addressed by performing the control-point calculation in multiple steps using a multi-resolution approach. In the first step, a few large image patches taken from downsampled detected images are used to calculate an approximate affine transformation warping function. This function is then employed to guide the placement of smaller, denser control-point patches in the next step. In this way, significant amounts of translation and rotation of the original images do not compromise the use of small, accurate, local control-point patches.

To begin our discussion of control-point generation, let us first consider the process of 2-D spatial correlation. We will then move on to algorithmic specifics. We will denote the target image as $g_1(x_1, y_1)$ and the source image as $g_2(x_1, y_1)$. That is, we wish to resample g_2 so that it matches g_1. Let $f_1(i,j) \subset g_1(x_1, y_1)$ be a subimage of g_1, and $f_2(i,j) \subset g_2(x_2, y_2)$ be a subimage of g_2. In general, these SAR image functions are complex. However, we will also have occasion to consider *detected*, or magnitude only, image functions. Let us now suppose for a moment that the source image is merely a translated version of the target image. In that case, we have:

$$f_2(i - s', j - t') = f_1(i,j) \ . \tag{5.16}$$

The 2-D spatial cross-correlation function of f_1 and f_2 (for zero-mean data) is defined to be

$$R_{f_1 f_2}(s,t) = E\left\{\sum_i \sum_j f_1(i,j)\, f_2^*(i-s, j-t)\right\} \ . \tag{5.17}$$

Use of Equation 5.16 then gives

$$\begin{aligned} E\{\sum_i \sum_j f_1(i,j) f_2^*(i-s, j-t)\} & \\ &= E\{\textstyle\sum_i \sum_j f_1(i,j)\, f_1^*(i-s+s', j-t+t')\} \\ &\triangleq R_{f_1 f_1}(s-s', t-t') \end{aligned} \tag{5.18}$$

where R is the (complex) autocorrelation function of f_1. Once again invoking the uncorrelated and wide-sense stationarity property of SAR image data (see Section

5.2.2), the autocorrelation function of f_1 is

$$R_{f_1 f_1}(s - s', t - t') = \delta(s - s', t - t') \; . \tag{5.19}$$

By computing the cross-correlation function of f_1 and f_2, we obtain a measurement of the image-to-image translation (s', t') as the position of the spike in $R_{f_1 f_2}(s, t)$.

It is often the case in SAR interferometry that fairly large correlation windows are required for control-point computations. Large correlation windows improve the peak-to-variance ratio in the function $R_{f_1 f_2}(s, t)$. Furthermore, the local translation uncertainty can also be large, necessitating knowledge of $R_{f_1 f_2}(s, t)$ over many lags. Under these circumstances, it is generally more efficient to employ transform-domain techniques than direct image-domain spatial correlations. If we take the two-dimensional Fourier transform of f_1 and f_2:

$$\begin{aligned} F_1(m, n) &= \mathcal{F}\{f_1(i, j)\} \\ F_2(m, n) &= \mathcal{F}\{f_2(i, j)\} \end{aligned} \tag{5.20}$$

and rewrite the spatial cross-correlation function as

$$\begin{aligned} R_{f_1 f_2}(s, t) &= \sum_i \sum_j f_1(i, j)\, f_2^*(i - s, j - t) \\ &= f_1(i, j) \otimes f_2(-i, -j) \end{aligned} \tag{5.21}$$

then by the convolution property of Fourier transforms, we have

$$\mathcal{F}\{R_{f_1 f_2}(s, t)\} \triangleq S(m, n) = F_1(m, n) F_2^*(m, n) \; . \tag{5.22}$$

Therefore, to obtain the cross-correlation matrix $[R_{f_1 f_2}]$ we must transform f_1 and f_2, conjugate multiply the transform-domain data, and inverse transform.

Having established the basis for estimating subimage displacements using 2-D spatial cross-correlation, let us now consider the specifics of a robust multi-resolution algorithm. From the original high-resolution, complex, target and source images g_1 and g_2, we form detected, downsampled, multi-look images $\tilde{g}_1$ and $\tilde{g}_2$. Next, we extract fairly large subimage patches from the detected images: $[\tilde{f}_1] \subset [\tilde{g}_1]$ and $[\tilde{f}_2] \subset [\tilde{g}_2]$. (In order to simplify the notation somewhat, we will no longer use the indices (i, j) and (m, n), unless they are required for clarity. We will use brackets $[\,\cdot\,]$ to denote a two-dimensional array of data.) These patches should be large enough to accommodate any unknown translation of the original images. This subimage data is transformed to give

$$\begin{aligned} [\tilde{F}_1] &= \mathcal{F}\{[\tilde{f}_1]\} \\ [\tilde{F}_2] &= \mathcal{F}\{[\tilde{f}_2]\} \end{aligned} \tag{5.23}$$

and conjugate multiplied, giving

$$[\tilde{F}_1 \tilde{F}_2^*] \,. \tag{5.24}$$

Because we are correlating detected data at this point, and because detected data is strictly non-negative, it has a non-zero mean. In order to avoid a bias in the cross-correlation function, we set to zero the dc coefficient of the conjugate multiplied array, i.e.

$$[\tilde{F}_1 \tilde{F}_2^*]_{0,0} = 0 \,. \tag{5.25}$$

Finally, we inverse transform to give

$$[R_{\tilde{f}_1 \tilde{f}_2}] = \mathcal{F}^{-1}\{[\tilde{F}_1 \tilde{F}_2^*]\} \tag{5.26}$$

and then locate the peak of $[R_{\tilde{f}_1 \tilde{f}_2}]$. The offset of this peak from the origin is an estimate of the local displacement (s', t') from g_1 to g_2 at the location of this subimage, properly scaled by the downsample ratio used to generate $\tilde{g}_1$ and $\tilde{g}_2$. The position of the correlation peak can be estimated accurately by a quadratic interpolation of the four nearest sample points.

Having obtained a set of control points in this manner, an affine transformation $A(x, y)$ is calculated from the displacements by linear regression (discussed in detail below). This transformation maps a location in the target image (x_1, y_1) into its corresponding location in the source image, (x_2, y_2). This mapping function is not particularly accurate, because it is calculated from the downsampled image control points. However, it is more than sufficient for placing subimage patches in the full-resolution imagery with adequate overlap to ensure good correlator performance. We then extract high-resolution, complex image patches from g_1 and g_2:

$$\begin{aligned} [f_1(x_1, y_1)] &\subset [g_1] \\ [f_2(x_2, y_2)] &\subset [g_2], \text{ where } (x_2, y_2) = A(x_1, y_1) \,. \end{aligned} \tag{5.27}$$

As before, we Fourier transform the subimage patches to get

$$\begin{aligned} [F_1] &= \mathcal{F}\{[f_1]\} \\ [F_2] &= \mathcal{F}\{[f_2]\} \,. \end{aligned} \tag{5.28}$$

At this point, we depart from the procedure for low-resolution control points. First of all, we are now correlating complex rather than detected data. Referring again to our backscatter model for rough terrain (Section 5.2.2), we note that the complex data are zero mean. It is not necessary, therefore, to set to zero the DC coefficient of the conjugate product of $[F_1]$ and $[F_2]$. More importantly, however, now we have to face the consequences of the phase terms in Equation 5.1. As we discussed in the introduction to this section, even if the two SAR collections are repeated with sufficient accuracy to have nearly identical aperture regions of support (and consequently the same apertured complex reflectivity functions), the complex image data for the two images will still not be identical, because of the phase terms. We noted in Section 5.3.1 that a difference in depression angle or radar center frequency between the two collections could introduce a linear phase difference between g_1 and g_2. We also noted that this term may be effectively eliminated by careful baseband translation during image formation. Equation 5.6 yields a phase difference that is proportional to terrain height and that represents the essential term leading to interferometric terrain mapping (Section 5.4.3). Thus, we cannot expect to correlate subimage patches $[f_1]$ and $[f_2]$ with these phase differences present. If one cycle of phase difference is present over the correlation window, the correlator output will be near zero. This problem may be circumvented in one of two ways: (1) the image data are detected first, before Fourier transformation, thus eliminating any effects of image-domain phase; or (2) if complex correlations are desired, we make use of the shift property of Fourier transforms, stated as

$$\mathcal{F}\{e^{jkx}\, f(x)\} = F(m-k)\,. \tag{5.29}$$

This property says that the Fourier transform of a function multiplied by a linear phase is a shifted version of the Fourier transform of that function. Consequently, if we assume the product of the above two phase-difference terms can be adequately modeled as a low-frequency plane wave in the local correlation window, we can remove this relative phase difference by shifting one of the transform arrays of Equation 5.28. Because the frequency and direction of this plane wave are unknown, we need to evaluate a number of cross-correlation arrays

$$\begin{aligned}[R_{f_1 f_2}]\,(k,l) = \mathcal{F}^{-1}\{[F_1(m,n)]\,[F_2^*(m+k,n+l)]\}\\ -p \le k,l \le p\,.\end{aligned} \tag{5.30}$$

We compute $(2p+1)^2$ cross-correlation arrays $[R_{f_1 f_2}]\,(k,l)$, each of which is compensated by a phase plane: $e^{j(kx+ly)}$. The parameter p is chosen to be commensurate with the terrain slope and the collection geometries. While demanding substantially more computation, correlating complex image data often yields a superior control point set. Complex correlations are particularly effective in eliminating false correlation peaks in imagery containing man-made, point-like targets.

Crossed-ground-track imaging geometries present another challenge to high-resolution control-point generation. From Equation 5.15, a substantial cross-range misregistration that is proportional to the terrain height may be present in these cases. For example, with $\rho = 1$ ft, $\psi = 30°$, $\theta_{g_1} = -30°$, $\theta_{g_2} = 30°$, and $h = 500$ ft, we calculate a differential cross-range position of

$$\frac{x_2 - x_1}{\rho} = 500(0.577)(0.577 + 0.577) = 333 \text{ pixels}\,. \tag{5.31}$$

This can be much larger than the subimage patches used for control-point correlations. Because $h(x, y)$ is *a priori* unknown, we must evaluate several cross correlations spaced in cross-range about the expected position. For $f_1(x_1, y_1) \subset g_1$, we cross correlate with the $(2q+1)$ image chips from g_2, given by

$$f_2(x_2 + k\frac{N}{2}, y_2) \subset g_2 \quad (x_2, y_2) = W(x_1, y_1)$$
$$-q \leq k \leq q \,. \tag{5.32}$$

This procedure of cross correlating with spaced, overlapped windows from the source image in the epipolar direction is well known in stereo matching algorithms [17].

Finally, we require a means for discriminating good control points from bad. Not every control-point displacement calculated in the low- and high-resolution steps is valid. There are a number of reasons for this. If we calculate a regular grid of control points, some correlation windows will invariably fall in radar-shadowed regions, in regions of low radar cross section, in regions covered by water, or in regions subjected to temporal changes. Because these areas represent an attempt to correlate only system noise or different reflectivity functions, failure is nearly certain. A simple but effective procedure for automatically editing out such bad control points is to calculate the peak-to-rms ratio of $[R_{f_1 f_2}]$:

$$\frac{\max_{s,t} |R_{f_1 f_2}(s,t)|}{\sqrt{\sum_s \sum_t |R_{f_1 f_2}(s,t)|^2}} \tag{5.33}$$

and apply a threshold test. Because of coherent speckle in interferometric pairs, good control points generally have an impulse-like correlation function with a very high peak-to-rms ratio. Therefore, it is not difficult to select a global threshold value. Reference [17] discusses some additional tests on control points that may be effective under some conditions.

Image Resampling

Using the above procedures, a dense set of accurate control-point displacements can be calculated

$$\left[D_i(x_1, y_1) = \begin{bmatrix} D_i^x(x_1, y_1) \\ D_i^y(x_1, y_1) \end{bmatrix} : (x_1, y_1) \in \mathcal{R}^2 \right] \tag{5.34}$$

where the subscript i refers to the ith control point, $D_i^x(x_1, y_1)$ is the displacement in x_1, and $D_i^y(x_1, y_1)$ is the displacement in y_1. It remains to calculate a warp function and to resample the source image to overlay the target image. The warp function is a mapping

$$W(x_1, y_1) : \mathcal{R}^2 \to \mathcal{R}^2 \tag{5.35}$$

that takes locations in g_1 into locations in g_2. A simple procedure for accomplishing this is to perform a global warp by modeling $W(x_1, y_1)$ as a 2-D polynomial

$$W(x_1, y_1) = \begin{bmatrix} W^x(x_1, y_1) \\ W^y(x_1, y_1) \end{bmatrix} = \begin{bmatrix} a_0 + a_1 x_1 + a_2 y_1 + a_3 x_1 y_1 + \cdots \\ b_0 + b_1 x_1 + b_2 y_1 + b_3 x_1 y_1 + \cdots \end{bmatrix} \tag{5.36}$$

where the coefficients $[a_i]$, and $[b_i]$ can be determined by linear regression. This is accomplished by solving the following two sets of linear equations:

$$\begin{bmatrix} 1 & x_{11} & y_{11} & x_{11}^2 & x_{11}y_{11} & y_{11}^2 \\ 1 & x_{12} & y_{12} & x_{12}^2 & x_{12}y_{12} & y_{12}^2 \\ 1 & x_{13} & y_{13} & x_{13}^2 & x_{13}y_{13} & y_{13}^2 \\ & & & \vdots & & \end{bmatrix} \begin{bmatrix} a_0 \\ a_1 \\ a_2 \\ a_3 \\ a_4 \\ a_5 \end{bmatrix} = \begin{bmatrix} x_{11} + D_1^x \\ x_{12} + D_2^x \\ x_{13} + D_3^x \\ \vdots \end{bmatrix} \tag{5.37}$$

$$\begin{bmatrix} 1 & x_{11} & y_{11} & x_{11}^2 & x_{11}y_{11} & y_{11}^2 \\ 1 & x_{12} & y_{12} & x_{12}^2 & x_{12}y_{12} & y_{12}^2 \\ 1 & x_{13} & y_{13} & x_{13}^2 & x_{13}y_{13} & y_{13}^2 \\ & & & \vdots & & \end{bmatrix} \begin{bmatrix} b_0 \\ b_1 \\ b_2 \\ b_3 \\ b_4 \\ b_5 \end{bmatrix} = \begin{bmatrix} y_{11} + D_1^y \\ y_{12} + D_2^y \\ y_{13} + D_3^y \\ \vdots \end{bmatrix} . \tag{5.38}$$

In these two equations, the notations x_{1i} and y_{1i} designate the particular values of the image coordinates (x_1, y_1) associated with the *i*th control point. Polynomial warps are excellent for accommodating global translation, rotation, skew, and change of scale. The procedure is efficient and sufficiently accurate provided that: (1) the collection geometries employ parallel ground tracks; and (2) the set of control points contains few "outliers." These restrictions are required because crossed-ground-track collections introduce height-dependent misregistration and because polynomial fitting using linear regression is somewhat intolerant of control-point outliers. The latter can be mitigated to a large extent by effective control-point editing and the use of a dense control-point set, producing a highly over-determined set of Equations 5.37 and 5.38.

Because crossed-ground-track collection geometries in the presence of terrain relief produce cross-range differential layover, a local rather than global warping function is required for these cases. One such procedure employs the 2-D spline warp. Here, we model $W^x(x, y)$ and $W^y(x, y)$ as surfaces. These surfaces are tessellated by a triangular grid in which the control points make up the vertices. The height of the surface represents displacement in x or y. Each triangle is a best spline fit of the displacements at its vertices subject to the constraint that the first partial derivatives are continuous at the boundaries (cubic spline fitting) [18]. Because of the continuity constraint, a given control point affects the warping function in its immediate neighborhood, but will not influence the function over the rest of the image. Local variations in cross-range layover are accommodated because we are locally interpolating between closely spaced control points.

While useful for tracking terrain-induced differential layover in crossed-ground-track collections, spline warping functions are not particularly good for rotations. Therefore, we can choose to combine the best attributes of the polynomial and spline warps. To accomplish this, we model $W(x_1, y_1)$ as the sum

$$W(x_1, y_1) = A(x_1, y_1) + S(x_1, y_1) \tag{5.39}$$

where $A(x_1, y_1)$ is a first order 2-D polynomial warp (an affine transformation) and $S(x_1, y_1)$ is a spline warp. $A(x_1, y_1)$ is first fit to the control points in the manner outlined above. Then, a new set of *residual* control-point displacements are calculated by subtracting the displacement predicted by A from the actual displacements:

$$\begin{aligned} R_i(x_1, y_1) &= \begin{bmatrix} R_i^x(x_1, y_1) \\ R_i^y(x_1, y_1) \end{bmatrix} \\ &= \begin{bmatrix} D_i^x(x_1, y_1) - A^x(x_1, y_1) \\ D_i^y(x_1, y_1) - A^y(x_1, y_1) \end{bmatrix} \end{aligned} \tag{5.40}$$

where

$$A(x_1, y_1) = \begin{bmatrix} A^x(x_1, y_1) \\ A^y(x_1, y_1) \end{bmatrix} \tag{5.41}$$

are the displacements predicted by the affine transformation. $S(x_1, y_1)$ is then fit to the $[R_i(x_1, y_1)]$. We note in passing that for crossed-ground-track geometries, the cross-range spline warp function $S^x(x_1, y_1)$ represents a scaled approximation to the terrain-height function $h(x, y)$. This follows from Equation 5.15, and forms the basis of stereoscopic terrain mapping (see Section 5.7).

Having computed a warping function W, we are ready to resample the source image g_2 to overlay the target image g_1. For every sample point of the target image (x_1, y_1) we calculate a location in g_2 as

$$(x_2, y_2) = W(x_1, y_1) . \tag{5.42}$$

Of course, the output point (x_2, y_2) generally does not fall on a sample location in g_2. Therefore, we must interpolate to solve for the value of $g_2(x_2, y_2)$ from its surrounding sample points. In the image domain, this can be accomplished efficiently by using the bilinear interpolator

$$\begin{aligned} g_2(x_2, y_2) = {} & (1-p)(1-q)g_2(i, j) + p(1-q)g_2(i+1, j) \\ & +(1-p)qg_2(i, j+1) + pqg_2(i+1, j+1) \end{aligned} \tag{5.43}$$

where p and q represent the x and y distances, respectively, of the point (x_2, y_2) from the nearest sample value of g_2 to the upper left (see Figure 5.5).

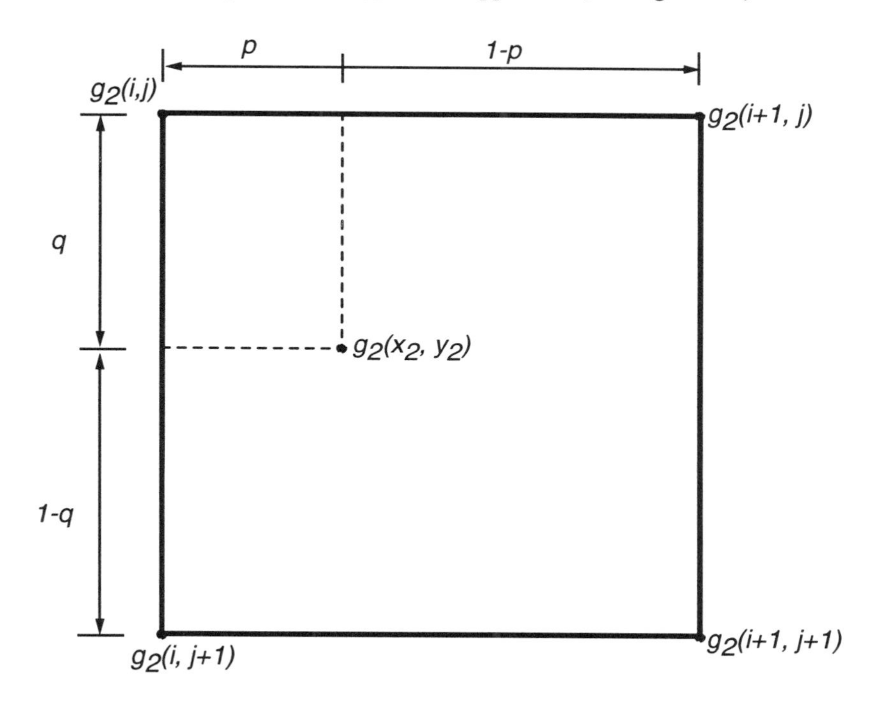

Figure 5.5 Bilinear interpolator geometry.

Summary of Critical Ideas from Section 5.3

- **Image-formation processors for interferometric applications should be phase preserving, should use a common baseband translation Y_0 for each image pair, should form the images in a common plane, and should use equal image-domain sampling rates, IPR functions, and oversample ratios.**

- **Registration of SAR image pairs is the first stage of interferometric processing and is accomplished in three steps: (1) generation of a set of control points or local image-to-image displacement vectors, (2) calculation of a warping function, and (3) image resampling.**

- **Two-dimensional correlation of small image patches is a robust means of generating registration control points. The method can take advantage of the phenomenon of coherent speckle in interferometric image pairs, the 2-D correlations can be efficiently calculated using FFTs, and a dense control-point set can be generated independent of image content.**

- **A multi-resolution control-point generation strategy can accommodate large initial offsets and rotations between two images, and still allow fine, sub-pixel accuracy from the final control-point set.**

- **Complex correlations can be used for greater accuracy in control-point generation, but allowance must be made for terrain dependent image-to-image linear phase terms.**

- **Where interferometric images are collected using crossing-ground-track geometries, terrain-height-dependent differential layover will be present in the images. This differential displacement will always be in the cross-range direction (the epipolar direction).**

- **Warping functions can be polynomial based (global), spline based (local), or a combination of the two. The latter is the best choice for accommodating rotation as well as terrain-height-dependent differential layover.**

- **A simple bilinear interpolator is sufficient for resampling the source image according to the warping function.**

5.4 INTERFEROMETRIC TERRAIN MAPPING

Having collected a pair of SAR images that meet the collection constraints of Section 5.2.2 and having registered them using the techniques of Section 5.3.2, we are now in a position to interfere the complex images. The two principal uses of IFSAR processing are for terrain mapping, discussed in this section, and for change detection, discussed in Section 5.5. Interferometric terrain mapping has received a great deal of attention in the literature. Our treatment of the subject, however, employs an approach quite unlike that of previous publications, in that it relies on the unifying theory of tomography. The major processing functions discussed here include parameter estimation, phase unwrapping, scaling, wavefront curvature correction, and orthorectification.

To begin, suppose that our two registered complex images (collected with overlapping spatial frequency domain aperture regions) are denoted f and g. We will model these two images as (see Equation 5.6)

$$\begin{aligned} f(x_1, y_1) &= r_A(x, y) \\ g(x_1, y_1) &= r_A(x, y)\ e^{j(\beta_f - \beta_g)Y_0 h(x,y)} \end{aligned} \tag{5.44}$$

where we assume that during the registration process, the second image has been resampled onto image coordinates (x_1, y_1). Recall that the image coordinates are related to the physical space coordinates via Equation 5.12. Here, several simplifying assumptions have been made: (1) the apertures were trimmed to a common aperture region A at image formation according to the procedure described in Section 5.3.1; (2) a baseband translation Y_0 was employed; and (3) the layover is sufficiently benign so as to avoid superposition of different physical locations in the imagery. Also, recall from the introductory discussion surrounding the derivation of Equation 5.6 that we require the height function $h(x, y)$ to be slowly varying with (x, y) so that the phase term $(\beta_f - \beta_g)Y_0 h(x, y)$ can be treated as constant over the support of the IPR. From Equation 5.44, we see that the terrain height function $h(x, y)$ modulates the image-domain phase difference between f and g. If we conjugate multiply the two image functions, we obtain

$$f^* g = |r_A(x, y)|^2\ e^{j(\beta_f - \beta_g)Y_0 h(x,y)} \tag{5.45}$$

we note that the phase of the product does *not* depend on the reflectivity function. *This is a direct result of collection geometries and aperture trimming that result in a common, aperture-domain region of support A. Without this overlap, the reconstructed images f and g would be uncorrelated, and their conjugate product would contain a rapidly varying reflectivity phase. Interferometry under these circumstances would be impossible.* From this point, the height function can be found by taking the argument of the conjugate product, performing phase unwrapping, and scaling. However, before examining those operations, we must return to the assumption of a common aperture support region.

5.4.1 Maximum-Likelihood Parameter Estimation

In reality, any two spatial-frequency-domain apertures cannot be precisely trimmed to their intersection, if only because of imprecision in the knowledge of the pointing vectors. As we know, any non-overlapping portions of the aperture regions contribute uncorrelated components to the reconstructed reflectivity functions. Furthermore, so far in this chapter we have not considered the effects of system noise on these processes. SAR radars, like all receivers, suffer from additive thermal noise, generally injected at the receiver front end. Under the assumption that these noise sources are independent of the radar waveform, additive, and Gaussian, a more complete model than Equation 5.44 for the registered images is

$$\begin{aligned} f(x_1, y_1) &= r_A(x,y) + n_f(x_1, y_1) \\ g(x_1, y_1) &= r_A(x,y)\, e^{j(\beta_f - \beta_g)Y_o h(x,y)} + n_g(x_1, y_1) \end{aligned} \tag{5.46}$$

where the noise terms are independent of each other and of the reflectivity function, and we denote

$$\begin{aligned} E\{|r|^2\} &= \sigma_r^2 \\ E\{|n_f|^2\} &= E\{|n_g|^2\} = \sigma_n^2 \\ E\{r^* n_f\} &= E\{r^* n_g\} = E\{n_f^* n_g\} = 0\,. \end{aligned} \tag{5.47}$$

The conjugate product of the two image functions is

$$\begin{aligned} f^* g = {} & |r|^2\, e^{jY_o(\beta_f - \beta_g)h(x,y)} \\ & + n_f^* r\, e^{-j(\beta_f - \beta_g)Y_o h(x,y)} \\ & \quad + r^* n_g + n_f^* n_g\,. \end{aligned} \tag{5.48}$$

Taking the expected value of this product, we have the important result

$$E\{f^* g\} = \sigma_r^2\, e^{jY_o(\beta_f - \beta_g)h(x,y)} \tag{5.49}$$

so that *on average*, the phase of $f^* g$ contains the information that we want. On a pixel-by-pixel basis, however, this product is noisy. The question naturally arises how best to estimate the phase of $f^* g$ in the presence of this noise. In [19], a maximum-likelihood estimate for the conjugate product phase is derived. This derivation closely follows that for the phase difference estimator used in maximum-likelihood SAR autofocus, as derived in [20] and as described in Chapter 4 and

Appendix F. The only real difference is that the noise terms in the interferometric case arise from receiver noise, whereas in the case of autofocus, it is the clutter surrounding point targets that is treated as the noise. The important aspects are summarized below.

At any given location (x_1, y_1) in f and g, we consider a local neighborhood of pixels, $k \in [1, N]$. We assume that: (1) the height function $h(x, y)$ is slowly varying, i.e., $h(x, y)$ can be considered constant over the local neighborhood; and (2) r_k, $n_{f,k}$, and $n_{g,k}$ are Gaussian, zero-mean, and mutually independent. The neighborhood of pixels f_k and g_k will be referred to as the observations. Our model for these observations becomes

$$\begin{aligned} f_k &= r_k + n_{f,k} \\ g_k &= r_k \, e^{j\phi} + n_{g,k}, k \in [1, N] \end{aligned} \tag{5.50}$$

where $\phi = (\beta_f - \beta_g)Y_o h$. In vector notation, this may be written as

$$\vec{X}_k = \begin{bmatrix} f_k \\ g_k \end{bmatrix} = r_k \begin{bmatrix} 1 \\ e^{j\phi} \end{bmatrix} + \begin{bmatrix} n_{f,k} \\ n_{g,k} \end{bmatrix} \quad k \in [1, N] \, . \tag{5.51}$$

Our problem, then, is this: Given the set of observations $\{\vec{X}_k, \quad k \in [1, N]\}$ from a local neighborhood with constant height h, we must estimate the value of ϕ. The observation covariance matrix can be computed as

$$\begin{aligned} Q &= E\{\vec{X}_k \vec{X}_k^H\} \\ &= E\left\{\begin{bmatrix} |f_k|^2 & f_k g_k^* \\ f_k^* g_k & |g_k|^2 \end{bmatrix}\right\} \\ &= \begin{bmatrix} \sigma_r^2 + \sigma_n^2 & \sigma_r^2 e^{-j\phi} \\ \sigma_r^2 e^{j\phi} & \sigma_r^2 + \sigma_n^2 \end{bmatrix} . \end{aligned} \tag{5.52}$$

Following the methodology presented in Chapter 4, we first need the probability function of the observations conditioned on the variable we wish to estimate. That is, we need to find $P(\vec{X}_1, \vec{X}_2, ..., \vec{X}_N | \phi)$. Letting

$$\vec{\mathbf{x}} = \begin{bmatrix} \vec{X}_1 \\ \vec{X}_2 \\ \vdots \\ \vec{X}_N \end{bmatrix} \tag{5.53}$$

and using the conditional probability of a single observation

$$P(\vec{X}_k | \phi) = \frac{1}{\pi^2 |Q|} \exp[-\vec{X}_k^H Q^{-1} \vec{X}_k] \quad k \in [1, N] \tag{5.54}$$

we can write the joint conditional probability of the set of observations as the product of the individual pdf's, because they are mutually independent:

$$P(\vec{\mathbf{x}}|\phi) = \frac{1}{\pi^2|Q|^N} \exp\left[-\sum_{k=1}^{N} \vec{X}_k^H Q^{-1} \vec{X}_k\right] . \tag{5.55}$$

Our job is to find the value of ϕ that maximizes $P(\vec{\mathbf{x}}|\phi)$. This is the maximum-likelihood estimate of ϕ, denoted $\hat{\phi}_{ML}$.

Taking the logarithm[4] of both sides of Equation 5.55, we have

$$\max_{\phi} \ln\{P(\vec{\mathbf{x}}|\phi)\} = \max_{\phi} \left[- \ln\{\pi^2|Q|^N\} - \sum_{k=1}^{N} \vec{X}_k^H Q^{-1} \vec{X}_k\right] . \tag{5.56}$$

But the modulus of Q is not a function of ϕ because

$$|Q| = \sigma_n^4 + 2\sigma_r^2\sigma_n^2 \tag{5.57}$$

so that we can perform the simpler maximization

$$\max_{\phi} \sum_{k=1}^{N} (f_k g_k^* e^{j\phi} + f_k^* g_k e^{-j\phi}) \tag{5.58}$$

or

$$\max_{\phi} \left|\sum_{k=1}^{N} f_k^* g_k\right| \cos\left(\phi - \angle\left(\sum_{k=1}^{N} f_k^* g_k\right)\right) . \tag{5.59}$$

The expression in Equation 5.59 obviously achieves its maximum over ϕ when the argument of the cosine is zero. Thus, the maximum-likelihood estimate for ϕ is simply

$$\hat{\phi}_{ML} = \angle\left(\sum_{k=1}^{N} f_k^* g_k\right) . \tag{5.60}$$

This is a significant result. It says that we can estimate the phase shift between two interferometric images at any location by calculating the above sum over pixels in a local neighborhood about that location. The estimate we compute is a maximum-likelihood estimate and, therefore, is at least as good as any other estimate in the sense of asymptotically meeting the Cramer-Rao lower bound (see Appendix F). The error in the estimate is a function of the input image phase noise and the neighborhood size. Larger neighborhoods yield lower noise estimates, but at the expense of spatial resolution. The only assumption required to arrive at Equation

[4] Because the logarithm is a monotonic function, it does not alter the maximization over ϕ.

5.60 is that the height function be slowly varying so as to be considered constant over the local neighborhood. From a computational point of view, we note that this sum can be computed recursively, because each product $f_k^* g_k$ has unit weighting in the sum.

In Appendix F we demonstrate that the maximum-likelihood (ML) phase estimator given in Equation 5.60 achieves the Cramer-Rao lower bound (CRLB) on estimator error variance for moderate-to-large values of signal-to-noise ratios. (Appendix F addresses the phase difference estimator as used in the autofocus scheme described in Chapter 4, but as we have suggested earlier, the interferometric phase difference estimator has the same mathematical form.) In typical SAR systems, the clutter-to-receiver noise ratio (which is the signal-to-noise quantity relevant to the IFSAR phase estimation problem) is in the range of 10 to 35 dB. As a result, we know that the ML interferometric phase difference estimates will have variance prescribed by the CRLB as derived in Appendix F, namely

$$\sigma_\phi^2 = \frac{1}{N\beta} \tag{5.61}$$

where β is the clutter-to-receiver noise ratio

$$\beta = \frac{\sigma_r^2}{\sigma_n^2} \,. \tag{5.62}$$

Note that Equation 5.61 only takes into account receiver noise. Other noise sources may also be present. These could include sidelobes from nearby, large cross-section targets, temporal decorrelation (see Section 5.5), and propagation effects.

Rodriquez and Martin [12] have computed a similar estimator for the interferometric phase difference. Their expression is

$$\hat{\phi} = \arctan\left[\frac{Im\sum_{n=1}^{N_L} v_1^{(n)} v_2^{*(n)}}{Re\sum_{n=1}^{N_L} v_1^{(n)} v_2^{*(n)}}\right] \tag{5.63}$$

where the sums are now over samples from multiple-look images generated from the radar phase history, and N_L is the number of looks to be averaged. While the noise performance for these two estimators should be similar, the Rodriquez and Martin estimator seems to be more awkward to use: changing the averaging sample size N_L involves starting over again with the image-formation process to compute the requisite multi-look images. In contrast, once the full-resolution images are initially formed and registered, different averaging windows can be computed with Equation 5.60. In fact, one could simultaneously compute the estimate for several different window sizes at once, and then choose that result with the best tradeoff between spatial resolution (size of N) and noise performance for a particular application.

Let us consider an example. In Figure 5.6, we show a pair of detected images that were collected with geometries meeting the collection constraints for performing interferometry. The images were collected in two passes using an airborne SAR test platform built and operated by Sandia National Laboratories. The system forms one-meter resolution imagery in spotlight mode, can be pointed over a large range of depression and squint angles, incorporates real-time digital synthesis of transmitted waveforms, motion compensation, image formation, and autofocus, and flies on a DH Twin Otter aircraft. The image pair in the figure shows an area just south of Albuquerque, New Mexico. Shown are several small hills, Tijeras Arroyo, Coyote Arroyo, a paved road, and a facility with a perimeter fence enclosing some storage bunkers. Each bunker has a dimension of approximately 40 ft x 25 ft x 15 ft.

These two images were subsequently registered using the techniques described in Section 5.3. The phase difference between the two images was then estimated at every location according to Equation 5.60. Figure 5.7 shows two examples of this estimation. The one on the left uses 3x3 neighborhoods; the one on the right uses 7x7 neighborhoods. Phase principal values are shown with the interval $[-\pi,\pi]$ mapped to the gray-scale brightness. This example clearly shows how larger estimation windows improve noise performance.

A crossed-ground-track example is shown in Figure 5.8. For this example, the same scene was again imaged in two passes. This time, however, one image was collected while squinting 10 degrees forward of broadside, and the second while squinting 10 degrees backward of broadside. The aircraft ground tracks likewise intersected in a 20 degree angle so that the boresight to the center of the scene was the same for each image. Because this imaging geometry introduces height-dependent differential layover (see Section 5.3.2), a combination polynomial/spline warp is required to locally register all portions of the two images. This is illustrated in the Figure. The left interference fringe image of Figure 5.8 was processed using only a global polynomial warp. Therefore, the height-induced differential layover caused local misregistration in the hilly region of the scene. The right fringe image was processed using a local spline warp in combination with the polynomial warp. This procedure accommodated the differential layover and high-quality interference fringes are present throughout the scene.

5.4.2 2-D Phase Unwrapping

Phase unwrapping problems have a long history in a number of scientific endeavors. They arise in situations where a physical quantity is transduced or related to the phase of a complex signal. Examples include optical interferometry, adaptive optics,

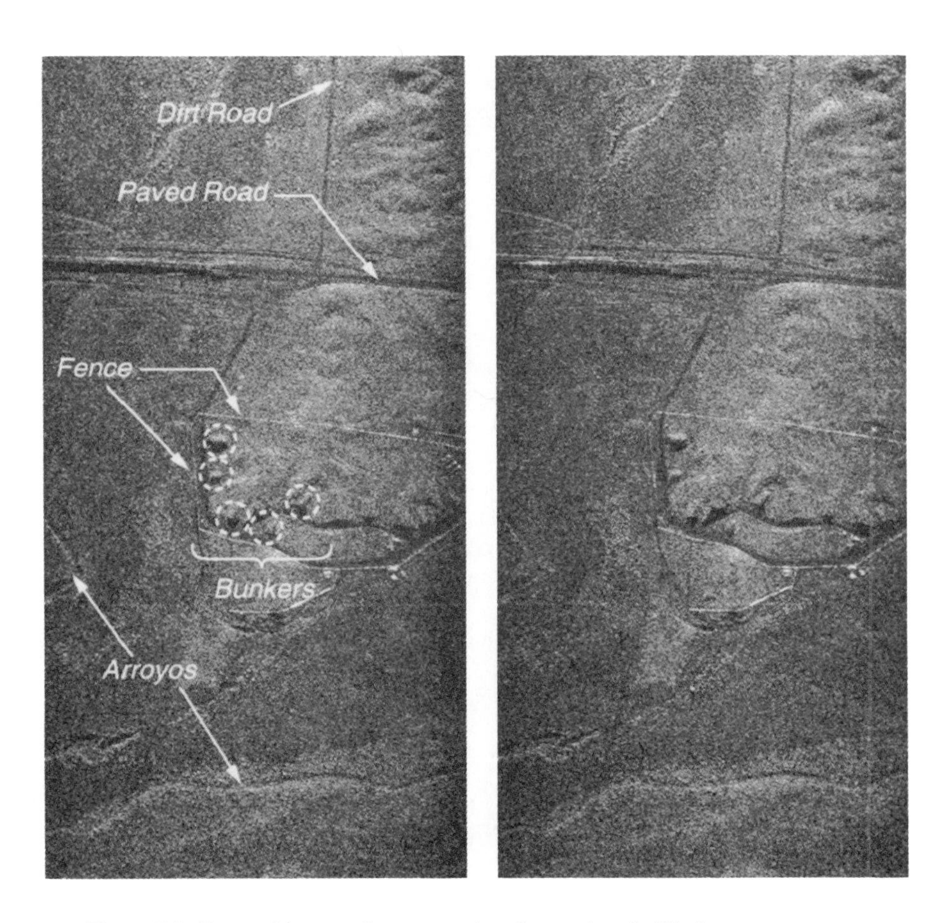

Figure 5.6 Detected images of a two-pass interferometric pair. The images encompass two arroyos,a paved road, and a fenced enclosure containing storage bunkers.

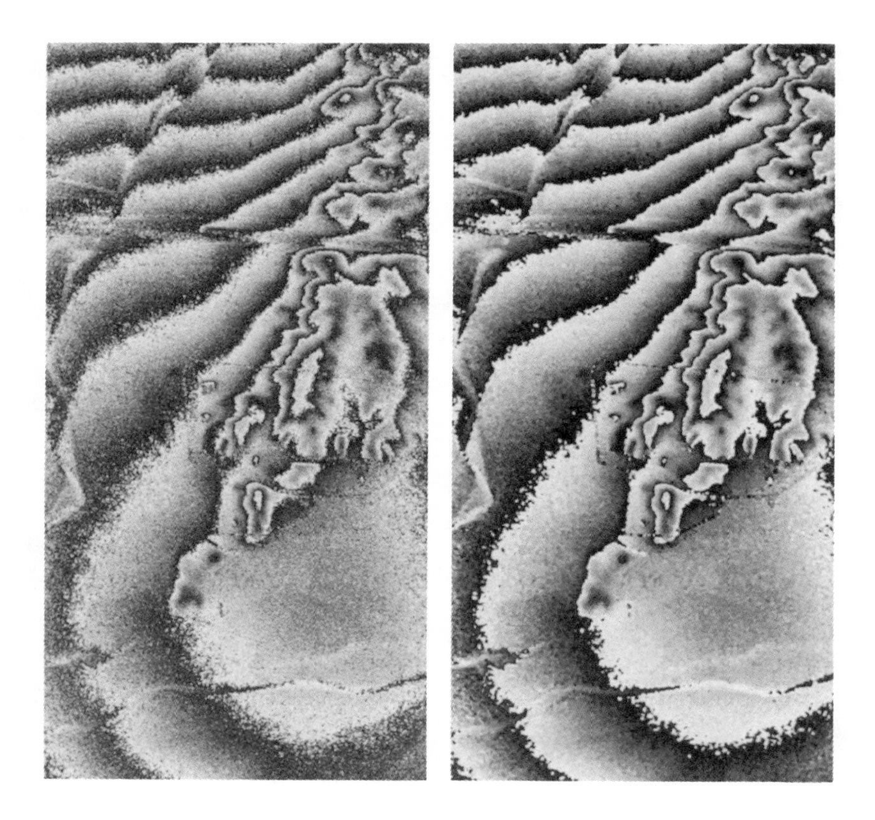

Figure 5.7 Interferometric phase difference principal values. The left image shows phase differences estimated using a 3x3 window. The right image results from a 7x7 estimation window. The decreased level of phase noise achieved by using the larger window is clearly evident. The price for reducing noise is decreased spatial resolution.

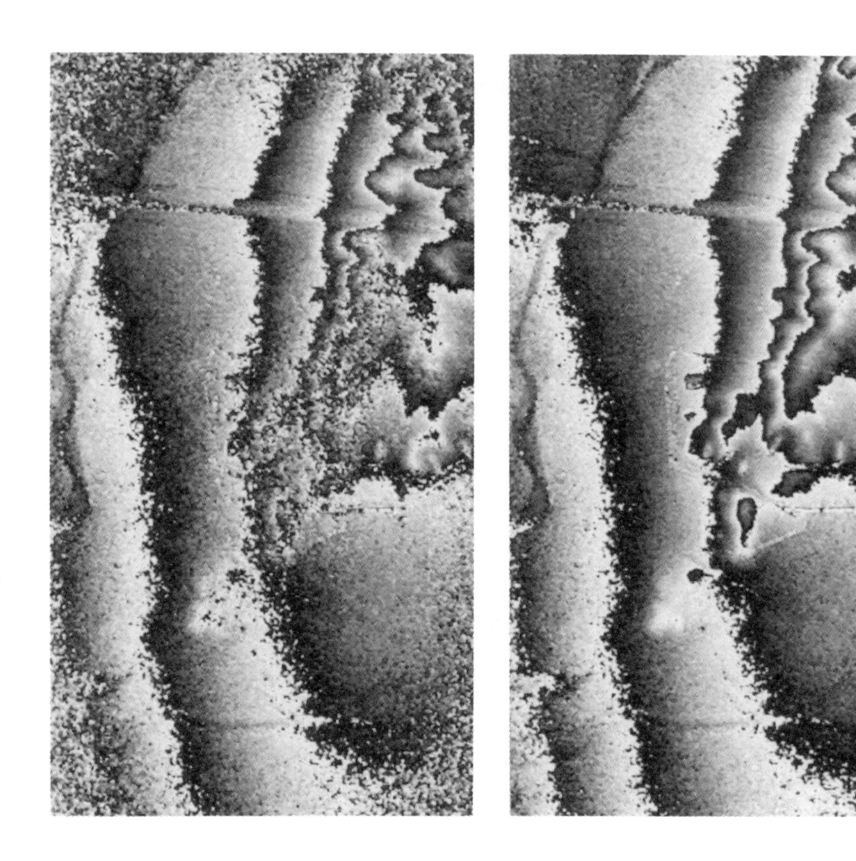

Figure 5.8 An example of crossed-ground-track interferometry. The pair of images used in this example were collected with a 20 degree difference in ground-track angles. The left image was processed without accommodating differential layover. Poor interference fringes in the hilly region of the scene result from local misregistration. The right image was processed using a local warp to accommodate the differential layover. High-quality interference fringes are evident throughout the scene.

speckle imaging, and SAR. Because the phase measurements in these problems usually result from an argument computation or from wavefront tilt sensing of phase differences, only principal values of phase are generally available.

In our problem, Equation 5.60 provides the maximum-likelihood estimate for the phase difference between two interferometric complex images at some location. However, because of the argument operation, only the principal value of the phase is obtainable. In fact, because of the model Equation 5.50, ϕ enters the problem as $e^{j\phi}$ and all we can say is that the estimate satisfies

$$E\{e^{j\hat{\phi}}\} = e^{jY_o(\beta_f-\beta_g)h(x,y)} \ . \tag{5.64}$$

In order to compute $h(x, y)$, we require not the principal values of ϕ, but the continuous function. The process of recovering the continuous ϕ from its principal values is known as phase unwrapping. Mathematically, we are presented with a two-dimensional array of sampled principal values of phase, $[\phi_{ij}]$. To unwrap this array, we need to find the integral number of 2π radians to add or subtract at every point (i,j) to obtain a continuous function $[\Psi_{ij}]$ such that

$$\Psi_{i,j} = \phi_{i,j} + 2\pi k_{i,j}, \quad -\pi \leq \phi_{i,j} \leq \pi, \quad i \in [0, M-1], \ j \in [0, N-1] \ . \tag{5.65}$$

There are two related but distinct issues associated with this task. The first is ensuring consistency and the second is accuracy. *These are not the same.* Consistency implies that in the unwrapped array $[\Psi_{ij}]$, the phase difference between any two arbitrary points is independent of the path from one point to the other. Accuracy implies that the unwrapped array faithfully reproduces the original phase function from which the principal value samples were obtained. We will begin by discussing the first point of consistency and then will return to the more difficult question of accuracy later in the section.

It is always possible to unwrap samples of wrapped phases consistently if they are considered a one-dimensional signal. One begins with the first sample and adds or subtracts a multiple of 2π radians to the next sample, so that the absolute value of the phase difference between the second and the first sample is less than π radians. This procedure continues for successive samples until the entire one-dimensional array is unwrapped:

$$\begin{aligned} \Delta_j &= \phi_{j+1} - \phi_j \\ \Psi_{j+1} &= \begin{cases} \Psi_j + \Delta_j \ , & |\Delta_j| < \pi \\ \Psi_j + \Delta_j - 2\pi \ , & \Delta_j > \pi \\ \Psi_j + \Delta_j + 2\pi \ , & \Delta_j < -\pi \ . \end{cases} \end{aligned} \tag{5.66}$$

This procedure is generically referred to as a linear path-following scheme. No matter how it is applied to a one-dimensional signal, it yields the same result, within an arbitrary constant offset. Consistency is ensured because there is only one possible path between two arbitrary points in the unwrapped array.

For signals of two or more dimensions, however, the problem becomes significantly more complicated. In higher dimensions, many possible unwrapping paths between any two points can be postulated. In the absence of any measurement noise or aliasing, path following may still be used to unwrap the array and to produce a consistent result. However, in the practical situation of noisy measurements, of absent data (e.g. shadowed or no-return areas), and of aliased data, it is generally impossible to obtain a consistent solution with path following. Techniques that use path-following structures generally apply heuristic rules that attempt to resolve or mitigate the resulting inconsistencies [21],[22],[23],[24].

Two-dimensional phase unwrapping methods that do not rely on path following have their roots in adaptive optics. The original literature cast the problem in the mathematical formalism of least-squares estimation. The papers by Fried [25], Hudgin [26], Noll [27], and Hunt [28] provide a thorough background. More recently, it has been shown that the least-squares solution to the phase unwrapping problem is mathematically identical to the solution of Poisson's equation on a rectangular grid with Neumann boundary conditions [29], [30]. Phase unwrapping on very large grids, making efficient use of memory and limited precision arithmetic, can be accomplished by solving Poisson's equation using a specific form of the fast cosine transform [31], or using FFTs [32]. This approach is numerically stable and robust, exactly solves Poisson's equation, automatically imposes the proper boundary conditions, can be posed as a separable process for computational efficiency, can be performed "in place" in memory, and is mathematically formal in construction. The salient parts of this algorithm are summarized below.

Let us begin with what we know. We know the phase, modulo 2π, of a function on a discrete grid:

$$\phi_{i,j} \;=\; \Psi_{i,j} - 2\pi k_{i,j}, \quad -\pi \leq \phi_{i,j} \leq \pi, \quad i \in [0, M-1],\ j \in [0, N-1]\,. \tag{5.67}$$

Given the wrapped values of phase $[\phi_{i,j}]$, we wish to determine the unwrapped phase values $[\Psi_{i,j}]$ at the same grid locations, with the requirement that the phase differences of the $\Psi_{i,j}$ agree with those of the $\phi_{i,j}$, *in the least-squares sense.* Let us define a wrapping operator that wraps all values of its argument into the range $(-\pi, \pi]$ by adding or subtracting an integral number of 2π radians:

$$W : \mathcal{R} \rightarrow (-\pi, \pi] : \; W(x) = x + 2\pi k\ . \tag{5.68}$$

With this notation we have, for example, that $W(\Psi_{ij}) = \phi_{ij}$. Next, we compute the first differences of phase in both cardinal directions and wrap those differences as

$$\Delta^x_{i,j} = \begin{cases} W\{\phi_{i+1,j} - \phi_{i,j}\}, & i \in [0, M-2],\ j \in [0, N-1] \\ 0 & \text{otherwise} \end{cases}$$

$$\Delta^y_{i,j} = \begin{cases} W\{\phi_{i,j+1} - \phi_{i,j}\}, & i \in [0, M-1],\ j \in [0, N-2] \\ 0 & \text{otherwise} \end{cases} \tag{5.69}$$

where the x and y superscripts refer to the differences in the i and j indices, respectively. The function $\Psi_{i,j}$ that minimizes the squared error metric

$$\begin{aligned} \min_{\Psi_{i,j}} \quad & \textstyle\sum_{i=0}^{M-2} \sum_{j=0}^{N-1} \left(\Psi_{i+1,j} - \Psi_{i,j} - \Delta^x_{i,j}\right)^2 \\ & + \textstyle\sum_{i=0}^{M-1} \sum_{j=0}^{N-2} \left(\Psi_{i+1,j} - \Psi_{i,j} - \Delta^y_{i,j}\right)^2 \end{aligned} \tag{5.70}$$

is the (global) least-squares solution to the 2-D phase unwrapping problem. The normal equations leading to the least-squares phase unwrapping solution can be summarized with the following simple equation (details can be found in Reference [31]):

$$\begin{aligned} & \Psi_{i+1,j} + \Psi_{i-1,j} + \Psi_{i,j+1} + \Psi_{i,j-1} - 4\Psi_{i,j} \\ & \quad = \Delta^x_{i,j} - \Delta^x_{i-1,j} + \Delta^y_{i,j} - \Delta^y_{i,j-1} \,. \end{aligned} \tag{5.71}$$

This equation gives the relationship between the wrapped phase differences (from the original wrapped phases via Equations 5.69) and the unwrapped phase values $\Psi_{i,j}$. A simple manipulation of Equation 5.71 yields

$$(\Psi_{i+1,j} - 2\Psi_{i,j} + \Psi_{i-1,j}) + (\Psi_{i,j+1} - 2\Psi_{i,j} + \Psi_{i,j-1}) = d_{i,j} \tag{5.72}$$

where

$$d_{i,j} = \left(\Delta^x_{i,j} - \Delta^x_{i-1,j}\right) + \left(\Delta^y_{i,j} - \Delta^y_{i,j-1}\right) \,. \tag{5.73}$$

This is nothing more than a discretization of Poisson's equation:

$$\frac{\partial^2}{\partial x^2}\Psi(x,y) + \frac{\partial^2}{\partial y^2}\Psi(x,y) = d(x,y) \tag{5.74}$$

on a rectangular M by N grid. If we take the 2-D discrete cosine transform (DCT) of both sides of Equation (5.72), apply the shift property of DCTs, and rearrange terms, the exact solution in the transform domain is

$$\bar{\Psi}_{i,j} = \frac{\bar{d}_{i,j}}{2\left(\cos \pi i/M + \cos \pi j/N - 2\right)} \tag{5.75}$$

where the overbar notation indicates the transformed function. The specific form of the 2-D discrete cosine transform pair is given in Reference [31]. It is important to note that the cosine expansion automatically imposes the Neumann boundary conditions.

In summary, the algorithm consists of these steps:

1. Compute the wrapped first differences via Equations 5.69.
2. Compute the driving function $d_{i,j}$ via Equation 5.73.
3. Compute $\bar{d}_{i,j}$, the DCT of $d_{i,j}$.
4. Modify $\bar{d}_{i,j}$ according to Equation 5.75.
5. Perform the 2-D inverse DCT of $\bar{\Psi}_{i,j}$ to obtain the least-squares unwrapped phase values $\Psi_{i,j}$.

When this unwrapping algorithm is applied to the wrapped phase data from the bunker site IFSAR image pair of Figure 5.7 (right image with 7x7 window), we obtain the unwrapped data shown in Figure 5.9. Here, the minimum to maximum brightness corresponds to 35.5 radians of phase. As Section 5.4.3 demonstrates, this image is but a scale change away from an elevation map. The hills, arroyos, and bunkers are quite evident in the image.

The discussion to this point has concerned itself with consistency. The unwrapping algorithm summarized in the preceding paragraphs guarantee a consistent result, regardless of the condition of the input data. Furthermore, this result is optimum in the sense of minimizing the squared error between the first-order differences of the input data and that of the output array. As such, it represents a considerable advance over heuristic path-following techniques, which can guarantee neither. However, a *consistent* solution is not necessarily an *accurate* one. In unwrapping SAR phase data to produce a terrain-elevation map, one generally has to deal with phase noise, no-return or low-return areas, and abrupt elevation discontinuities that alias the phase measurements. The phase unwrapping algorithms, whether path following or based on Poisson's equation, attempt to reconstruct a phase function whose differences match the differences in the wrapped measurements *and that are less than π radians in absolute value*. Where the phase values have been aliased, however, erroneous results occur. The unwrapper, furthermore, has no way of "knowing" the number of phase cycles that have slipped in the aliasing process. These problems can be partially resolved by developing the ability to classify phase values into high-confidence versus low-confidence categories. Aliased and noisy phase data ought

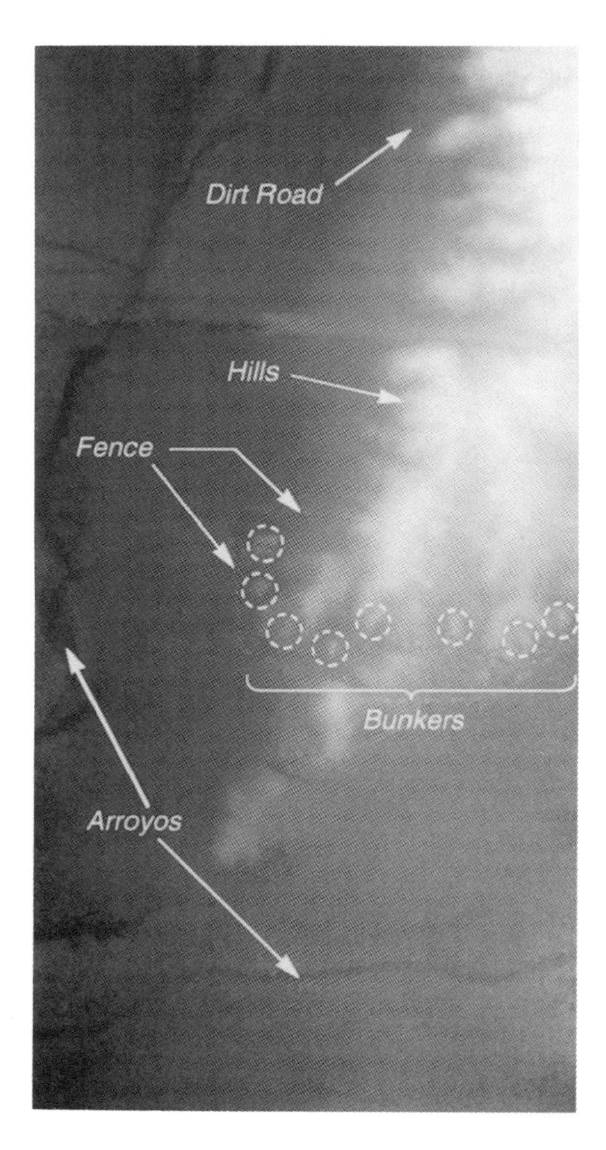

Figure 5.9 Unwrapped phase difference image. Phase unwrapping was accomplished using the 2-D fast transform solution to Poisson's Equation.

to be discounted relative to clean, non-aliased data. Path-following methods attempt to do this by identifying singularities and branch cuts in the data array. It is always problematic to tie such points correctly and unambiguously together.

Fortunately, the mathematical formalism that led to the unweighted least-squares phase unwrapper can be extended to a weighted phase unwrapper [31]. Also fortunate is the fact that we have at our disposal a rather explicit weighting function, namely the sample complex correlation coefficient $\alpha(i, j)$ that is developed in Section 5.5.1. For the present, suffice it to say that this coefficient is precisely what is required of a weighting function. It is normalized in the interval [0,1] and is high in those regions of the interfered images in which the interferometric phase is consistent over a local neighborhood; otherwise, it is small in value. This weighting function[5] can be used with the iterative, weighted phase unwrapper described in Reference [31]. This technique provides a robust, consistent, and computationally efficient approach to the difficult problem of phase unwrapping in the presence of aliased data. Of course, it must be understood that where a region of the interfered images is completely cut off from its surroundings by missing or aliased data (such as the top of a building), no unwrapper can arrive at the correct values of phase in that region relative to the surroundings. In such cases, care must be taken to choose the imaging geometry and consequent topographic scale factor in such a manner as to avoid the aliasing in the first place.

5.4.3 Topographic Scale Factor and Orthorectification

The tomographic formulation of SAR has led us to a mathematical model of the image function Equation 5.1 in which terrain height directly modulates the phase. The speckle model of terrain reflectivity along with careful imaging geometry, aperture trimming (see Section 5.3.1), and registration allows us to cancel the apertured complex reflectivity function between a pair of images. Maximum-likelihood estimation of the phase difference between the images yielded Equation 5.60. Equation 5.64 then dictated that the 2-D unwrapped phase function Ψ be proportional to the terrain height; specifically

$$\Psi(x_1, y_1) = Y_o(\beta_f - \beta_g)h(x, y) \, . \tag{5.76}$$

In our quest for $h(x, y)$, two processing steps remain: (1) removing the scale factor $Y_o(\beta_f - \beta_g)$; and (2) taking into account the layover equations, which relate the image coordinates (x_1, y_1) to the physical space coordinates (x, y). This latter process is sometimes known as *orthorectification*. Beginning with the scale factor, we can

[5] In practice, the correlation coefficient is first smoothed to reduce noise and impose some blending of "speckled weights;" then, it is thresholded to produce a binary weight array. Zero-valued weights impose a total "no-confidence" vote in regions of poor correlation.

gain some insight by substituting for Y_0, β_f, and β_g, giving

$$\begin{aligned}\Psi(x_1,y_1) &= \frac{4\pi}{\lambda}\cos\psi_g\ [\tan\psi_f - \tan\psi_g]\,h(x,y) \\ &= \frac{4\pi}{\lambda}\cos\psi_g\ [\tan(\psi_g+\Delta\psi) - \tan\psi_g]\,h(x,y) \qquad (5.77)\end{aligned}$$

where $\Delta\psi = \psi_f - \psi_g$. (We have taken Y_0 to be $(4\pi/\lambda)\cos\psi_g$.) Using the fact that collection constraints dictate $\Delta\psi \ll 1$, a little trigonometry yields

$$\begin{aligned}\Psi(x_1,y_1) &= \frac{4\pi}{\lambda}\cos\psi_g\left[\frac{\tan\psi_g+\tan\Delta\psi}{1-\tan\psi_g\tan\Delta\psi} - \tan\psi_g\right]h(x,y) \\ &= \frac{4\pi}{\lambda}\cos\psi_g\ \tan\Delta\psi\left[\frac{1+\tan^2\psi_g}{1-\tan\psi_g\tan\Delta\psi}\right]h(x,y) \\ &\cong \frac{4\pi}{\lambda}\cos\psi_g\ (\Delta\psi)\left[\frac{\sec^2\psi_g}{1-\tan\psi_g(\Delta\psi)}\right]h(x,y) \\ &= \frac{4\pi}{\lambda}\frac{\Delta\psi}{\cos\psi_g}\,h(x,y)\left[\frac{1}{1-\tan\psi_g(\Delta\psi)}\right] \\ &\cong \frac{4\pi}{\lambda}\frac{\Delta\psi}{\cos\psi_g}\,h(x,y)\,. \qquad (5.78)\end{aligned}$$

This equation reveals that the scale factor that relates unwrapped phase with terrain height depends only on the radar wavelength and on the depression angles of the two collections. Note especially that range does not appear anywhere. This result reinforces a principal idea expressed at the outset of this chapter: tomographic development stresses that the radar merely transduces spatial-frequency-domain data about a scene. We can then forget about the radar *per se*. The subsequent information that we extract from the data is determined by the position and size of the spatial-frequency-domain apertures. We see that the topographic scale factor is completely determined by the angles ψ_i of the apertures, and their distance from the origin $4\pi/\lambda$. In particular, we need not consider range, timing, Doppler, or any other radar concept.

To illustrate this point, Figure 5.10 shows the wrapped interferometric phase differences of the bunker site from two pairs of collections. The image on the left resulted from interfering two images collected with a depression angle difference of 0.08 degrees. The image on the right was produced from a second image pair whose depression angles differed by 0.30 degrees. Note how the larger angular difference yields a larger scale factor and, hence, a higher rate of change of phase with terrain height. Because these are wrapped phase values, the right image possesses more cycles of phase.

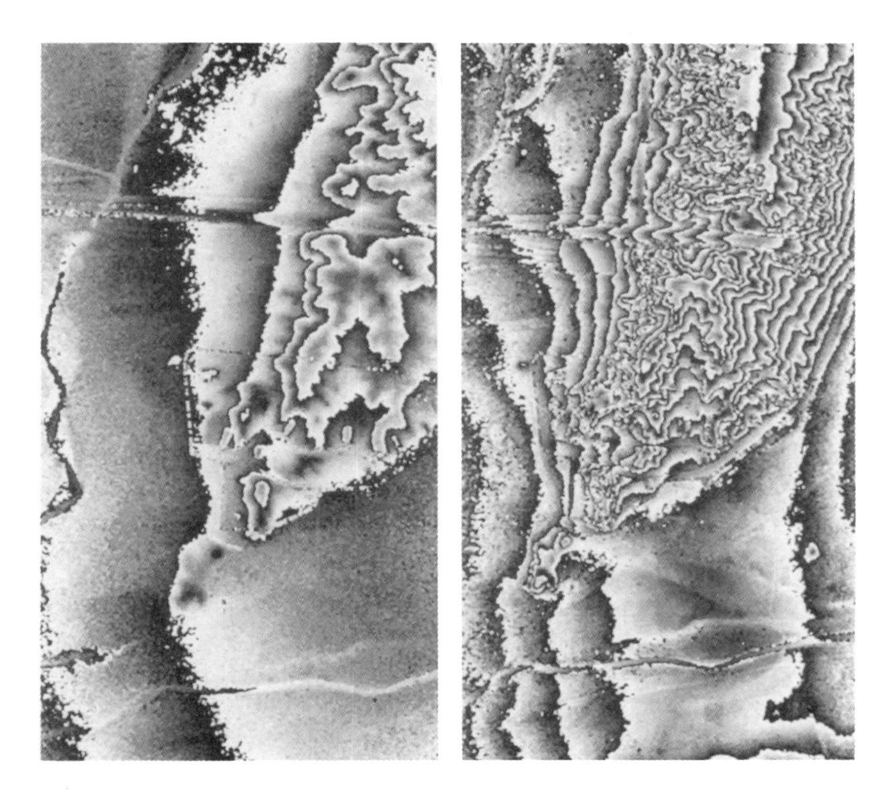

Figure 5.10 A comparison of interferometric phase images from two collection pairs illustrating the relationship between depression angle differences and terrain height-to-phase scale factor. The left image results from an image pair with a depression angle difference of 0.08 degrees. The right image was obtained from another pair with a depression angle difference of 0.30 degrees.

We are still not finished with the tomographically derived image function (Equation 5.2). The layover transformation (Equation 5.3) relates image coordinates with three-space physical coordinates. In the registration process, we accounted for differential layover and resampled the source image g to overlay the target image f. After phase estimation and unwrapping, the function $\Psi(x_1, y_1)$ is still in image coordinates. Scaling this function by Equation 5.78 yields a distorted terrain model $(x_1, y_1, h(x, y))$ with layover foreshortening. Assuming, again, that no superposition has occurred, we have

$$\begin{aligned} x_1 &= x + \tan \eta_g \; h(x, y) \\ y_1 &= y + \tan \psi_g \; h(x, y) \end{aligned} \tag{5.79}$$

and, from Equation 5.78,

$$h(x, y) = \frac{\lambda}{4\pi} \frac{\cos \psi_g}{\Delta \psi} \; \Psi(x_1, y_1) \; . \tag{5.80}$$

The above three equations can be solved simultaneously for the true three-space terrain model $(x, y, h(x, y))$ and we are finished. Once again, the layover transformation that proceeds directly from the tomographic formulation is seen to be only a function of the aperture position angles η and ψ.

When this last step of processing is performed on our bunker-site interferometric data, the terrain elevation model of Figure 5.11 is produced. These data are portrayed as a 3-D surface rendering using a ray tracing of the elevation data. This rendering clearly shows the arroyos, hills, bunkers, and even the perimeter fence, although the fence appears more like a block wall due to the spatial averaging of the phase estimation step. The rms height error due to phase noise (described in Section 5.4.4) averages six inches in this dataset. The fence is eight feet high, the bunkers are 15 feet high, and the arroyos vary from two to six feet in depth.

5.4.4 Error Sources

We complete this section on interferometric terrain mapping with a discussion of the various sources of error in the process. A short-wavelength SAR with the appropriate imaging geometry can yield extraordinary terrain height accuracy, but only if these error sources are properly controlled. While many subsidiary error sources exist, we will focus on four primary ones: (1) wavefront curvature, (2) system phase noise, (3) navigation errors, and (4) spatial phase aliasing.

The question of wavefront curvature in interferometric processing has not yet surfaced, because the entire development has been based on Equation 2.72. That equa-

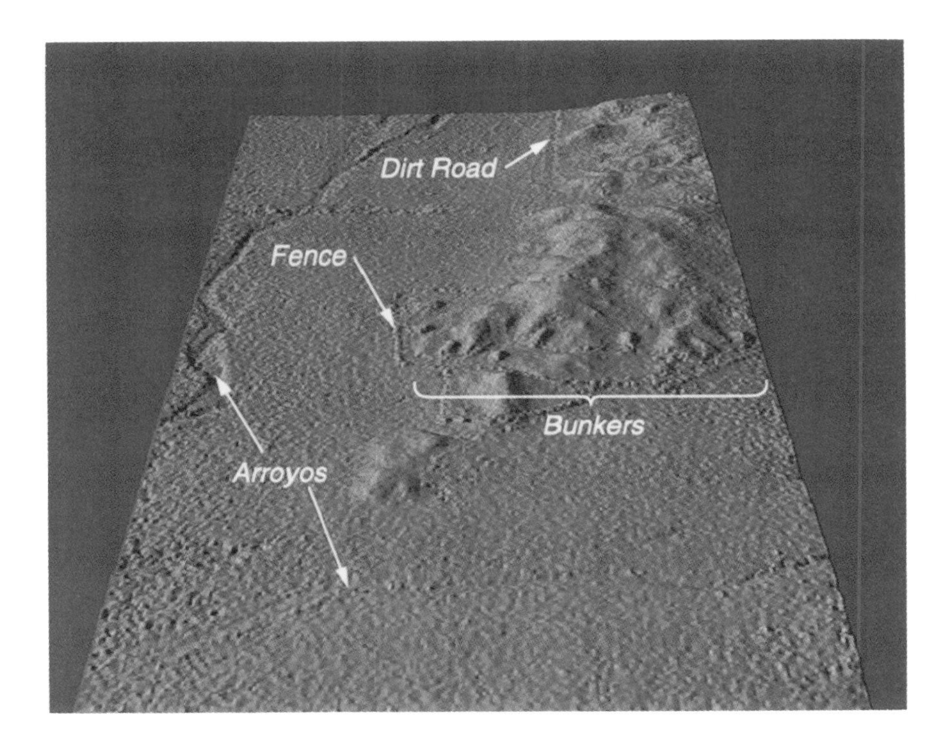

Figure 5.11 A 3-D surface rendering of the terrain elevation model generated by interferometric processing of the bunker-site image pair. Note that the arroyos, hills, bunkers, and perimeter fence are all visible. The rms height error due to phase noise is about one-half foot.

tion was derived using tomographic principles in Chapter 2. Although the tomographic formulation embodies many effects seen in SAR imagery in one compact theory, it does not, in fact, deal with wavefront curvature. The projection functions of Equation 2.40 are assumed to be taken by integration across planes, not sections of spherical wavefronts. This discrepancy led to some additional analysis where we derived the amounts of wavefront curvature that could be tolerated without impacting image quality. (See Section 2.6 and Appendix B.)

In the context of interferometry, we again cannot ignore the effects of wavefront curvature, except perhaps at extreme ratios of image range to size. At nearer ranges, there will be appreciable curvature of the wavefronts across the image patch of interest. With two collections, the corresponding curvatures are not identical. The difference in curvatures gives rise to path-length phase differences that are not due to terrain height. This distortion can be removed by calculating the differential range from the two imaging apertures to a hypothetical flat earth and then subtracting the corresponding phase from one of the complex images. If we denote the mid-aperture position of the sensor for image f as (x_f, y_f, z_f), then the range to a point (x, y, z) on the ground[6] is

$$R_f(x, y, z) = [(x_f - x)^2 + (y_f - y)^2 + (z_f - z)^2]^{1/2} \tag{5.81}$$

and similarly for image g. During image formation, both images are phase stabilized to the same ground reference point. Both images contain residual image-domain phase due to the difference between the assumed planar projections and the actual spherical wavefronts. However, the *phase difference between the two images* is simply due to the path-length difference

$$\phi_{rc}(x, y, z) = \frac{4\pi}{\lambda}[R_g - R_f] \, . \tag{5.82}$$

Assuming a hypothetical flat earth (or perhaps an earth geoid model if the image size is large enough to warrant it) and good platform position data, it is straightforward to use Equation 5.82 to calculate a two-dimensional phase function to be subtracted from image g. The remaining phase difference between f and g can then be attributed solely to terrain modulation. As an aside, when images f and g have slightly different depression angles, Equation 5.82 also contains a linear term in the y direction. This is due to a plane wave phase term that is normally eliminated with proper aperture trimming.

This processing step is easily visible in Figure 5.12. Our familiar bunker site was again imaged by the SAR with a two-pass interferometric geometry. On this occasion, however, the slant range was shortened to exaggerate the wavefront curvature

[6] This is one of few times in this book the reader will encounter a range equation.

over the image patch. The theoretical flat-earth correction phase of Equation 5.82 for the collection geometry of the pair is shown on the left side of the Figure. The interferometric phase of the actual data is shown on the right side. Of course, the real terrain modulates the phase data on the right, but the effect of the wavefront curvature is clearly seen in the similarity of the two images. This is especially evident in the flatter areas of terrain (left and lower left).

The topic of system phase noise is a diverse one. There are many design considerations, system tradeoffs, and hardware characteristics that result in a level of complex system noise in formed SAR imagery. Because it is not the intent of this book to provide an in-depth treatment of this subject, the interested reader is referred to Reference [33] not only for its exposition of this topic, but for its excellent bibliography as well. References [13] and [34] provide a more specific treatment of the various sources of phase noise in interferometric pairs. Radar-receiver and data-processing phase noise sources, as well as temporal scene decorrelation and baseline decorrelation are treated there. Our principal concern here focuses on how image-domain phase noise affects terrain-height estimation, independent of the sources of that noise. We note, however, that SAR systems are designed to a prescribed system noise level in absolute power. The observed clutter-to-system noise ratio in a given scene also depends on the backscatter coefficient (σ_o) of the terrain. Therefore, target areas with a high σ_o have a lower phase noise for the purposes of interferometry [12].

Once a value for the image-domain phase noise has been determined, either through calculation or measurement, its effect on terrain-height estimation is quite straightforward. Earlier, we derived a simple proportionality relationship between the (unwrapped) image-domain phase difference and the terrain-height estimate, given by

$$h = \frac{\lambda}{4\pi} \frac{\cos\psi_g}{\Delta\psi} \Psi \, . \tag{5.83}$$

If we assume an additive and independent phase noise term with variance σ_ϕ^2 (see Equation 5.61), then the variance of the height measurement is

$$\sigma_h^2 = \left[\frac{\lambda}{4\pi} \frac{\cos\psi_g}{\Delta\psi} \right]^2 \sigma_\phi^2 \, . \tag{5.84}$$

This equation is only realistic when the major contributors to image-domain phase noise can be assumed to be independent from image to image. Contributors such as thermal noise certainly fall in this regime. There may exist, however, some non-negligible effects, e.g., the antenna-beam pattern, which are systematic in nature and correlated between two images. Instead of contributing zero mean noise as Equation 5.84 does, this type of source would bias the height measurement.

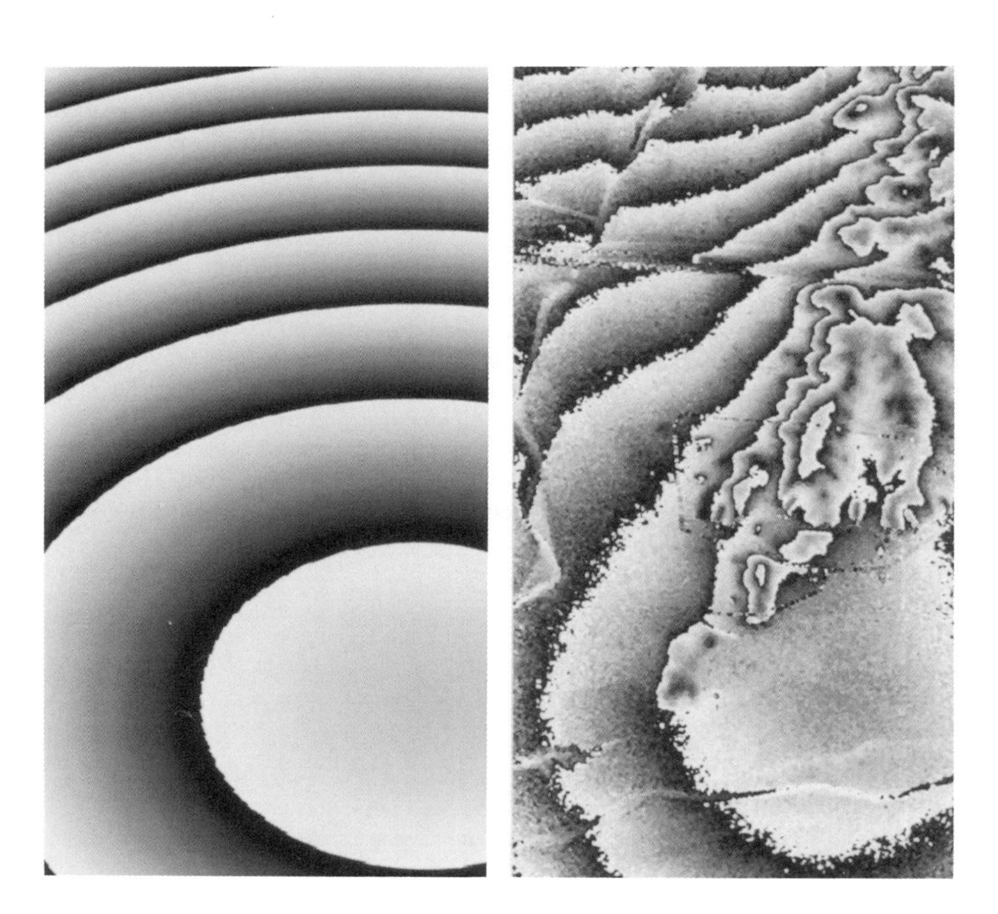

Figure 5.12 The effect of wavefront curvature on interferometric phase differences. The left image shows the theoretical flat-earth phase resulting from wavefront curvature for a specific collection geometry at close slant range. The image on the right shows the interferometric phase from an actual collection pair with the same geometry.

A third and extremely important source of terrain-height measurement errors can be attributed to the SAR platform navigator. These errors are particularly troublesome because they generally result in bias errors. While it may be quite possible to control uncorrelated phase noise and to use an imaging geometry that yields rms errors from Equation 5.84 on the order of inches, it is much harder to reduce bias errors from platform position and attitude to less than one-meter accuracy over large images. By navigation errors, we mean uncertainties in the absolute position of the antenna phase center for each image of an interferometric pair. (Even this may not be enough. Electrical delays in the SAR hardware beyond the antenna phase center may also be important. Such *system design* considerations, while important, will not be treated here.)

Inaccuracies in this positional information affect terrain mapping in two ways. First, any *common* offset in the positional data from the pair of collections will introduce the same offset in the reconstructed terrain function. After all, the SAR interferometer is measuring the terrain relative to its own position. Unless one admits the possibility of using ground-reference points that may be distinguishable in the image, the only source of absolute position information is the SAR navigator itself. Second, an unknown *differential* position error between the pair of collections introduces a tilt in the terrain data. This is easily seen in Figures 2.18 and 5.4. A position error in the direction normal to the slant plane of one aperture changes the depression angle and, therefore, introduces an error into the Y offset used during image formation. This, in turn, imposes a linear-phase term (plane wave) in the phase difference image, which, after phase unwrapping, manifests itself as an unknown tilt (plane) in the y direction of the estimated height map. A similar argument holds in the X direction. An unknown error in the cross-range position of one aperture changes the squint angle and offsets the spatial frequency data in the X direction (see Figures 2.21 and 5.4), resulting in a linear-phase term in the image-domain x direction when the two images are interfered. Interestingly, uncertainties in the ranges of the two collections would seem to have no effect at all, because they change neither the depression nor the squint angles. This is because the tomographic formulation does not account for range-curvature effects, as discussed above. At nearer ranges, unknown offsets in range do, in fact, affect the interfered phase function, but the result is a generalized conic error instead of a linear one, as we discussed earlier with Equation 5.82.

The sources described above are the first-order effects. Higher-order effects of navigation errors also exist. For example, unknown errors in depression angles affect the phase-to-height scale factor, as prescribed by Equation 5.80. Generally speaking, however, the first-order effects are usually the most damaging. An unknown linear ramp quickly builds up a substantial height bias in a large image. Simple calculations will convince the system designer that certain navigation errors must be held to strict limits.

One final note concerning navigation errors. It seems that one-pass collections wherein two antennas are used to obtain simultaneous images with slightly different depression angles would be immune to differential errors. This is not true. The exact position of the antenna phase centers is determined from the *orientation* of the baseline between the phase centers. As the aircraft or spaceborne platform rolls, pitches, or yaws, this orientation with respect to line-of-sight vectors changes. Because the phase centers are not very far apart in this case, the difference in depression angles to the scene is small. Hence, the phase-to-height scale factor is correspondingly small. Even small orientation angle changes can lead to fairly large linear-phase functions *compared to this scale factor.* When the height function is eventually computed using Equation 5.80, the bias term has a significant influence. This is also affected by the static orientation of the baseline. A baseline that is nominally at right angles to the line-of-sight vector introduces the smallest out-of-slant-plane motion with rotation. Depending on a number of such considerations for short interferometer baselines and long ranges, the orientation error budget can fall into the region of arc-seconds.

A final error source to be discussed here is that of spatial phase aliasing. In Section 5.4.2, we discussed the implications of aliased phase data on the 2-D phase unwrapping process. We pointed out that the weighted least-squares phase unwrapper described in reference [31] shows good promise in resolving the ambiguous problem of unwrapping data with regions of aliased phase by deweighting those areas in favor of regions with a consistent phase. However, it was also clear from that discussion that regions of imagery that are *surrounded* by a band of aliased phase are a particular problem. In this case, the cutoff region can never be correctly unwrapped vis-a-vis its surroundings. A building for which the height exceeds the aliasing distance given by

$$h_a = \pi \frac{\lambda}{4\pi} \frac{\cos \psi}{\Delta \psi} = \frac{\lambda}{4} \frac{\cos \psi}{\Delta \psi} \tag{5.85}$$

will exhibit aliased phase around all four of its walls. This is the height given by Equation 5.80 with Ψ equal to one-half cycle of phase. Even if the top surface of the building has consistent phase values that may be unwrapped, the unwrapper cannot "walk" up any side of the building to yield the correct height at the top (after scaling), in reference to the ground elevation. This condition can be *detected* by a weighting function with low values around all four walls of the building, but it cannot be *corrected.* The only solution here is to collect the interferometric imagery with a smaller baseline so that the wrapping height exceeds the height of the building. This example points out the need for detecting aliased data. A phase unwrapper will always produce an output array, even if one or more cycles of phase have been slipped. Thus, without detection, it is common to produce an incorrect terrain height map in the presence of spatially aliased data.

We note that regions of aliased data, e.g., radar shadow regions, must be treated as areas of missing data. Even if a weighted-phase unwrapper can work around the aliased data to provide consistent, accurate terrain height, the aliased and shadow regions are interpolated from the surrounding good data. In these interpolated regions, we have no actual data to support any local structure, i.e., they are in fact *holes* in the data. Finally, we note that it might be possible to obtain better results using more than two mutually interfering images. Interfering multiple observations could combine the high relative accuracy of large baselines with the low ambiguity characteristics of small baselines. This is a topic of ongoing research [35].

Summary of Critical Ideas from Section 5.4

- **Using the tomographically derived image function for spotlight-mode SAR images, we have shown that two properly collected and formed images may be conjugate multiplied, pixel-by-pixel, to yield**

$$f^*(x_1, y_1)g(x_1, y_1) = |r_A(x, y)|e^{jY_o(\beta_f - \beta_g)h(x,y)}$$

where (x, y) and (x_1, y_1) are related by

$$\begin{aligned} x_1 &= x + \tan\eta\, h(x, y) \\ y_1 &= y + \tan\beta\, h(x, y) \, . \end{aligned}$$

The phase of the above product is a wrapped and scaled function of the terrain height $h(x, y)$.

- **After taking into consideration the effects of receiver noise, we have derived a maximum-likelihood estimator for the phase angle**

$$\phi(x_1, y_1) = Y_o(\beta_f - \beta_g)h(x, y)$$

which is

$$\hat{\phi}_{ML} = \angle\left(\sum_{k=1}^{N} f_k^* g_k\right)$$

calculated over a local neighborhood of N pixels surrounding (x_1, y_1).

- **The two-dimensional function of phase principal values $\hat{\phi}(x_1, y_1)$ may be unwrapped using a fast transform-based solver of Poisson's equation on a rectangular grid. This unwrapping technique can be used in an unweighted sense, or can be employed in an iterative, weighted algorithm to mitigate the effects of phase aliasing and low radar return areas. Other phase unwrapping algorithms could also be candidates.**

Summary of Critical Ideas from Section 5.4 (cont'd)

- **The unwrapped 2-D phase function $\Psi(x_1, y_1)$ can be scaled to terrain height using the relationship**

$$h(x,y) = \frac{\lambda}{4\pi} \frac{\cos\psi}{\Delta\psi} \Psi(x_1, y_1)$$

which depends *only* on the radar wavelength and the depression angles of the two collections.

- **The scaled unwrapped phase function can be solved simultaneously with the layover equations for the true three-space (orthorectified) terrain model.**

- **Error sources in interferometric terrain mapping include: a) wavefront curvature, b) system phase noise, c) navigation errors, and d) spatial phase aliasing.**

5.5 INTERFEROMETRIC CHANGE DETECTION

The second branch of SAR interferometry is change detection. While this application has not received as much attention in the literature as terrain mapping, it is nonetheless an extremely useful and sensitive technique [36]. Interferometric change detection is, of course, a two-pass application. As discussed in Section 5.2.1, two (or more) collections of the same terrain are made, satisfying the collection constraints of Section 5.2.2. We are then interested in learning about temporal changes in the scene reflectivity function that have occurred in the interval between collections. Generically, SAR change detection is similar to change detection techniques for optical or other types of imagery. However, interferometric change detection has several important advantages over traditional incoherent methods. First, the requisite processing step of image registration is considerably easier to accomplish. The phenomenon of coherent speckle discussed in Section 5.3.2 ensures that SAR interferometric image pairs contain correlated, high-frequency information to assist the registration process. Second, the change statistic involves the complex reflectivity function. Thus, subtle changes can be transduced. An example of such a change is the rearrangement of some of the scatterers that make up a resolution cell. These changes may not affect to an appreciable extent the magnitude of the backscattered energy (and therefore do not show up in the detected imagery), but they do affect the phase. Thus, interferometry presents a method for automated processing and an extremely sensitive measure of change in SAR imagery. It has much potential for such applications as environmental or activity monitoring, land-use surveys, etc. We develop this method by building a mathematical model for change in the scene reflectivity function, by deriving a maximum-likelihood change statistic, and finally by examining the peculiar consequences of the statistic's probability density function.

5.5.1 A Complex Reflectivity Change Model

We again return to Equation 5.1. Let us imagine that we have collected a pair of images of the same terrain, separated by some interval of time. We will assume, for the present, that the aperture regions of support A_i are sufficiently coincident that they can be considered identical. Interferometric terrain mapping requires an angular separation in the apertures in order to generate a non-zero, phase-to-height scale factor (Equation 5.78), but change detection performs best with identical aperture regions. This minimizes the uncorrelated components arising from non-overlapping portions of the apertures (Section 5.2.2). We will also assume, as we have previously, that the terrain is sufficiently smooth to let us ignore the superposition of multiple resolution cells due to layover.

Following the logic of Section 5.4.1, we include in our image model the presence of independent additive Gaussian noise. Therefore, our two reconstructed image functions f and g, (see Equation 5.46) after registration, can be written as

$$\begin{aligned} f(x_1, y_1) &= r_f(x, y) + n_f(x_1, y_1) \\ g(x_1, y_1) &= r_g(x, y)\, e^{j(\beta_f - \beta_g) Y_0 h(x,y)} + n_g(x_1, y_1) \end{aligned} \tag{5.86}$$

where

$$\begin{aligned} E\{|r_f|^2\} &= E\{|r_g|^2\} = \sigma_r^2 \\ E\{|n_f|^2\} &= E\{|n_g|^2\} = \sigma_n^2 \\ E\{r_f n_f^*\} &= E\{r_g n_g^*\} = 0 \\ E\{r_f^* n_g\} &= E\{r_g n_f^*\} = E\{n_f^* n_g\} = 0\,. \end{aligned} \tag{5.87}$$

In these equations, we assume that the second image of the pair has been registered and resampled onto the image coordinates (x_1, y_1). The only difference between this equation and Equation 5.46 involves the subscripts for the reflectivity functions r. In Section 5.4.1, we assumed that the spatial-frequency-domain data was trimmed to the aperture intersection A, but that no physical change occurred in the terrain reflectivity functions. Here, the assumption of identical aperture regions is implicit, but we admit the possibility of different terrain reflectivities r_f and r_g, due to physical changes in the time interval between collections. Furthermore, we have retained the subscripts on the collection angle β. Even though we have proposed identical collection geometries, no two-pass SAR platform is sufficiently accurate in navigation to assume the phase terms of Equation 5.86 will be identical everywhere.

What about the relationship between r_f and r_g? As before, for any given location (x_1, y_1), we consider a local neighborhood of pixels, $k \in [1, N]$. If there have been no physical changes in this local neighborhood in the time interval between the collections of f and g, the reflectivity functions should, in fact, be identical, because the same frequency-domain aperture region A was assumed. On the other hand, this patch of ground may have been disturbed, completely randomizing the multitude of scattering locations that make up the resolution cells of the neighborhood. In this case, the complex reflectivity functions might become completely uncorrelated. Therefore, we will model the function r_g as composed of two parts:

$$r_{g,k} = \alpha\, r_{f,k} + \left(\sqrt{1-\alpha^2}\right) z_k \tag{5.88}$$

where $\alpha \in [0, 1]$, $k \in [1, N]$, and the z_k are zero-mean, Gaussian, uncorrelated with $r_{f,k}$, $n_{f,k}$ and $n_{g,k}$, and $E\{|z_k|^2\} = \sigma_r^2$. In this model, α becomes the "change parameter." For example, if $\alpha = 1$, then $r_{g,k} = r_{f,k}$, and we say that no change has occurred in the scene reflectivity. If $\alpha = 0$, then $r_{g,k} = z_k$, and the reflectivities of f and g are completely uncorrelated; this suggests a very significant change. Rewriting Equation 5.86 using Equation 5.88 gives

$$\begin{aligned} f_k &= r_k + n_{f,k} \\ g_k &= \alpha\, r_k\, e^{j\phi} + \left(\sqrt{1-\alpha^2}\right) z_k + n_{g,k} \end{aligned} \tag{5.89}$$

with $\phi = Y_o(\beta_f - \beta_g)h$, and $k \in [1, N]$. In Equation 5.89, we have dropped the superfluous subscript f in $r_{f,k}$, and invoked the same assumption used earlier, namely that the height function is slowly varying and may be considered constant over the neighborhood. Also, because r_k and z_k are uncorrelated, and z_k is Gaussian (with random phase), we can drop the phase term $e^{j\phi}$ multiplying z_k without loss of generality. Our problem can now be stated succinctly. Given the set of image observations f_k and g_k in a local neighborhood $k \in [1, N]$ and the model of Equation 5.89, we must estimate the change parameter α.

For notational brevity, we write Equation (5.89) in matrix form as

$$\vec{X}_k = \begin{bmatrix} f_k \\ g_k \end{bmatrix} = r_k \begin{bmatrix} 1 \\ \alpha e^{j\phi} \end{bmatrix} + z_k \begin{bmatrix} 0 \\ \sqrt{1-\alpha^2} \end{bmatrix} + \begin{bmatrix} n_{f,k} \\ n_{g,k} \end{bmatrix} \tag{5.90}$$

and calculate its probability density function conditioned on not only α but also ϕ, which we will now treat simply as an unknown nuisance parameter. The probability density is:

$$P(\vec{X}_k|\phi, \alpha) = \frac{1}{\pi^2|Q|}\; e^{-\vec{X}_k^H Q^{-1} \vec{X}_k} \tag{5.91}$$

where the covariance matrix Q is given by

$$Q = E\{\vec{X}_k \vec{X}_k^H\} = \begin{bmatrix} \sigma_r^2 + \sigma_n^2 & \alpha\sigma_r^2 e^{-j\phi} \\ \alpha\sigma_r^2 e^{j\phi} & \sigma_r^2 + \sigma_n^2 \end{bmatrix}. \tag{5.92}$$

As before, we will denote by $\vec{\mathbf{x}}$ the vector of observations in a local neighborhood:

$$\vec{\mathbf{x}} = \begin{bmatrix} \vec{X}_1 \\ \vec{X}_2 \\ \vdots \\ \vec{X}_N \end{bmatrix}. \tag{5.93}$$

Invoking the mutual independence of the observations, we have a joint conditional density function given by

$$P(\vec{\mathbf{x}}|\phi, \alpha) = \frac{1}{\pi^2|Q|^N}\; e^{-\sum_{k=1}^{N} \vec{X}_k^H Q^{-1} \vec{X}_k}. \tag{5.94}$$

The maximum-likelihood estimate of α is that value of α that maximizes

$$\begin{aligned} & - N ln\{(1-\alpha^2)\sigma_r^4 + 2\sigma_r^2\sigma_n^2 + \sigma_n^4\} \\ & - \sum_{k=1}^{N} \left(\delta|f_k|^2 - \beta f_k g_k^* e^{j\phi} - \beta f_k^* g_k e^{-j\phi} + \delta|g_k|^2\right) \end{aligned} \tag{5.95}$$

where

$$\begin{aligned} \delta &= \frac{1}{\sigma_r^2(1-\alpha^2)} \\ \beta &= \frac{\alpha}{\sigma_r^2(1-\alpha^2)} \ . \end{aligned} \tag{5.96}$$

To find the extrema of this equation over α, we differentiate with respect to that variable, set the result equal to zero, and solve for α. Thus, we must solve

$$\left(\hat{\alpha}_{ML}^2 + 1\right)\left(\hat{\alpha}_{ML} - \frac{2\sum_{k=1}^{N} \mathbf{Re}(f_k^* g_k e^{-j\phi})}{\sum_{k=1}^{N}|f_k|^2 + \sum_{k=1}^{N}|g_k|^2}\right) = 0 \tag{5.97}$$

to determine the optimal estimate $\hat{\alpha}$. According to our interpretation of α, the parameter cannot be imaginary; the only root to the above equation that is real is that of the second term, giving

$$\hat{\alpha}_{ML} = \frac{2\sum_{k=1}^{N} \mathbf{Re}(f_k^* g_k e^{-j\phi})}{\sum_{k=1}^{N}|f_k|^2 + \sum_{k=1}^{N}|g_k|^2} \ . \tag{5.98}$$

Differentiation of Equation 5.95 with respect to ϕ gives the same ML estimate of ϕ that we determined in Section 5.4.1 for a somewhat simpler problem, namely

$$\hat{\phi}_{ML} = \angle\left(\sum_{k=1}^{N} f_k^* g_k\right) \ . \tag{5.99}$$

Substituting this into Equation 5.98 yields the maximum-likelihood estimator for α:

$$\hat{\alpha}_{ML} = \frac{2\left|\sum_{k=1}^{N} f_k^* g_k\right|}{\sum_{k=1}^{N}|f_k|^2 + \sum_{k=1}^{N}|g_k|^2} \ . \tag{5.100}$$

Calculating $\hat{\alpha}_{ML}$ over local neighborhoods, we obtain a local estimate, normalized to the interval [0,1]. Displaying these values in a two-dimensional grid yields a *change map* that is registered with image f. The change map has values near unity when images f and g have highly correlated reflectivity functions (no change), and values near zero when they are uncorrelated (i.e., when they change).

The $\hat{\alpha}$ just derived is normalized by the arithmetic mean of the clutter power in the two images given by

$$1/2\left[\sum_{k=1}^{N}|f_k|^2+\sum_{k=1}^{N}|g_k|^2\right] \tag{5.101}$$

Under the assumption that image f and g have the same average clutter power, this indeed is the maximum-likelihood estimator. This condition was ensured in the model of Equation 5.89 by the α and $\sqrt{1-\alpha^2}$ coefficients, and by the fact that $E\{|r_k|^2\} = E\{|z_k|^2\}$. While mathematically convenient, this estimator suppresses the estimate of α in the practical case of image power variations. A simple example will illustrate the point. Let $g_k = f_k/2$. Even though these two images are perfectly correlated, Equation 5.100 results in $\hat{\alpha}_{ML} = 0.8$. If, instead, we normalize by the geometric mean of the clutter power, we achieve just such an accommodation:

$$\hat{\alpha}_{ML(gm)} = \frac{|\sum_{k=1}^{N} f_k^* g_k|}{\sqrt{\sum_{k=1}^{N}|f_k|^2\sum_{k=1}^{N}|g_k|^2}}. \tag{5.102}$$

Three comments are in order. First, we note that $\hat{\alpha}_{ML(gm)} \geq \hat{\alpha}_{ML(am)}$ with equality if and only if $\sum_{k=1}^{N}|f_k|^2 = \sum_{k=1}^{N}|g_k|^2$, the equal power condition. Here, *gm* denotes geometric mean and *am* denotes arithmetic mean. Second, the $\hat{\alpha}_{ML(gm)}$ of Equation 5.102 is a well-known quantity; it is the sample complex correlation coefficient. This relationship will be exploited in Section 5.5.2. Finally, Equation 5.102 can be computed recursively because the weights of the summed quantities are all unity.

To illustrate the power of this technique, let us consider a simple demonstration. In Figure 5.13, we show a pair of images collected using the same SAR platform that provided the terrain-mapping examples of Section 5.4. These one-meter resolution spotlight images were collected as a two-pass interferometric pair, and show a scene with a landfill, several vehicles, equipment, and an excavated trench. The near-side vertical wall of the trench gives rise to the shadowed region at image center. Close inspection of these detected images does not reveal any difference in the appearance of the landfill, except for the position of two vehicles. In fact, during the interval of time between the two collections, a self-loading earthmover made a circuit of the site and a bulldozer was busy pushing dirt into the trench from the right. After the second image was collected, the bulldozer continued to work near the right side of the trench and the unpaved road running across the upper half of the scene was graded. Finally, a third image (not shown in Figure 5.13) was collected. This image again matched the interferometric constraints of the first two images.

After image registration, the sample complex correlation coefficient of Equation 5.102 was calculated over 7x7 windows for two sets of interferometric pairs: the

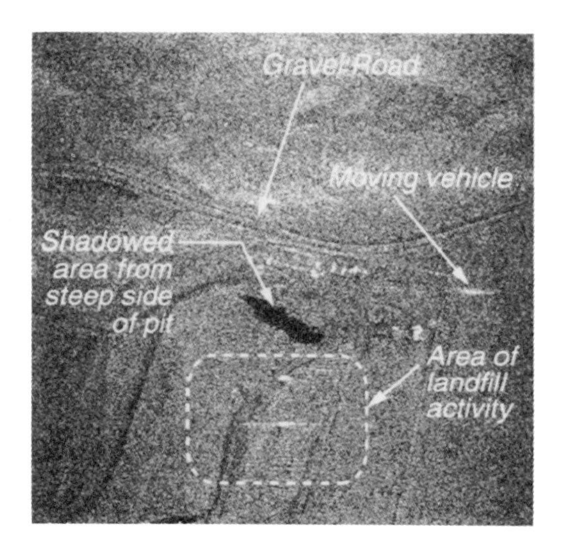

Figure 5.13 Detected images of landfill site used for the interferometric change detection experiment. The images show a landfill, several vehicles, equipment, and an excavated trench. Two earthmoving vehicles were at work between the collections of these images.

first to the second, and the first to the third image. These change maps are shown in Figure 5.14. In these maps, low correlation values (changes) have been mapped into dark pixels, while high correlation values (no changes) are shown in light pixels. The activity of the earthmovers is readily apparent in these maps. Because the reflectivity changes are *cumulative*, the lower map shows all the changes of the upper map plus the additional activity that occurred between the second and third collection. In particular, the bulldozed area is larger, and the effect of the road grading is evident. (The careful observer will note the existence of two horizontal streaks in the upper image that cannot be seen in the lower image. These are a result of sidelobes of vehicles captured in motion during the second collection. These streaks do not correlate with the first image.)

5.5.2 Statistical Considerations

The sample complex correlation coefficient estimator given by Equation 5.102 possesses some interesting statistical properties. While α is a parameter with a specific value, $\hat{\alpha}$ is an *estimate* of that value and, in fact, is a random variable. For any given local neighborhood, the value of the estimator is a function of the observations in that neighborhood. One may naturally ask how good an estimator of the true α it is. Under the condition that, as we have assumed, the observations are jointly normal, it can be shown (see Reference [37]) that the density function of the sample correlation coefficient depends only on the parameters α and N. An important aspect of this density function, however, is that its variance is a function of the mean. At a given signal-to-noise ratio, the variance for small values of α is much larger than the variance for values of α near 1. Therefore, our change map will appear to be much noisier in regions of change ($\alpha \approx 0$), than in regions of no change ($\alpha \approx 1$). This assumes a relatively constant clutter-to-system noise ratio. This is readily apparent in typical change maps. The white areas of such maps generally appear rather uniform, whereas the darker areas exhibit rapid variations in the value of $\hat{\alpha}$.

A second statistical property of sample correlation coefficients is also manifested in change maps. That is, $\hat{\alpha}$ is very unreliable as an estimate of α unless the sample size N is large. This is a well-known fact in regression analysis. Thus, to improve the statistical reliability of $\hat{\alpha}$, we should use a large local neighborhood. On the other hand, we cannot make the neighborhood too large. One assumption used in this development was that the height function $h(x, y)$ could be considered constant over the neighborhood. This results in a constant phase rotation angle ϕ for all observations. Also, a large value for N results in considerable spatial averaging; this causes reduced resolution of the final change map. Therefore, we have a classical

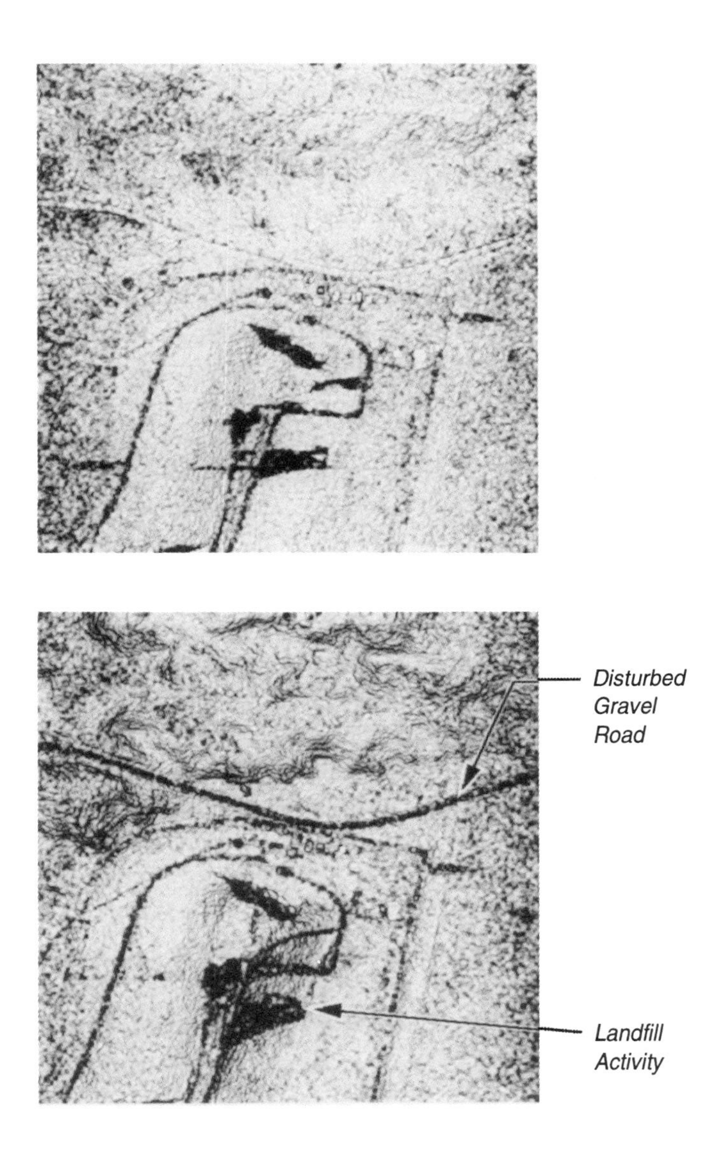

Figure 5.14 Two interferometric change-detection maps. The upper image shows changes over a short interval of time (about five minutes); the lower image shows changes at the same site over a longer interval (fifteen minutes). Changes are mapped into dark areas; light areas indicate no change.

tradeoff between statistical noise and resolution. Note that in developing the $\hat{\alpha}$ estimator, the noise process of concern is system noise. It follows that the higher the clutter-to-system noise ratio, the better the performance of the estimator. Therefore, if we increase the irradiated power density on the terrain or image terrain with a higher average backscatter coefficient, we can use a a smaller computation window.

This dependence of the estimator noise performance on computation window size is demonstrated in Figure 5.15. These two interferometric change maps were generated from the same image pair of the landfill site. The upper map was produced using a 3x3 correlation window, whereas a 7x7 window was used for the lower map. The 3x3 neighborhood provides a very unreliable estimate of the true value of α, unless that value is very close to unity. The larger window provides a much better result at the clutter-to-system noise ratio of the input images used in this experiment.

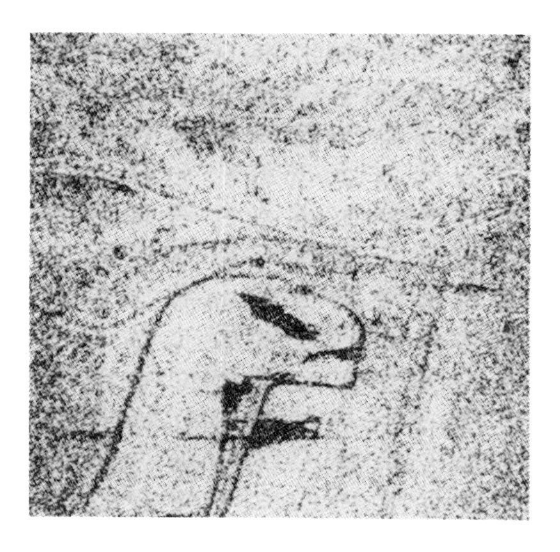

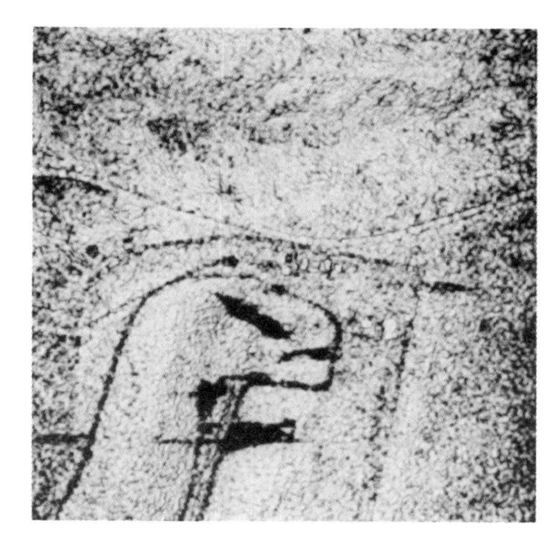

Figure 5.15 A comparison of two correlation window sizes for the landfill interferometric change-detection images. The upper image was correlated using 3x3 windows. The lower image used 7x7 correlation windows. The larger sample size (the 7x7 windows) gives a much better estimate of the true values of α.

Summary of Critical Ideas from Section 5.5

- **Interferometric change detection far outperforms incoherent change detection with SAR imagery. It allows completely automated data processing and offers a much more sensitive change statistic.**

- **Using the tomographically derived image function for spotlight-mode SAR images, a normalized change parameter, $\alpha \in [0, 1]$, between two interferometric SAR images f and g can be estimated at every pixel location as**

$$\hat{\alpha}_{ML}(x_1, y_1) = \frac{|\sum_{k=1}^{N} f_k^* g_k|}{\sqrt{\sum_{k=1}^{N} |f_k|^2 \sum_{k=1}^{N} |g_k|^2}}$$

calculated over a local neighborhood of N pixels surrounding (x_1, y_1).

- **This estimator is, in fact, the sample complex correlation coefficient between image functions $f(x_1, y_1)$ and $g(x_1, y_1)$ over the N samples. It shares two characteristics with all such statistics: its variance is larger for small values of α (large degree of change between the images) than it is for values near unity, and it is a rather unreliable estimator for α, unless N is large and/or the signal-to-noise ratio is high.**

5.6 INTERFEROMETRIC TERRAIN MOTION MAPPING

Before leaving the topic of interferometric processing, we will briefly discuss the subject of terrain *motion* mapping. As the name suggests, this technique addresses the subject of dynamic, rather than static, terrain. To begin, consider collecting two SAR images of a scene over an interval of time during which the reflectivity function $r(x, y)$ has not changed. Further, let these collections be accomplished with identical imaging geometries. Finally, let us suppose the height function $h(x, y)$ has undergone a slight change over the interval. What do the fundamental SAR imaging equations say about this situation? Let us rewrite them slightly to reflect the change in the height function, much as we did with Equation 5.6 for the case of IFSAR static-terrain mapping. As usual, we will postulate the absence of superposition in the images. The relevant equations are

$$\begin{aligned} f(x_1, y_1) &= s_A(x, y) \otimes \left[r(x, y) e^{-j\beta Y_0 h_f(x,y)} e^{-jyY_0} \right] \\ g(x_1, y_1) &= s_A(x, y) \otimes \left[r(x, y) e^{-j\beta Y_0 h_f(x,y)} e^{j\beta Y_0 (h_f(x,y) - h_g(x,y))} e^{-jyY_0} \right] . \end{aligned} \tag{5.103}$$

Here, we have assumed that the imaging geometries for the two collections are essentially identical, so that $\beta_f = \beta_g = \beta$, and that the apertures have been appropriately trimmed to retain only the common region A. Assuming that the difference in height varies slowly enough that the term $e^{j\beta Y_0 (h_g(x,y) - h_f(x,y))}$ may be considered constant over the width of the IPR, we can rewrite Equations 5.103 in a fashion exactly analogous to Equation 5.6:

$$\begin{aligned} f(x_1, y_1) &= r_A(x, y) \\ g(x_2, y_2) &= r_A(x, y) \; e^{j\beta Y_o (h_f(x,y) - h_g(x,y))} . \end{aligned} \tag{5.104}$$

The conjugate product of the above image expressions (after image registration) is then given by

$$f^*(x_1, y_1) g(x_1, y_1) = |r_A(x, y)|^2 \; e^{j\beta Y_o (h_f(x,y) - h_g(x,y))} \tag{5.105}$$

where the phase depends only on the change in the height function between the two collections. Of course on a sample-by-sample basis, the conjugate product is very noisy as we saw in Section 5.4.1. Fortunately, the information we seek is in the phase rotation angle between f and g. Therefore, we use precisely the same maximum-likelihood estimator here as we used for the static-terrain problem.

The scale factor relating change in height at a location and the phase angle $\Delta\phi$ of the conjugate product is easily found by substituting for βY_o to be

$$\Delta\phi = \beta Y_o \left(h_f(x, y) - h_g(x, y) \right)$$

$$
\begin{aligned}
&= \frac{4\pi}{\lambda} \cos\psi \tan\psi \, (h_f(x,y) - h_g(x,y)) \\
&= \frac{4\pi}{\lambda} \sin\psi \, (h_f(x,y) - h_g(x,y))
\end{aligned} \tag{5.106}
$$

so that

$$
\Delta\phi = \frac{4\pi}{\lambda} \Delta h \sin\psi \, . \tag{5.107}
$$

This equation has a very simple physical interpretation. The phase change observed is equal to the two-way, line-of-sight change in range of the terrain scaled by the wavelength. Because the phase noise of the maximum-likelihood estimator is usually on the order of a few degrees, it can be appreciated that (for a short wavelength SAR), extremely subtle terrain movements could be transduced. This opens the possibility of observing and mensurating earth movements (such as fault creep, ground subsidence, ice movement, and magma bulges) using SAR interferometry. See [38] and [39] for two examples of this technique. Even more impressive, with large-cross-section targets (such as corner reflectors) one can do even better. A large-cross-section target exhibits a much higher signal-to-system noise ratio than does terrain clutter. This results in a much smaller error in the phase estimate. Indeed, Gray, et al. have reported observing millimeter movements of corner reflectors using this method [40]. It must be stressed that identical imaging geometries for the two collections are required in order to suppress any phase change due to *static* terrain undulations. If the depression angles are not matched exactly, then the image-domain, conjugate-product phase contains a term proportional to $h(x,y)$. There is no *a priori* way to separate the contributions to the phase of static terrain versus terrain motion. Reference [39], for example, used a terrain digital elevation model to back out the small, static-terrain phase terms in that study.

Summary of Critical Ideas from Section 5.6

- **Land-form movements (such as subsidence, fault movement, or ice flow motion) can be mapped interferometrically. Again, the tomographic formulation of the image function points the way. Assuming no change in the reflectivity functions over a time interval, the conjugate product of a pair of registered spotlight-mode SAR images with identical collection geometries is given by**

$$f^*(x_1, y_1)g(x_1, y_1) = |r_A(x, y)|^2 \, e^{j\beta Y_o(h_f(x,y) - h_g(x,y))} .$$

The phase of this product depends only on the difference in terrain-height functions over the time interval.

- **The scale factor relating the change in terrain height to the phase of the conjugate product is equal to the round-trip change in the line-of-sight range of the terrain scaled by the wavelength:**

$$\phi = \frac{4\pi}{\lambda} \Delta h \sin \psi .$$

- **If the collection geometries are not identical, phase modulation of the conjugate product due to *static* terrain undulations will also be present. There is no *a priori* way to separate the contributions of static terrain versus terrain motion in the phase data.**

5.7 STEREOSCOPIC TERRAIN MAPPING

Finally, in this section, we consider the method of determining terrain height by stereoscopy. While this technique is, in general, quite distinct from interferometry, the two come together in an interesting way for crossed-ground-track imaging geometries. We have already seen evidence of this in Section 5.3.2 where we noted that the cross-range spline warp function $S^x(x, y)$ (used to register crossed-track interferometric pairs) is a scaled approximation to the terrain height function $h(x, y)$. Let us examine this relationship a little more closely.

5.7.1 Layover Equations and Scale Factor

Consider again the image Equation 5.2. In Chapter 2, we learned that the SAR image-formation process effects a projection of the three-dimensional reflectivity surface onto the two-dimensional image. A scatterer at three-space location $(x, y, h(x, y))$ will end up in the image at location (x_1, y_1). These latter coordinates are given by Equation 5.3. If we denote the layover at (x, y) as $\vec{l}(x, y)$, then substitution for α and β yields a total layover magnitude (see Figure 2.28) of

$$|\vec{l}(x, y)| = h(x, y) \ \tan \varsigma \tag{5.108}$$

where ς is the slope angle, i.e., the angle between the slant plane and the ground plane.[7] As was demonstrated in Chapter 2, layover always occurs in the direction normal to the line of intersection between the slant plane and the ground plane. For level flight of the SAR platform, this is also in the direction normal to, and toward, the ground track (see Figure 2.28). Using the law of cosines, we have for the total differential layover distance between two images:

$$|\vec{l}_2(x, y) - \vec{l}_1(x, y)| =$$
$$h(x, y) \ \left[\tan^2 \varsigma_1 + \tan^2 \varsigma_2 - 2 \tan \varsigma_1 \tan \varsigma_2 \cos(\theta_{g_2} - \theta_{g_1})\right]^{1/2} . \tag{5.109}$$

From this equation, it is easily seen how the terrain height function $h(x, y)$ could be inferred from stereo pairs of images. As long as the two images are collected with different grazing angles and/or different squint angles, the differential change in position of scatterers due to layover is proportional to the height of those scatterers.

In practice, scatterers are brought into stereo correspondence using automated or interactive procedures. Then, the imaging geometry is used to calculate height.

[7] The slope angle and depression angle, ς and ψ, are related by: $\tan \varsigma = \tan \psi / \cos \theta_g$. See Appendix C.

Usually with SAR stereo, the terrain is imaged in overlapping strips. In the regions of overlap, the geometry from two adjacent strips differ in grazing angles, so the predominant differential layover is in the range direction. If the strips overlap by half their width, a continuous survey can be accomplished.

Virtually any pair of images of the same terrain can be used for this purpose. The difference in their grazing or squint angles will determine the direction and magnitude of the differential layover with terrain height. However, the reflectivity of many scenes depends much more on aspect angle for SAR than for optical images. Consequently, the angle between the line-of-sight vectors for SAR stereo images tends to be much smaller than is typical for optical images. There is one imaging geometry available to SAR that is impossible to duplicate optically, however. That is the crossed-ground-track geometry discussed in Section 5.3.2. As we demonstrated in Equation 5.15, at the point where the ground tracks cross, two images can be made of the same terrain with identical boresight angles but different squint angles, because the platform velocity vectors are not the same. In optical systems, if the boresight angles are identical, the stereo convergence angle is zero, and one's ability to infer terrain height disappears altogether. With SAR, the different squint angles still result in differential layover in the cross-range direction. The important point, however, is that the identical boresight angles yield stereo images that are highly correlated even in open, clutter-dominated terrain. Indeed, we have seen examples where such images have been interfered with each other (see Figure 5.8). Because of this high degree of correlation, these images can be registered (brought into stereo correspondence) using the completely automated techniques discussed earlier in this chapter. Furthermore, the presence of interferometric fringes from conjugate multiplication of the registered images constitutes independent proof of proper registration. Therefore, such stereo pairs can be processed automatically and can be visually inspected without the tedious manual task of checking correspondence on a post-by-post basis.

This collection scenario may prove useful in terrain-motion mapping as well. As we pointed out in Section 5.6, collections for such studies should be made with depression angles that are as similar as possible. This is necessary in order to make the static-terrain scale factor small. Of course, in any practical two-pass collection, a scale factor of zero cannot be guaranteed; a small influence of static terrain on the interferometric phase must be assumed. If the collection is made using crossed ground tracks, however, a rough estimate of the static-terrain elevation may be accomplished using the stereo correspondence techniques just outlined. This elevation model can then be used to back out the static-terrain phase modulation. This will leave only the phase term due to terrain motion. Because the static-terrain scale factor will generally be very small, even a rough elevation model would be sufficient for the purpose.

5.7.2 Error Sources

Errors associated with stereo terrain mapping are quite different from those that arise in interferometry. On the one hand, stereoscopy does not suffer from ambiguities, system phase noise effects, or wavefront curvature. Small navigation errors do not have as large an impact. On the other hand, this technique cannot approach the spatial resolution of interferometry and is subject to errors due to poor correlation of two-pass data.

For small images, there are three principal sources of error. First, because layover is a result of projecting three-space terrain onto the ground plane, an accurate ground-plane normal is essential. The ground-plane normal vector is generally taken to be the normal to the local geoid at the ground reference point (grp). During image formation, the accuracy of the pointing vector set completely determines the accuracy of the projection. Fortunately, at long ranges, even moderately large navigation errors of tens of meters have a small effect on the calculated values of α and β. This is true because the angular error is on the order of $arcsin(\Delta/R)$, where R is the range and Δ is the position error of the platform. The same ground-plane normal should be used for both stereo images. The measured differential layover gives a height measurement relative to this ground plane. If desired, the resulting elevation data can be referenced to the local geoid.

The second source of error is the scale factor computed from Equation 5.109. Here again, the angles involved, ς_i and θ_{g_i}, are determined by the pointing vectors. The angular errors are on the order of $arcsin(\Delta/R)$. It is quite reasonable to collect image pairs where the slant plane or squint angle differences are much larger than the errors. This is in contrast to interferometry where the all-important grazing angle difference is constrained to be very small and the effects of angular errors in imaging geometry are much more noticeable.

Finally, a third source of error must be postulated due to the question of correlation accuracy when measuring the differential layover. This is somewhat harder to quantify. Because two-dimensional correlations are performed, this appears to be a classic tradeoff between accuracy and spatial resolution by varying the size of the correlation window used. However, the implicit assumption leading to Equation 5.19 is that the translation of the source image relative to the target image is constant over the correlation window. This further implies that the elevation is constant over this area. Obviously, one cannot increase the size of the correlation window indefinitely. We can say, however, that at least for the case of crossed-ground-track geometries with equal boresight angles, the correlation measurements one obtains by taking advantage of the coherent speckle effect (Section 5.3.2) are accurate to within a fraction of an impulse-response width. Were this not so, interference fringes would not appear in the conjugate product image. Thus applying the scale factor in Equation 5.109 to an impulse-response width overbounds the error due to registration.

Summary of Critical Ideas from Section 5.7

- **The layover equations resulting from the tomographic formulation of spotlight-mode SAR capture the stereoscopic effect in its most general form. The total differential layover distance of a corresponding scatterer in two images is related to the height of the scatterer above the processing ground plane by**

$$|\vec{l}_2(x,y) - \vec{l}_1(x,y)| =$$
$$h(x,y)\ \left[\tan^2\varsigma_1 + \tan^2\varsigma_2 - 2\tan\varsigma_1\tan\varsigma_2\cos(\theta_{g_2} - \theta_{g_1})\right]^{1/2} .$$

  The scale factor contains only the slope angles and ground-plane squint angles.

- **Any pair of SAR images of the same terrain can be used for stereoscopic terrain mapping using this equation, so long as the correspondence problem can be solved.**

- **Stereoscopic *interferometric* image pairs, imaged with the same boresight angles but with different ground-track angles, can be easily registered (brought into correspondence) by automated techniques due to their coherent speckle. Such pairs transduce terrain height by two independent means: differential layover and interferometric phase.**

REFERENCES

[1] L. Graham, "Synthetic Interferometer Radar for Topographic Mapping," *Proceedings of the IEEE*, Vol. 62, No. 6, pp. 763-768, June 1974.

[2] F. T. Ulaby, R. K. Moore, and A. K. Fung, *Microwave Remote Sensing*, Vol. II, Artech House, Norwood, MA, 1982.

[3] J. W. Goodman, *Statistical Optics*, Wiley and Sons, New York, 1985.

[4] F. W. Leberl, *Radargrammetric Image Processing*, Artech House, Norwood, MA, 1990.

[5] N. Levanon, *Radar Principles*, Wiley and Sons, New York, 1988.

[6] A. W. Naylor and G. R. Sell, *Linear Operator Theory in Engineering and Science*, Springer-Verlag, New York, 1982.

[7] C. Pratti, R. Rocca, and M. Guarmieri, "SAR Interferometry Experiments With ERS-1," *Proceedings of the Interferometric SAR Technology and Applications Symposium*, ARPA and US Army TEC, April 13-14, 1993.

[8] P. H. Eichel, C. V. Jakowatz, Jr., P. A. Thompson, and D. C. Ghiglia, "Interferometric Processing of Coherent Spotlight-Mode SAR Images," *Proceedings of the Interferometric SAR Technology and Applications Symposium*, ARPA and US Army TEC, April 13-14, 1993.

[9] C. Prati, F. Rocca, A. M. Guarnieri, and E. Damonti, "Seismic Migration for SAR Focusing: Interferometric Applications," *IEEE Transactions on Geoscience and Remote Sensing*, Vol. 28, No. 4, pp. 627-640, July 1990.

[10] R. K. Raney and P. W. Vachon, , "A Phase-Preserving SAR Processor," *Proceedings IGARSS '89*, Vancouver, British Columbia, 1989.

[11] F. Li and R. M. Goldstein, "Studies of Multi-baseline Spaceborne Interferometric Synthetic Aperture Radars," *IEEE Transactions on Geoscience and Remote Sensing*, Vol. 28, No. 1, pp. 88-97, January 1990.

[12] E. Rodriques and J. Martin, "Theory and Design of Interferometric Synthetic Aperture Radars," *Proceedings of the IEEE*, Vol. 139, No. 2, pp. 147-159, April 1992.

[13] D. Just and R. Bamler, "Phase Statistics of Interferograms With Applications to Synthetic Aperture Radar," *Applied Optics*, Vol. 33, No. 20, pp. 4361-4368, July 1994.

[14] S. Alliney and C. Morandi, "Digital Image Registration Using Projections," *IEEE Transactions on Pattern Analysis and Machine Intelligence*, Vol. PAMI-8, pp. 222-233, March 1986.

[15] D. Lee, S. Mitra, and T. Krile, "Analysis of Sequential Complex Images Using Feature Extraction and Two-dimensional Cepstrum Techniques," *J. Opt. Soc. Am. A*, Vol. 6, No. 6, pp. 863-870, June 1989.

[16] A. K. Gabriel and R. M. Goldstein, "Crossed Orbit Interferometry: Theory and Experimental Results from SIR-B," *International Journal of Remote Sensing*, Vol. 9, No. 5, pp. 857-872, 1988.

[17] M. D. Pritt, "Image Registration With Use of the Epipolar Constraint for Parallel Projections," *J. Opt. Soc. Am. A*, Vol. 10, No. 10, pp. 2187-2192, October 1993.

[18] W. H. Press and S. A. Teukolsky, *Numerical Recipes in C: The Art of Scientific Computing*, Cambridge University Press, 1992.

[19] P. H. Eichel, D. C. Ghiglia, C. V. Jakowatz, Jr., P. A. Thompson, and D. E. Wahl, "Spotlight SAR Interferometry for Terrain Elevation Mapping and Interferometric Change Detection," Sandia National Labs Tech. Report, SAND93-2072, 1993.

[20] C. V. Jakowatz, Jr. and D. E. Wahl, "An Eigenvector Method for Maximum-Likelihood Estimation of Phase Errors in SAR Imagery," *J. Opt. Soc. Am. A*, pp. 2539-2546, December 1993.

[21] R. Cusack, J. M. Huntley, and H. T. Goldrein, "Improved Noise-immune Phase-unwrapping Algorithm," *Applied Optics*, Vol. 34, No. 5, pp. 781-789, February 1995.

[22] R. M. Goldstein, H. A. Zebker, and C. L. Werner, "Satellite Radar Interferometry: Two-dimensional Phase Unwrapping", *Radio Sci.*, 23, pp. 713-720, 1988.

[23] J. M. Huntley, "Noise-immune Phase Unwrapping Algorithm," *Applied Optics*, 28, pp. 3268-3270, 1989.

[24] D. J. Bone, "Fourier Fringe Analysis: The Two-dimensional Phase Unwrapping Problem," *Applied Optics*, 30, pp. 3627-3632, 1991.

[25] D. L. Fried, "Least-squares Fitting a Wave-front Distortion Estimate to an Array of Phase-difference Measurements," *J. Opt. Soc. Am.*, Vol. 67, pp. 370-375, 1977.

[26] R. H. Hudgin, "Wave-front Reconstruction for Compensated Imaging," *J. Opt. Soc. Am.*, Vol. 67, pp. 375-378, 1977.

[27] R. J. Noll, "Phase Estimates from Slope-type Wave-front Sensors," *J. Opt. Soc. Am.*, Vol. 68, pp. 139-140, 1978.

[28] B. R. Hunt, "Matrix Formulation of the Reconstruction of Phase Values from Phase Differences," *J. Opt. Soc. Am.*, Vol. 69, pp. 393-399, 1979.

[29] R. L. Frost, C. K. Rushforth, and B. S. Baxter, "Fast FFT-Based Algorithm for Phase Estimation in Speckle Imaging," *Applied Optics*, Vol. 18, pp. 2056-2061, June 1979.

[30] D. C. Ghiglia and L. A. Romero, "Direct Phase Estimation from Phase Differences Using Fast Elliptic Partial Differential Equation Solvers," *Optics Letters*, Vol. 14, No. 20, pp. 1107-1109, October 1989.

[31] D. C. Ghiglia and L. A. Romero, "Robust Two-dimensional Weighted and Unweighted Phase Unwrapping That Uses Fast Transforms and Iterative Methods," *J. Opt. Soc. Am. A*, Vol. 11, No. 1, pp. 107-117, January 1994.

[32] M. D. Pritt and J. S. Shipman, "Least-squares Two-dimensional Phase Unwrapping Using FFT's," *IEEE Transactions on Geoscience and Remote Sensing*, Vol. 32, No. 3, pp. 706-708, May 1994.

[33] J. C. Curlander and R. N. McDonough, *Synthetic Aperture Radar, Systems and Signal Processing*, Wiley and Sons, New York, 1991.

[34] H. A. Zebker, C. L. Werner, P. A. Rosen, and S. Hensley, "Accuracy of Topographic Maps Derived from ERS-1 Interferometric Radar," *IEEE Transactions on Geoscience and Remote Sensing*, Vol. 32, No. 4, July 1994.

[35] D. C. Ghiglia and D. E. Wahl, "Interferometric Synthetic Aperture Radar Terrain Elevation Mapping from Multiple Observations," *Proceedings IEEE 6th DSP Workshop*, Yosemite National Park, CA, October 2-5, 1994.

[36] J. Villasenor and H. A. Zebker, "Studies of Temporal Change Using Radar Interferometry," *Proceedings SPIE*, Vol. 1630, pp. 187-198, 1992.

[37] P. G. Hoel, *Introduction to Mathematical Statistics*, Fourth Edition, Wiley and Sons, New York, 1971.

[38] A. K. Gabriel, R. M. Goldstein, H. A. Zebker, "Mapping Small Elevation Changes Over Large Areas: Differential Radar Interferometry," *Journal of Geophysical Research*, Vol. 94, No. B7, pp. 9183-9191, July 1989.

[39] D. Massonnet et al, "The Displacement Field of the Landers Earthquake Mapped by Radar Interferometry," *Nature*, Vol. 364, pp. 138-142, July 1993.

[40] A. L. Gray and P. J. Farris-Manning, "Repeat-pass Interferometry with Airborne Synthetic Aperture Radar," *IEEE Transactions on Geoscience and Remote Sensing*, Vol. 31, No. 1, pp. 180-191, January 1993.

A

ROTATION PROPERTY OF FOURIER TRANSFORMS

A proof of the theorem stating that the Fourier transform of a rotated function is equal to a rotated version of the Fourier transform of that function follows.

Let $g(\mathbf{x})$ be a function of $\mathbf{x}$, where $\mathbf{x}$ is a column vector with arbitrary dimension. Denote the Fourier transform of g by

$$\mathcal{F}[g(\mathbf{x})] = G(\mathbf{X}) = \int g(\mathbf{x}) \exp\{-j\mathbf{x}^T\mathbf{X}\}d\mathbf{x} \, . \tag{A.1}$$

Consider the Fourier transform of a rotated version of g. The linear transformation describing the rotation is

$$\mathbf{u} = \mathbf{A}\mathbf{x} \tag{A.2}$$

where $\mathbf{A}$ is the orthonormal matrix such that

$$\mathbf{x} = \mathbf{A}^{-1}\mathbf{u} = \mathbf{A}^T\mathbf{u} \, . \tag{A.3}$$

We then have for the transform of the rotated version of g

$$\mathcal{F}[g(\mathbf{A}\mathbf{x})] = \int g(\mathbf{u}) \exp\{-j(\mathbf{A}^T\mathbf{u})^T\mathbf{X}\}|\mathbf{J}|d\mathbf{u} \tag{A.4}$$

where $|\mathbf{J}|$ is the determinant of the Jacobian of the transformation of Equation A.3. It can be easily shown that $|\mathbf{J}| = 1$, so that we obtain

$$\mathcal{F}[g(\mathbf{A}\mathbf{x})] = \int g(\mathbf{u}) \exp\{-j\mathbf{u}^T\mathbf{A}\mathbf{X}\}d\mathbf{u} = \mathbf{G}[\mathbf{A}\mathbf{X}] \, . \tag{A.5}$$

The final expression represents a rotated version of $\mathbf{G}(\mathbf{X})$. This completes the proof.

B

PHASE ANALYSIS FOR LIMITATIONS OF THE TOMOGRAPHIC PARADIGM

B.1 THE PHASE OF THE RETURN FROM A POINT TARGET

The tomographic paradigm states that the demodulated echo of each transmitted pulse evaluates the three-dimensional Fourier transform of the scene reflectivity function along a particular line in the Fourier domain. It has been argued that the combined effect of many such line traces (from many demodulated pulses) is to sample a two-dimensional slice through the Fourier domain. We will formalize these ideas somewhat to arrive at an expression for the phase of the two-dimensional signal predicted by the tomographic model.

Consider the SAR image collection scenario shown in Figure B.1. Imagine a scene to be imaged that lies nominally in the x-y plane of Figure B.1(a), and consider the three-dimensional Fourier transform of the scene function existing in the X-Y-Z domain shown in Figure B.1(b). Let the vector $\mathbf{p} = x_0\bar{\mathbf{x}} + y_0\bar{\mathbf{y}} + z_0\bar{\mathbf{z}}$ denote the location of a certain point target in the scene. Let $r_0\mathbf{u}_\theta$ denote the position of the SAR platform at a particular point in the synthetic aperture, where r_0 is the distance of the platform from scene center, and $\mathbf{u}_\theta$ is a unit-length *pointing* vector that denotes the direction from the center of the scene to the SAR platform. The subscript θ denotes aperture position in terms of the angle of the pointing vector relative to the center of the aperture.

The contribution of the point target at $\mathbf{p}$ to the scene reflectivity can be represented in terms of a delta function as follows:

$$s_{\mathbf{p}}(x, y, z) = A_{\mathbf{p}}\delta(x - x_0, y - y_0, z - z_0) \tag{B.1}$$

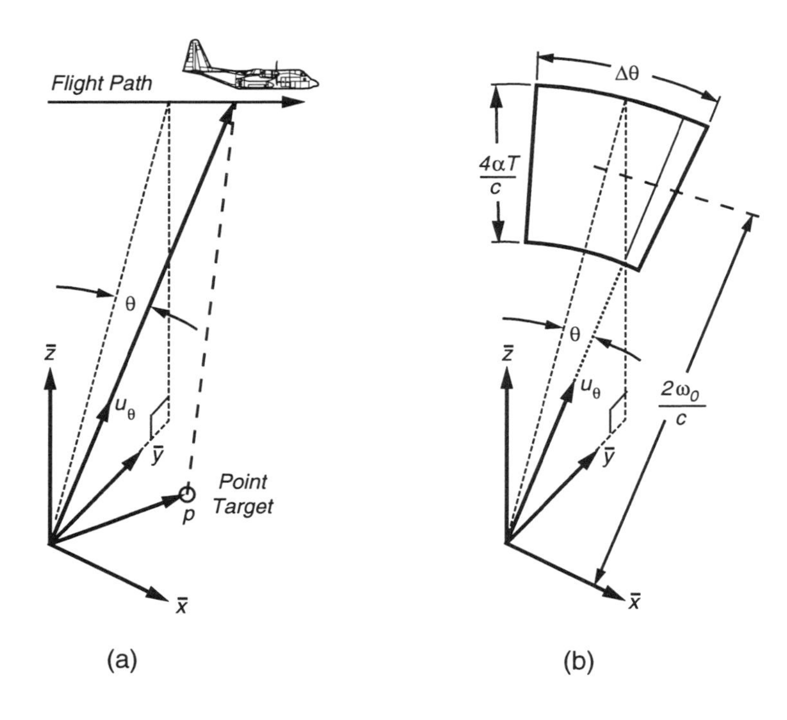

Figure B.1 Three-dimensional SAR imaging geometry

where $A_{\mathbf{p}}$ is a complex-valued scalar denoting the amplitude and phase of the radar return. In the Fourier domain of Figure B.1(b), this point target contributes a three-dimensional complex sinusoid of the form

$$S_{\mathbf{p}}(X, Y, Z) = A_{\mathbf{p}} e^{j(x_0 X + y_0 Y + z_0 Z)} . \tag{B.2}$$

An equivalent vector expression for this Fourier-domain point-target response is

$$\bar{S}_{\mathbf{p}}(\mathbf{k}) = A_{\mathbf{p}} e^{j\mathbf{p}\cdot\mathbf{k}} \tag{B.3}$$

where $\mathbf{k} = X\bar{\mathbf{x}} + Y\bar{\mathbf{y}} + Z\bar{\mathbf{z}}$ denotes position in the Fourier domain.

Let us return now to the notion that each demodulated pulse consists of evaluating the Fourier transform of the scene function along a line in three dimensions. A vector denoting a position in the Fourier domain on the line in question can be written in the form

$$\mathbf{k}_{\theta t} = k_t \mathbf{u}_\theta \tag{B.4}$$

where the subscript t denotes time variation within a given return pulse. The magnitude assumed by this vector (see Equation 1.41) at a particular intra-pulse time t

is

$$k_t = \frac{2\omega_0}{c} + \frac{4\alpha}{c}(t - \tau_0), \qquad |t - \tau_0| \leq T/2 \tag{B.5}$$

where ω_0 is the radar center frequency, α is the chirp rate parameter, and τ_0 is the time delay used to demodulate the return signal. It follows that the demodulated return signal sampled at aperture position θ and time t due to a point target at scene location $\mathbf{p}$ is of the form

$$d_{\mathbf{p}}(\theta, t) = A_{\mathbf{p}} e^{j\mathbf{p}\cdot\mathbf{k}_{\theta t}} \ . \tag{B.6}$$

During the SAR collection, the vector $\mathbf{k}_{\theta t}$ sweeps out a two-dimensional surface in the three-dimensional Fourier domain as its angle θ and magnitude k_t vary. This Fourier *collection surface* is shown in Figure B.1(b). Generally, the SAR flight path is close to a straight line. In this case, the collection surface lies in a *slant plane* of the Fourier domain analogous to the spatial slant plane defined by the flight line and a point at the scene center. The two-dimensional collection geometry and corresponding Fourier domain for this situation are shown in Figure B.2. Here, the target position vector $\mathbf{p}' = \mathbf{x}'_0\hat{\mathbf{x}} + \mathbf{y}'_0\hat{\mathbf{y}}$ is the projection of $\mathbf{p}$ into the slant plane. Because any component of $\mathbf{p}$ lying outside the slant plane is orthogonal to the vector $\mathbf{k}$, we can rewrite Equation B.6 as

$$d_{\mathbf{p}}(\theta, t) = A_{\mathbf{p}} e^{j\mathbf{p}'\cdot\mathbf{k}'_{\theta t}} \ . \tag{B.7}$$

Here, $\mathbf{k}'_{\theta\mathbf{t}}$ represents the same vector as $\mathbf{k}_{\theta\mathbf{t}}$, but it is expressed in the two dimensions of the slant plane instead of the original three dimensions. By expressing $\mathbf{p}'$ and $\mathbf{k}'$ in terms of their respective x-y coordinates, we can write the point-target response as

$$d_{x'_0, y'_0}(X', Y') = A_{\mathbf{p}} e^{j(x'_0 X' + y'_0 Y')} \ . \tag{B.8}$$

This result shows that the phase function is linear in each of the slant- plane spatial-frequency coordinates X' and Y' and that it is proportional to target location (x'_0, y'_0) in the corresponding slant-plane coordinates of the scene. Thus, according to the tomographic paradigm, the SAR data from a point target consists of a complex sinusoid whose frequency in two dimensions corresponds to the location of the target projected into the slant plane. It follows that a simple inverse Fourier transform will produce an image of this target and, by extension, the entire scene. (This assumes that the data have been resampled from their original polar coordinates to the Cartesian coordinates X' and Y' indicated in Equation B.8, a process referred to as *polar-to-rectangular reformatting* discussed in detail in Chapter 3.)

The point-target phase history expressed by Equation B.8 conveys in a simple and concise way the essence of the SAR imaging process, but recall that it is based on a tomographic development that relies on certain approximations. In particular, the tomographic paradigm first approximates spherical wavefronts as planar, and

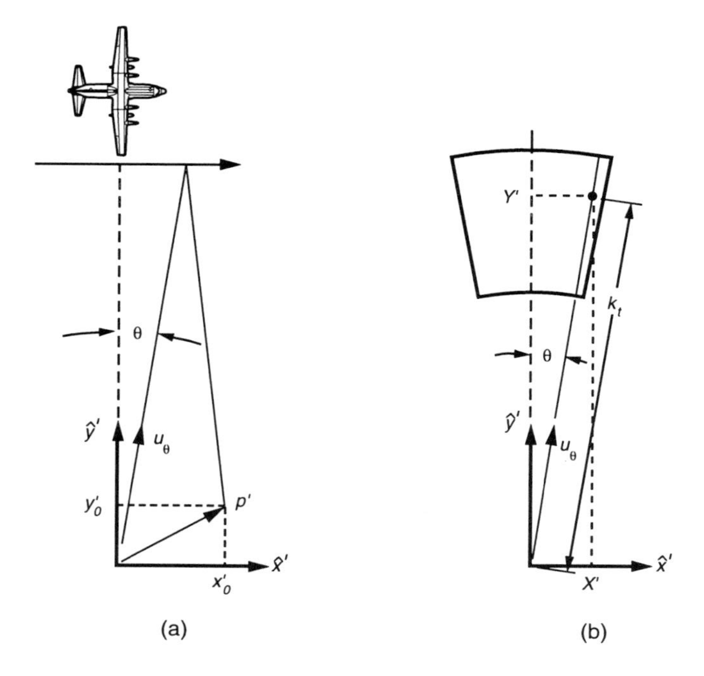

Figure B.2 Slant-plane SAR imaging geometry

secondly it ignores a troublesome quadratic phase term in the demodulated return (see Equation 1.37). To determine the significance of these two unaccounted-for effects, let us now re-derive the point-target phase history from the precise radar echo in the context of the actual geometry involved. Consider the slant-plane imaging geometry shown in Figure B.3. The target location designated by p' is again the projection of the actual target location in three-space into the slant plane. (It should be noted here that this projection is not an orthogonal projection as we implied earlier, but rather a projection along a circular arc in three dimensions centered on and normal to the flight path, i.e., centered along a contour of constant range and range rate. The consequence of assuming an orthogonal projection and interpreting the slant-plane image accordingly is a slight geometric distortion that can be corrected after the image is formed.)

Assume the radar transmits a linear FM chirp pulse of duration T represented by

$$s_x(t) = Re\{e^{j\phi_x(t)}\}, \qquad |t| \leq T/2 \tag{B.9}$$

whose phase function is a quadratic of the form

$$\phi_x(t) = \omega_0 t + \alpha t^2 \; . \tag{B.10}$$

Consider a point target in the scene at distance r from the radar at a certain point in the aperture, denoted by θ, as shown in Figure B.3(a). The phase of the radar return

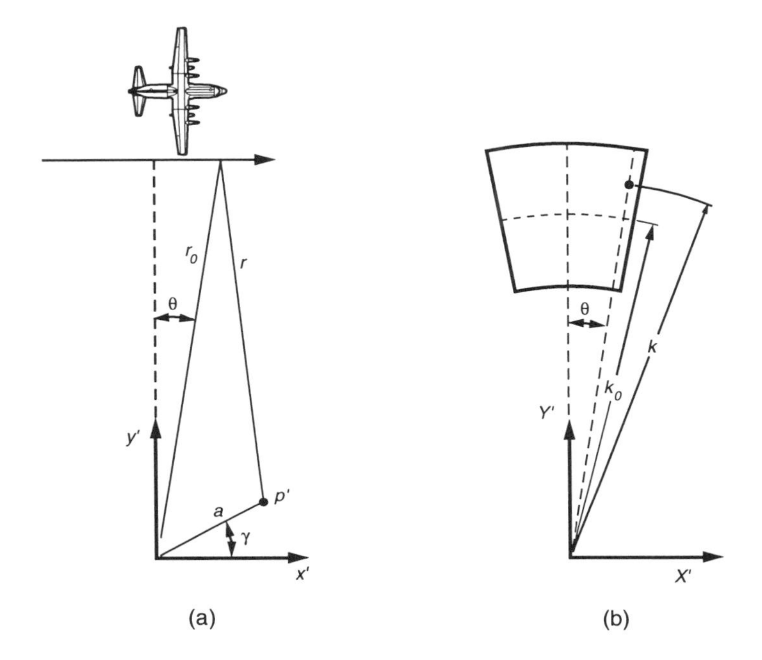

Figure B.3 Slant-plane geometry for deriving phase

or echo from this target at aperture position θ is simply an appropriately delayed replica of the transmitted phase:

$$\phi_r(t,\theta) = \omega_0(t - 2r/c) + \alpha(t - 2r/c)^2 \, . \tag{B.11}$$

The local reference used to demodulate the return signal has exactly the same phase function as the return from a hypothetical target at the center of the scene. With r_0 denoting the distance of the radar platform from scene center at aperture position θ, as shown in Figure B.3(a), the reference phase is expressed by

$$\phi_0(t,\theta) = \omega_0(t - 2r_0/c) + \alpha(t - 2r_0/c)^2 \, . \tag{B.12}$$

The process of quadrature demodulation forms a complex video signal represented in separate I and Q components. The phase of this signal is the difference of the return and the reference signal phases:

$$\begin{aligned} \phi(t,\theta) &= \phi_r(t) - \phi_0(t) \\ &= -\frac{2}{c}(\omega_0 + 2\alpha t)(r - r_0) + \frac{4\alpha}{c^2}(r^2 - r_0^2) \\ &= -\frac{2}{c}\left[\omega_0 + 2\alpha\left(t - \frac{2r_0}{c}\right)\right](r - r_0) + \frac{4\alpha}{c^2}(r - r_0)^2 \, . \end{aligned} \tag{B.13}$$

Equation B.13 is an exact expression for the phase of the complex video signal obtained at aperture angle θ in terms of the radar parameters and the relative geometry of the target and platform. To see how this function describes a two-dimensional phase history that can be processed into an image of the scene, let us rewrite it in the form

$$\bar{\phi}(\theta, k) = -k(r - r_0) + \frac{4\alpha}{c^2}(r - r_0)^2 \tag{B.14}$$

where

$$k = \left[\frac{2\omega_0}{c} + \frac{4\alpha}{c}\left(t - \frac{2r_0}{c}\right)\right] \tag{B.15}$$

denotes a scaled and offset measure of intra-pulse time t. Now suppose that a certain sample of the video signal, with phase $\bar{\phi}(\theta, k)$, at time index k in a return pulse at aperture angle θ is laid down in a two-dimensional array at *angular coordinate* θ and *radial position* k, as shown in Figure B.3(b). As θ spans the extent of the aperture and as k varies throughout the duration of each return pulse, a polar annulus is swept out in a two-dimensional *phase space*. At this point we have done nothing more than to arranged the data in a particular way in two dimensions without yet defining this so-called phase space relative to the SAR imaging process. We will eventually identify the phase space with the Fourier transform domain corresponding to the scene, because, as we will demonstrate, a two-dimensional Fourier transform of the data laid down as described will reconstruct an image of the scene. But we will also see that the phase history contains certain components in addition to those associated with the Fourier transform of the scene function. So to be precise, the space in which we have placed the data is not, as we stated in the tomographic development, the Fourier domain of the scene. It is, nonetheless, close enough to being so to allow us to use the mathematically convenient and numerically efficient Fourier techniques to form an acceptable image within certain limits that we will identify. Thus in the subsequent development, we will assume that a two-dimensional Fourier transform is used to process the phase history into an image. We will then examine the image-domain effect for each of the components in the phase function.

The precise variation of phase $\bar{\phi}$ with angle θ in the phase history is implied by the geometry of the collection as represented in the relationship (equivalent to the Law of Cosines)

$$r^2 = r_0^2 + a^2 - 2ar_0 \sin(\theta + \gamma) \tag{B.16}$$

where a and γ specify the polar-coordinate location of the target in the scene. If we assume that the dimensions of the scene being imaged are small compared to the

SAR standoff distance, then $a \ll r_0$ and we can write

$$r - r_0 \approx -a \sin(\theta + \gamma) + \frac{a^2}{2r_0} \cos^2(\theta + \gamma) \tag{B.17}$$

and

$$(r - r_0)^2 \approx a^2 \sin^2(\theta + \gamma) . \tag{B.18}$$

With these approximations, the video phase function (Equation B.14) can be written

$$\begin{aligned}
\bar{\phi}(\theta, k) \quad \approx \quad & k a \sin(\theta + \gamma) && (1) \\
& - \frac{k a^2}{2r_0} \cos^2(\theta + \gamma) && (2) \\
& + \frac{4\alpha a^2}{c^2} \sin^2(\theta + \gamma) . && (3)
\end{aligned} \tag{B.19}$$

The first term in Equation B.19 constitutes the ideal phase history for SAR imaging. To see that this is the case, let us rewrite it as follows:

$$\begin{aligned}
\phi_1 &= k a \sin(\theta + \gamma) \\
&= k a (\sin\theta \cos\gamma + \cos\theta \sin\gamma) \\
&= (k \sin\theta)(a \cos\gamma) + (k \cos\theta)(a \sin\gamma) \\
&= x_0' X' + y_0' Y' .
\end{aligned} \tag{B.20}$$

The last line here is identical to the phase term in Equation B.8. This then is the same phase function predicted by the tomographic formulation, which will reconstruct an image of the target when the inverse Fourier transform is computed. If this were the only phase term present, the tomographic paradigm would be completely accurate and Fourier processing could be used without restriction. The presence of the two other terms in Equation B.19 in the phase history, however, impose certain limitations as we will see below. They constitute phase-error terms that have the effect of both defocusing and distorting the formed image.

B.2 THE RANGE CURVATURE EFFECT

The second term in Equation B.19 is a phase-error term attributable to range curvature. Notice that it vanishes in long-range imaging scenarios where $r_0 \rightarrow \infty$ and the

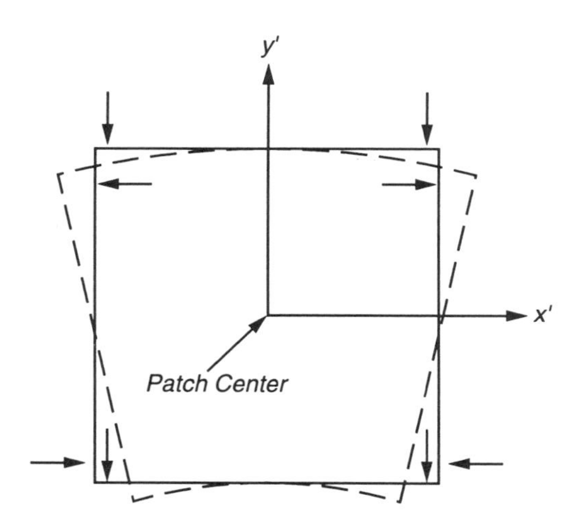

Figure B.4 Wavefront curvature causes image distortion. Because of the Fourier transform shift property, linear phase terms correspond to image translations. The $(x'_0 y'_0 / r_0)X'$ term of Equation B.21 produces a cross-range offset proportional to $x'_0 y'_0$, whereas the $(x'^2_0/(2r_0))Y'$ term produces a range target offset proportional to x'^2_0. The net effect is to distort a square image patch into an "annulus", as shown.

spherical wavefronts become linear. Conversely, it becomes increasingly significant in close-range situations. The effect of this term on the image can be seen by using a Taylor's series expansion in θ, assuming that $\theta \ll \pi/2$ and $k - k_0 \ll k_0$, and converting to rectangular coordinates (X', Y'):

$$\begin{aligned} \phi_2 &= -\frac{ka^2}{2r_0}\cos^2(\theta+\gamma) \\ &\approx -\frac{a^2}{2r_0}\left[\cos^2\gamma - 2\theta\sin\gamma\cos\gamma - \theta^2\cos(2\gamma)\right] \\ &\approx -\frac{x_0^2}{2r_0}Y' + \frac{x'_0 y'_0}{r_0}X' + \frac{x'^2_0 - y'^2_0}{2r_0 k_0}X'^2 . \end{aligned} \tag{B.21}$$

The two terms in Equation B.21 that are linear in X' and Y' give rise to spatial offsets of a point target in the formed image because a linear phase term corresponds to a translation in the Fourier transform domain. Because the magnitudes of these terms depend on target location (x'_0, y'_0), the amounts of the offsets will vary with position in the image and the result will be a geometric distortion of the image, as shown in Figure B.4.

Some amount of geometric distortion can be corrected quite easily once an image is formed, so this phenomenon is not of particular consequence to us in the current development. A potentially more serious degradation of the image results from the quadratic-in-X' term in Equation B.21. To minimize the effect of this phase-error term, we must restrict its amplitude to be no greater than, say, $\pi/4$ radians (see Figure 2.29) for any point in the image. Specifically, the requirement is

$$\frac{x_0'^2 - y_0'^2}{2r_0 k_0} X_1'^2 \leq \frac{\pi}{4} \tag{B.22}$$

where X_1' defines the maximum extent of the aperture ($|X'| \leq X_1'$). Recall now the nominal cross-range resolution of the image formed from such an aperture is $\rho_{x'} = \pi/X_1'$. (This is the half-power width of the ideal response function in Figure 2.29.) Using this relationship and recalling that $k_0 = 2\omega_0/c = 4\pi/\lambda$, we obtain from Equation B.22 the conditions

$$\begin{aligned} \frac{x_0'}{\rho_{x'}} &\leq \sqrt{\frac{2r_0}{\lambda}} \\ \frac{y_0'}{\rho_{x'}} &\leq \sqrt{\frac{2r_0}{\lambda}} \end{aligned} \tag{B.23}$$

for a well-focused image. Thus, the maximum image dimensions in both range y_0' and cross range x_0', relative to the azimuth resolution $\rho_{x'}$, must be restricted to maintain good focus. Notice that these restrictions are most severe at close range (small r_0) and at low frequency (large λ).

B.3 DERAMP RESIDUAL PHASE ERROR

The third term in Equation B.19 represents a phase error resulting from a residual of the deramp process. Recall that this is the term we ignored in the tomographic development. It closely resembles the form of the range-curvature phase-error term ϕ_2, but it depends on different radar and geometric parameters. Therefore, our analysis of this term parallels that in the previous section. By the same methods used to produce Equation B.21 we obtain

$$\begin{aligned} \phi_3 &= \frac{4\alpha a^2}{c^2} \sin^2(\theta + \gamma) \\ &\approx \frac{4\alpha a^2}{c^2} \left[\sin^2 \gamma + 2\theta \sin \gamma \cos \gamma + \theta^2 \cos(2\gamma)\right] \\ &\approx \frac{4\alpha y_0'^2}{c^2} + \frac{8\alpha x_0' y_0'}{k_0 c^2} X' + \frac{4\alpha(x_0'^2 - y_0'^2)}{k_0^2 c^2} X'^2 . \end{aligned} \tag{B.24}$$

As was the case in Section B.2 dealing with range curvature, the linear term in X' gives rise to geometric distortion in the image and the X'^2 term will cause image

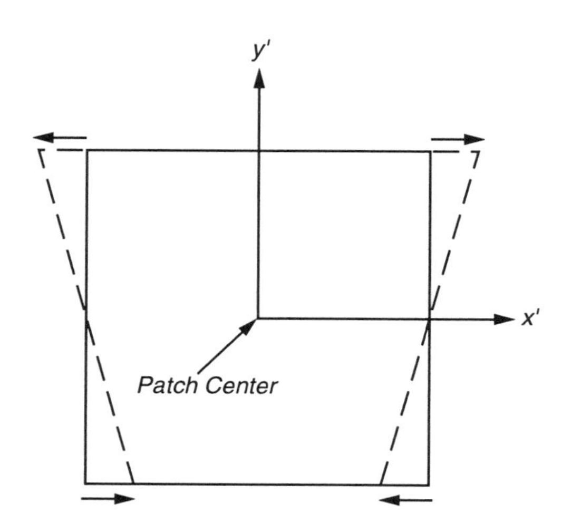

Figure B.5 A deramp residual phase error causes geometric image distortion. The linear phase term $(8\alpha x_0' y_0'/(k_0 c^2))X'$ of Equation B.24 introduces a cross-range target offset proportional to $x_0' y_0'$. Targets in the first and third image quadrants are shifted to the right, while targets in the second and fourth quadrants are shifted left. The net effect is to distort a square image patch into a "keystone" shape as shown.

defocus in the cross-range dimension. The distortion effect from the linear term is illustrated in Figure B.5.

If we again restrict the quadratic phase-error term to an amplitude of $\pi/4$ radians, we must require that

$$\frac{4\alpha(x_0'^2 - y_0'^2)}{k_0^2 c^2} X_1'^2 \leq \frac{\pi}{4} \,. \tag{B.25}$$

With the relationships $\rho_{x'} = \pi/X_1'$ and $k_0 = 4\pi/\lambda$, this inequality leads to the following constraint on image dimensions relative to cross-range resolution for the image to be well-focused:

$$\begin{aligned} \frac{x_0'}{\rho_{x'}} &\leq \frac{\omega_0}{2\sqrt{\pi\alpha}} = \frac{f_0}{\sqrt{\dot{f}}} \\ \frac{y_0'}{\rho_{x'}} &\leq \frac{\omega_0}{2\sqrt{\pi\alpha}} = \frac{f_0}{\sqrt{\dot{f}}} \,. \end{aligned} \tag{B.26}$$

Here, $f_0 = \omega_0/2\pi$ is RF center frequency in Hertz and $\dot{f} = \alpha/\pi$ is chirp rate expressed in Hertz per second.

Finally, we note that a processing step known commonly as " deskew processing" or "deskewing" can be performed to mitigate the phase-errors and distortions described above. However, we will not describe the details of the procedure here.

C

DEVELOPMENT OF A PRACTICAL IMAGE FORMATION GEOMETRY

C.1 INTRODUCTION

We begin with a discussion of a typical SAR imaging geometry and the definition of relevant terms. The geometry is vital because it defines how the SAR "sees" the terrain, which ultimately determines the information content of the imagery. The geometry also establishes a basis for the practical implementation of an image-formation algorithm and makes the calculation of important image phenomena (such as layover direction, shadow angles, grazing and squint angles, etc.), straightforward and quantitative.

C.2 IMAGING GEOMETRY AND COORDINATE SYSTEM DEVELOPMENT

The fundamentals of spotlight-mode SAR and the essence of image formation are easily understood when we view the SAR collection as a tomographic process. The theoretical development of such a view was presented in the Chapter 2. In summary, the tomographic interpretation uses signal processing methods to arrive at the same result others have obtained using traditional radar terminology. For example, each demodulated return of a linear FM (chirp) waveform evaluates the three-dimensional Fourier transform of the scene function along a line in Fourier space. (The scene function referred to here is actually the complex radar-backscatter function for the scene.) A series of such returns spanning the synthetic aperture then evaluates the Fourier transform on a surface (typically a plane) in the three-dimensional Fourier space (Figure C.1). It follows that a two-dimensional inverse Fourier transform of

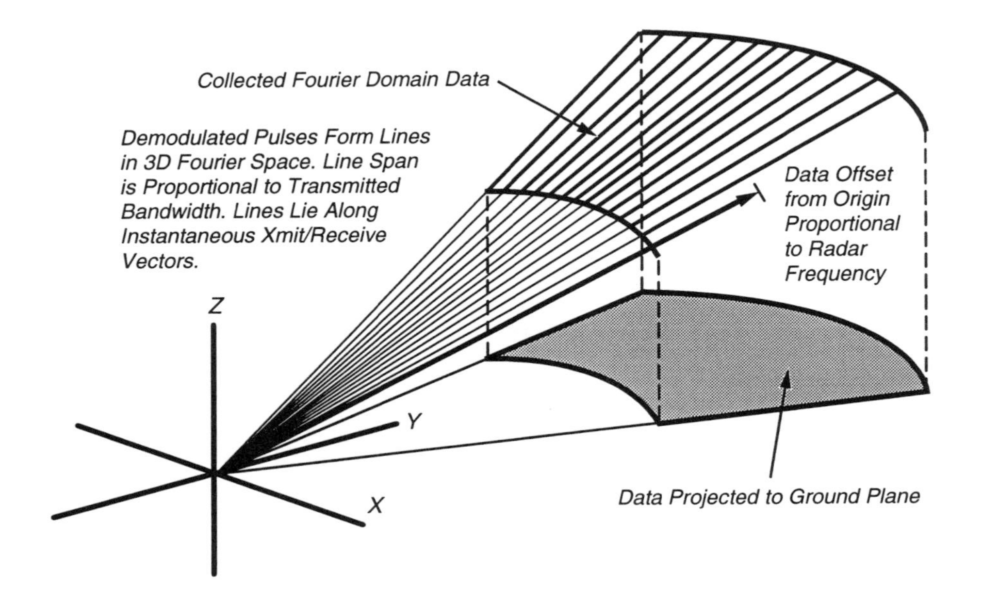

Figure C.1 A series of demodulated spotlight SAR returns spanning the synthetic aperture evaluates the Fourier transform on a surface in 3-D Fourier space.

the collected data will constitute an image of the scene. A further application of tomographic principles establishes that the two-dimensional image obtained here is an orthogonal projection of the three-dimensional scene function onto a plane that contains both the center of the scene and the flight path of the radar platform.

The projection plane described above is commonly called the slant plane, and the image formed in it is a slant-plane image. A slant-plane image is analogous to an optical image obtained by viewing the scene normal to the slant plane, because this involves the same orthogonal projection. If this slant-plane image is back-projected (orthogonal to the slant plane) onto a plane parallel to the nominal plane of the ground surface, the result is a ground-plane image. The ground-plane image is orthographically correct in the case where the scene is perfectly flat. It is equivalent to a view of the plane surface from a vantage point directly above the center of the scene. In the context of interferometry, where it is necessary to produce two essentially identical images from possibly different collection geometries (different slant planes), this orthographic property of ground-plane images has the advantage of making the images independent of the collection geometry, at least for flat terrain.

Real terrain, of course, is not perfectly flat, which makes the situation slightly more complicated. Targets in the scene that lie outside the ground plane appear offset from their true orthographic position, because instead of being projected vertically into the ground plane, they have been projected along lines orthogonal to the slant plane. This phenomenon is commonly referred to as target layover. Layover is an unavoidable phenomenon in all SAR images, regardless of the plane in which they are projected. It presents a significant problem in interferometry when the two images involved exhibit differing amounts of layover, as is the case when the two slant planes are inclined at different angles. Even relatively small differences in flight trajectories can introduce enough differential layover to require some compensation of the effect in the images. Large differences in flight path and/or steeply varying terrain can make SAR interferometry tediously difficult.

One important point to remember, however, is that no matter how data or imagery are projected into various planes, no more nor less information is available because of those projections.

It might be apparent from the previous discussion that one of the most significant and fundamental two-dimensional planes is the nominal *ground plane.* The ground plane (or focus plane) is defined as the plane containing the complex terrain scatterers that the radar ultimately illuminates. For perfectly flat terrain, this definition is unambiguous. For non-planar terrain there is, of course, some ambiguity in this definition. In practice, we choose to specify the ground plane by selecting a surface normal unit vector $\bar{z}$. This unit vector is orthogonal to the ground plane of interest, has an origin at the desired spotlight patch we wish to illuminate, and conveniently defines the z axis of our developing coordinate system.

With a little thought, it should become apparent that the unit normal to the ground plane must be defined in terms of a coordinate system that simultaneously defines the desired imaging location and the instantaneous position of the SAR. For example, an earth-centered coordinate system that encompasses an earth model might be appropriate. In other cases, simple latitude, longitude, and altitude might be sufficient. In many respects, it is irrelevant what this universal coordinate system is because our useful imaging and processing coordinate system will be derived from it. Therefore, we will dispense with further discussion of this universal coordinate system and consider it "hidden" in the mathematics. Just be aware that any coordinate system that we derive will, in some way, relate back to the universal system.

It is fundamentally important that we assume we have at our disposal a collection of *pointing vectors*, $\hat{R}_i$, that define the instantaneous position of the SAR (at each effective transmit/receive location) with respect to the imaged patch center. These pointing vectors are directed from the imaged patch center to the instantaneous position of the SAR at the time the pulse is transmitted. We will use these pointing vectors to establish formally the image-formation geometry and the relevant properties that result.

We must now orient the x and y axes of our coordinate system in the ground plane. As might be expected, the orientation depends precisely on how the radar illuminates the terrain.

Let us begin by specifying a simple but typical SAR collection geometry and see how this geometry helps us define the remainder of our coordinate system. It will also show how other useful coordinate system parameters are developed and related. As the imaging geometry becomes more complex, simple extensions of these concepts allow important acquisition and imaging properties (such as nominal slant plane, depression or grazing angle, squint angle, slope angle, tilt angle, range and cross-range resolutions, scale factors, height-dependent layover, shadow angle, etc.) to be developed.

One of the simplest and most common spotlight SAR acquisition geometries is called a broadside collection and is shown in Figure C.2. The SAR platform is typically carried in an aircraft flying a constant velocity straight line path parallel to the ground patch of interest. The aircraft must illuminate the terrain with sufficient angular diversity to obtain the desired cross-range resolution. During this acquisition, the aircraft traverses the distance L and synthesizes an aperture of this length. The ground-plane projection of the vector pointing from the patch center to the SAR at the midpoint of the aperture defines the direction of the y axis. It is now obvious why this is called a broadside collection. The aircraft velocity vector is parallel to the x axis and the SAR illuminates the terrain in a direction nominally broadside to the direction of travel. Even though the SAR antenna must be physically or electronically steered to illuminate the desired patch center, the fact that it is nominally pointed 90 degrees to the velocity vector qualifies this collection as broadside. This type of collection is also commonly known as zero-degree squint because the antenna is nominally pointed orthogonal to the velocity vector. Nominal forward and backward squint acquisitions are easily imagined from this definition. We will see further ramifications of squinted acquisitions later in this appendix, and we will develop a formal definition of squint angle that does not depend on the flight path.

From this simple broadside acquisition, it is straightforward to develop a useful coordinate system. From our knowledge of where the imaging took place (i.e. patch center), we define (or select) an appropriate ground-plane unit normal vector $\bar{z}$. The unit vector $\bar{y}$ is established by projecting the pointing vector directed from the patch center to the midpoint of the collection ($\hat{R}_{mid}$) into the ground plane. Simple vector

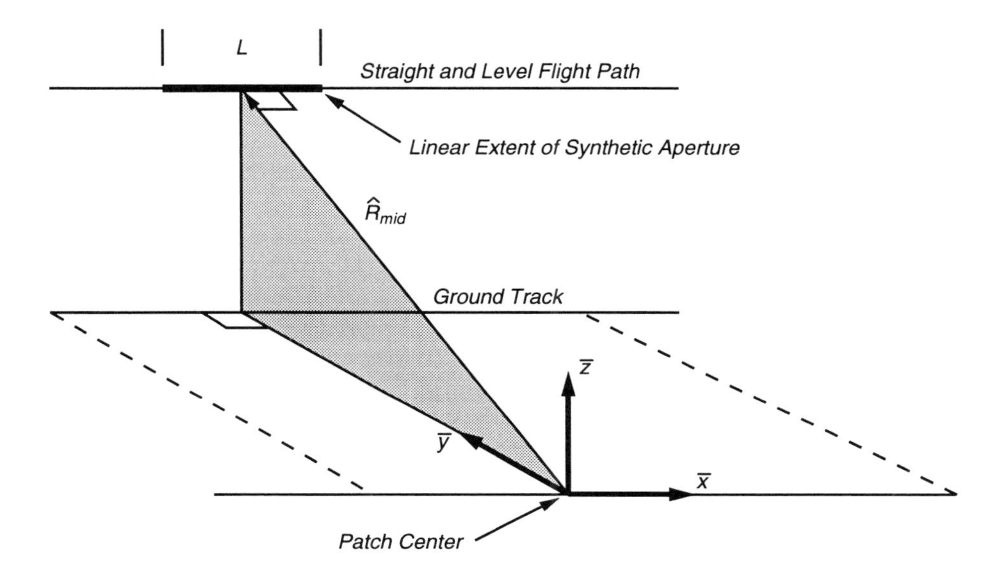

Figure C.2 A simple broadside imaging geometry.

operations show that

$$\bar{y} = \frac{\hat{R}_{mid} - (\hat{R}_{mid} \cdot \bar{z})\bar{z}}{|\hat{R}_{mid} - (\hat{R}_{mid} \cdot \bar{z})\bar{z}|} \tag{C.1}$$

where $\hat{R}$ and $\bar{z}$ are defined in the "hidden" coordinate system units. The unit vector $\bar{x}$ is readily established from the vector cross product of $\bar{y}$ and $\bar{z}$:

$$\bar{x} = \bar{y} \times \bar{z} \ . \tag{C.2}$$

These relations are used whether or not the collection occurred at broadside. Broadside imaging initially makes it easier to see how some simple relationships develop. As a useful mental aid, we associate the "flat" overbar on $\bar{x}, \bar{y}, \bar{z}$ with coordinates in the "flat" ground plane. In the following paragraphs, we will establish the $\hat{x}, \hat{y}, \hat{z}$ unit slant-plane vectors. (The circumflex symbol on these vectors refers to the slant plane.)

We see that with a straight-line flight path, all instantaneous pointing vectors lie in a common plane called the *slant plane*. We can establish a unit normal to this slant plane with a simple procedure. Take any two sufficiently separated pointing vectors and form their cross product. Normalization of this vector and assignment of the proper sign establishes the slant-plane unit normal,

$$\hat{z} = \pm \frac{\hat{R}_i \times \hat{R}_j}{|\hat{R}_i \times \hat{R}_j|} \tag{C.3}$$

where the sign is chosen so that $\hat{z}$ has some component in the same direction as $\bar{z}$. In other words, the sign is chosen to force $\hat{z} \cdot \bar{z} > 0$. Because all pointing vectors lie in the same plane in this case, any two different vectors could be selected and the resulting unit normal would be the same.

In practice, the SAR platform does not usually follow a straight-line path (as is certainly the case with satellite systems or aircraft in turbulent flight). Therefore, we must define a nominal slant plane to approximate the non-planar collection surface. We typically use the pointing vectors that coincide with 20% and 80% of the way through the synthetic aperture to establish $\hat{z}$. (All the pointing vectors can be used in a more sophisticated eigenvector solution to obtain the least-squares unit normal, if desired.)

The slant-plane y-axis unit vector is simply computed from the normalized mid-aperture pointing vector:

$$\hat{y} = \frac{\hat{R}_{mid}}{|\hat{R}_{mid}|} \tag{C.4}$$

while $\hat{x}$ is obtained by the vector cross product,

$$\hat{x} = \hat{y} \times \hat{z} \; . \tag{C.5}$$

Figure C.3 shows the relationship between these unit normal vectors in the broadside imaging condition.

C.3 THE FUNDAMENTAL ANGLES: GRAZING (ψ), SLOPE (ς), AND TILT (φ)

There are three fundamental and useful angles that can now be defined from our established unit vectors. These angles relate important image features and phenomena as we shall see later.

In Figure C.3, the nominal grazing angle ψ is obtained from the unit vectors as

$$\psi = \arccos\left(\hat{y} \cdot \bar{y}\right) , \tag{C.6}$$

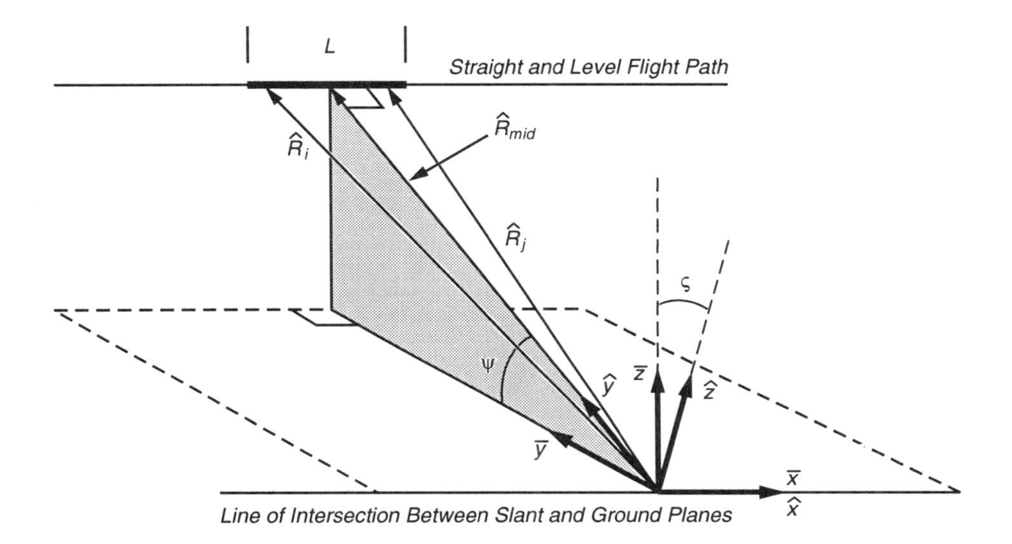

Figure C.3 Slant and ground-plane axes.

which is nothing more than the angle between the ground (focus) plane and the mid-aperture pointing vector (i.e., mid-aperture line-of-sight). The grazing angle affects the image scale factor and resolution in the "range" dimension.

Similarly, we can define a slope angle ς which is the angle between the slant-plane and ground-plane normals:

$$\varsigma = \arccos\left(\hat{z} \cdot \bar{z}\right) . \tag{C.7}$$

The slope angle affects elevated-target layover magnitude. For level-flight broadside imaging the grazing angle and slope angle are the same. We will soon see how these angles change and how other angles develop when the imaging geometry becomes more complex. For example, if the flight path was not level or not broadside, the grazing and slope angles might not be equal. In addition, another important angle called tilt, denoted by φ, comes into play.

Let us consider the slightly more complex imaging geometry shown in Figure C.4. We still have straight and level flight, but now the radar is squinted forward (off broadside) by several degrees. If we establish the two sets of coordinate axes exactly as we did in the broadside case, we see that a rotation exists between the slant-plane and ground-plane axes. Specifically, the slant-plane axes are rotated about the mid-

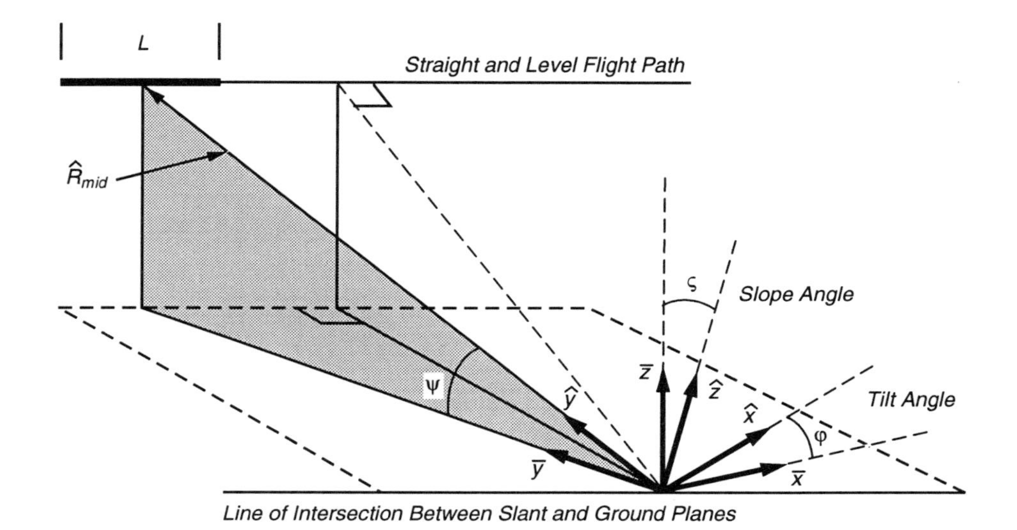

Figure C.4 Squinted imaging causes a rotation of the slant-plane axes around the line-of-sight mid-aperture pointing vector.

aperture pointing vector (which establishes $\hat{y}$) by the angle φ. Mathematically, this tilt angle is easily computed as

$$\varphi = \arccos(\hat{x} \cdot \bar{x}) \tag{C.8}$$

or (as we shall derive later)

$$\cos\varphi = \frac{\cos\varsigma}{\cos\psi} . \tag{C.9}$$

Because tilt can be positive or negative, we establish a consistent sign (while looking in the direction of $\hat{y}$, counterclockwise is positive) with the following equation:

$$(\hat{x} - \tan\varphi\hat{z}) \cdot \bar{y} = 0 . \tag{C.10}$$

Thus

$$\tan\varphi = \frac{\hat{x} \cdot \bar{y}}{\hat{z} \cdot \bar{y}} . \tag{C.11}$$

The tilt angle affects the image scale factor and resolution in the "cross-range" dimension. In addition, we see that the grazing angle ψ and the slope angle ς defined by Equations C.6 and C.7, respectively, are no longer equal. Notice that even though the radar is looking at the patch center with a slight squint angle, the

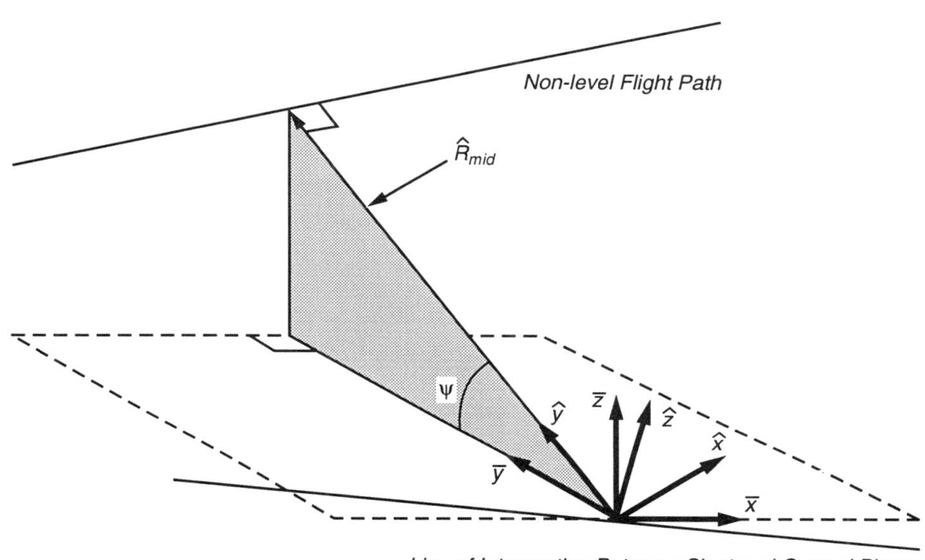

Figure C.5 Effect of non-level flight on coordinate axes.

flight path is the same as in the broadside case. Therefore, the slant plane has not changed. The slope angle and the line of intersection between the slant and ground planes also remain unchanged.

Figure C.5 shows the imaging geometry for an inclined flight path. Note that this situation results in a non-zero tilt angle even at broadside. Thus it can be seen that any combination of non-level-flight and/or squinted geometry generally causes coordinate system tilt and rotation that have direct bearing on the properties of the images produced. In summary, the fundamental angles between the respective slant and ground-plane axes are given by Equations C.6, C.7, and C.8 (or C.11).

C.4 ESTABLISHING THE GENERALIZED SQUINT ANGLE (θ)

It is easy to imagine that the same slant plane can be swept out by an infinite number of flight paths. Therefore, it is important to decouple the flight path (velocity vector) from a formal definition of squint angle. We accomplish this objective by defining a squint angle θ separately in the slant plane and ground plane as follows. Let us first define a vector $\bar{i}$ that specifies the line-of-intersection between the slant and ground planes. From our established slant-plane and ground-plane unit vectors, we obtain

$$\bar{i} = \frac{\bar{z} \times \hat{z}}{|\bar{z} \times \hat{z}|} . \tag{C.12}$$

Now, we establish two unit vectors, $\hat{l}$ and $\bar{l}$, in the slant and ground planes, respectively, that are directed perpendicular to the line-of-intersection of the slant and ground planes. Mathematically, $\hat{l}$ and $\bar{l}$ are perpendicular to $\bar{i}$ and are obtained from

$$\bar{l} = \bar{z} \times \bar{i} \tag{C.13}$$

$$\hat{l} = \hat{z} \times \bar{i} . \tag{C.14}$$

Finally, we define the slant-plane squint angle θ_s as the angle between $\hat{y}$ and $\hat{l}$ (clockwise is positive as viewed from above):

$$\tan \theta_s = -\frac{\hat{y} \cdot \bar{i}}{\hat{y} \cdot \hat{l}} . \tag{C.15}$$

Similarly, the ground-plane squint angle θ_g is obtained from

$$\tan \theta_g = -\frac{\bar{y} \cdot \bar{i}}{\bar{y} \cdot \bar{l}} . \tag{C.16}$$

Note that the flight path is not involved in the squint-angle definitions in Equations C.15 and C.16. Any flight path in a particular slant plane will exhibit the same squint angle as any other flight path in the same plane that subtends the same aperture. A positive-valued, generalized squint angle as defined here is consistent with an actual straight and level acquisition when the SAR is illuminating off the right side of the platform and ahead of the nominal broadside direction.

The squint angle defined above is equal to the *layover angle*, defined as the angle at which out-of-plane targets are projected in the formed image. Thus, a positive squint angle implies a positive layover angle, whereby an elevated target projects positive x and y layover components. Figure C.6 summarizes some of the important vectors and angles discussed above. A simulation involving an example of layover resulting from positive squint is shown in Figure C.9 described in Section C.7 below.

C.5 COMPUTING THE SHADOW ANGLE (ζ)

With our now formally defined coordinate axes, we can calculate another useful angle in image analysis called the shadow angle. We will define the shadow angle

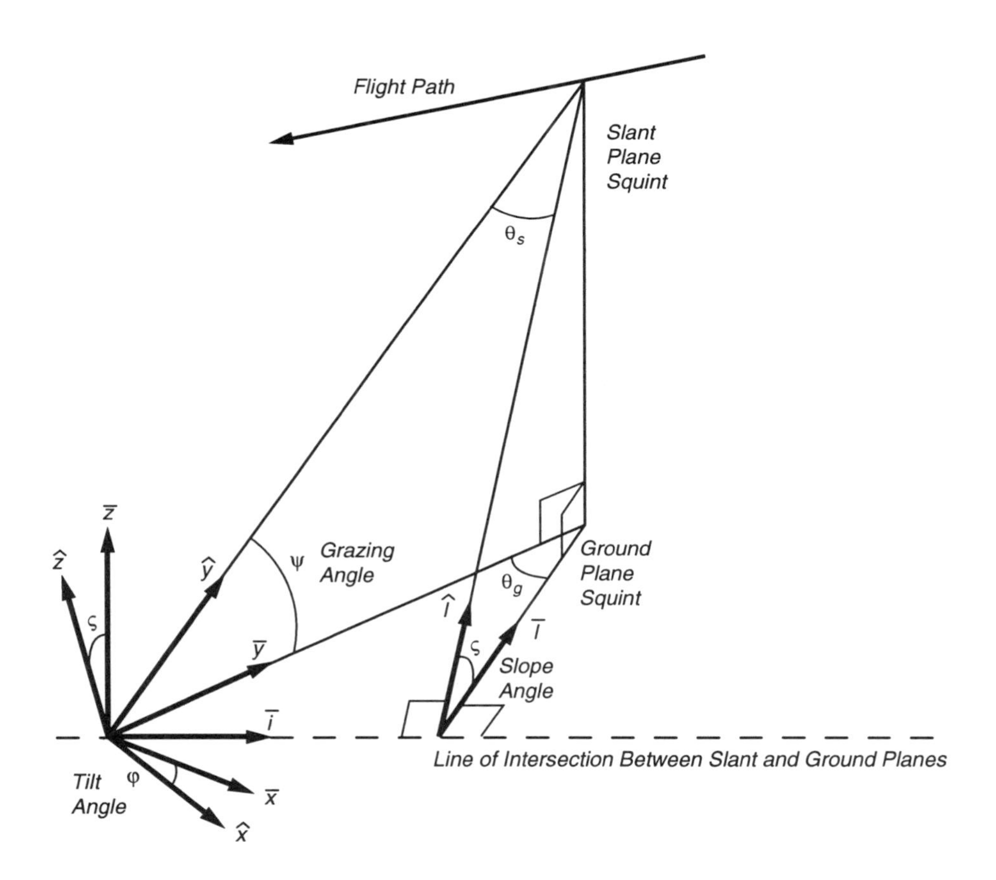

Figure C.6 Summary of important unit vectors and angles.

ζ as the angle in the image (with respect to the x axis) one must follow from the tip of a shadow to intersect the object that caused the shadow. Vector analysis makes the development straightforward. At first, this may appear to be a somewhat esoteric concept. However, it can certainly be imagined that with squinted imagery and seemingly strange projection effects of elevated targets, it might be useful to know the direction from the tip of a shadow one must go to find the "laid-over" object that caused it. We will illustrate the utility of our vector analysis in solving this seemingly difficult problem with ease.

We will develop the shadow angles for both slant-plane and ground-plane imagery.[1] Without loss of generality, let an elevated object be located at the tip of the ground-plane unit normal vector $\bar{z}$. We now ask: where does such an elevated object project into the slant plane? With the help of Figure C.7(a), we find a vector $\hat{p}$ that points to the projected position of the elevated target:

$$\hat{p} = \bar{z} - (\bar{z} \cdot \hat{z})\hat{z} \ . \tag{C.17}$$

Similarly, we can find where $\bar{z}$ projects into the ground plane (see Figure C.7b). For some value of k, the following must be true:

$$\bar{p} = \bar{z} - k\hat{z} \tag{C.18}$$

therefore

$$[\bar{z} - k\hat{z}] \cdot \bar{z} = 0 \ . \tag{C.19}$$

Expanding the above equation we find

$$(\bar{z} \cdot \bar{z}) - k(\bar{z} \cdot \hat{z}) = 0 \ . \tag{C.20}$$

Thus

$$k = \frac{1}{(\bar{z} \cdot \hat{z})} \ . \tag{C.21}$$

Finally

$$\bar{p} = \bar{z} - \frac{\hat{z}}{\bar{z} \cdot \hat{z}} \ . \tag{C.22}$$

We now have the vectors $\hat{p}$ and $\bar{p}$ that locate the projected position of the elevated target in the slant and ground planes, respectively. The components of $\hat{p}$ or $\bar{p}$ with respect to the slant or ground-plane x and y axes, respectively, are the fractional cross-range and range components of height-dependent layover. We will return to this point later. We now must find the corresponding location of the shadow of the

[1] Equal range and cross-range image scale factors are assumed in this analysis. If imagery is processed with unequal range and cross-range scale factors (i.e., non-square pixels), appropriate scale factors must be incorporated in the vector analysis.

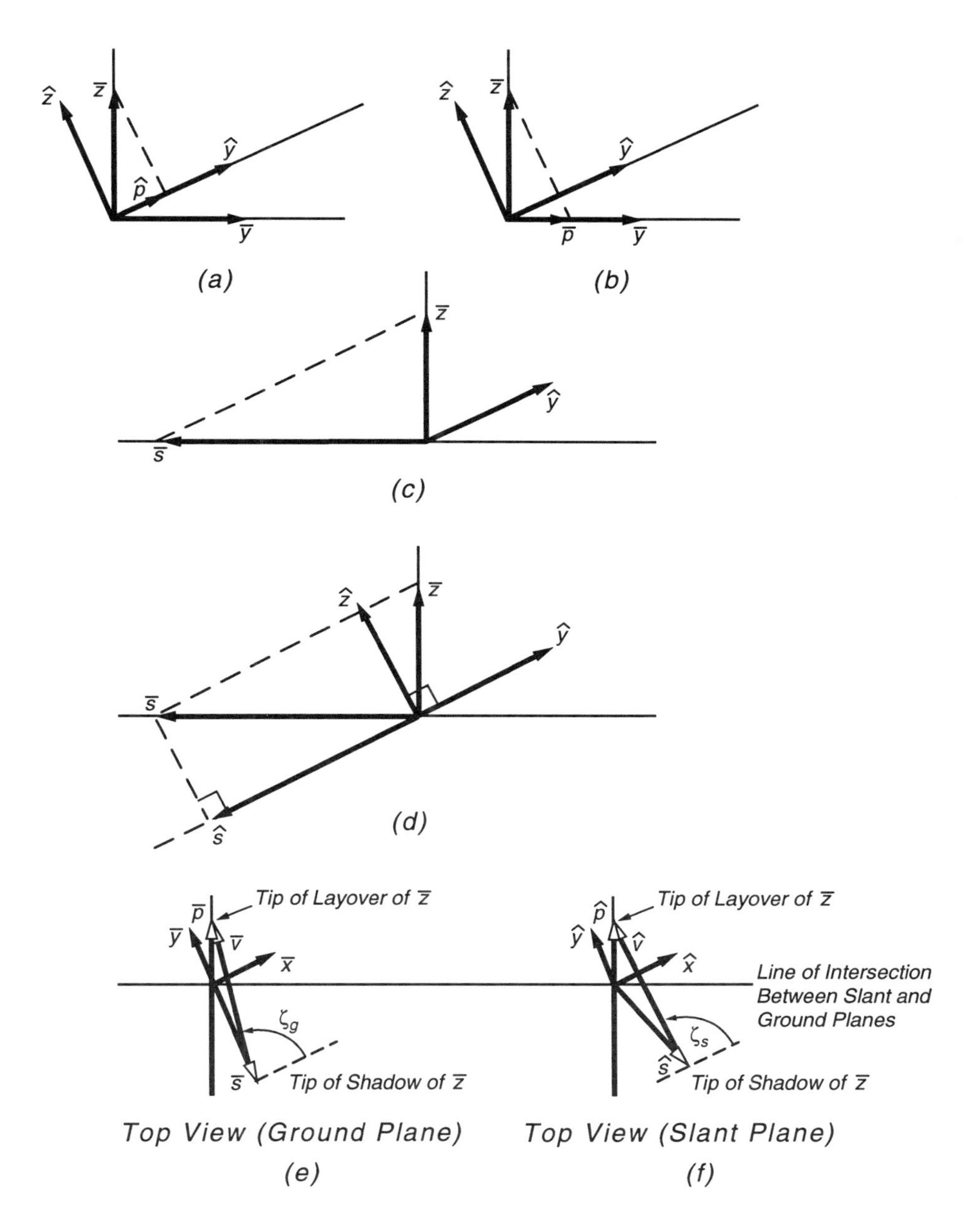

Figure C.7 Shadow geometry sketches for computing slant-plane and ground-plane shadow angles. See text for explanation.

elevated object. With the help of Figure C.7(c), we find the shadow vector $\bar{s}$ in the ground plane is of the form

$$\bar{s} = \bar{z} - k\hat{y} \ . \qquad (C.23)$$

(Remember that shadows are cast in the direction of $-\hat{y}$ in 3−space. Where the shadow falls on the ground plane depends on the geometry.) Because, for some value of k,

$$[\bar{z} - k\hat{y}] \cdot \bar{z} = 0 \qquad (C.24)$$

we find that

$$k = \frac{1}{(\bar{z} \cdot \hat{y})} \qquad (C.25)$$

and

$$\bar{s} = \bar{z} - \frac{\hat{y}}{\bar{z} \cdot \hat{y}} \ . \qquad (C.26)$$

Similarly, with reference to Figure C.7(d), we see that

$$\hat{s} \cdot \hat{z} = 0 \ . \qquad (C.27)$$

For some value of k we have

$$\hat{s} = \bar{s} - k\hat{s} \ . \qquad (C.28)$$

Carrying out the math we see that

$$(\bar{s} - k\hat{z}) \cdot \hat{z} = 0 \qquad (C.29)$$

and

$$k = \bar{s} \cdot \hat{z} \ . \qquad (C.30)$$

Therefore, we obtain the shadow vector in the slant plane:

$$\hat{s} = \bar{s} - (\bar{s} \cdot \hat{z})\hat{z} \ . \qquad (C.31)$$

Finally, (see Figures C.7e and f) we compute the shadow-angle direction unit vectors $\bar{v}$ and $\hat{v}$ and the shadow angles ζ_g and ζ_s for the ground and slant planes, respectively:

$$\bar{v} = \frac{\bar{p} - \bar{s}}{|\bar{p} - \bar{s}|} \qquad (C.32)$$

$$\hat{v} = \frac{\hat{p} - \hat{s}}{|\hat{p} - \hat{s}|} \qquad (C.33)$$

$$\zeta_s = \arccos(\hat{v} \cdot \hat{x}) \qquad (C.34)$$

$$\zeta_g = \arccos(\bar{v} \cdot \bar{x}) \ . \qquad (C.35)$$

It is easily verified from the above mathematics that the *slant*-plane shadow angle ζ_s is *always* 90 degrees.

Figure C.8 shows two SAR images of the Solar Power Tower at Sandia National Laboratories gathered with squinted geometry. One is processed in the slant plane and the other is processed in the ground plane. The array of heliostats is evident along with the shadow cast by the 60-meter-high tower containing the solar-energy heat-exchange apparatus. The graphic overlay indicates the tip of the shadow cast by the tower. It also shows the shadow angle, which indicates the direction that must be followed to intersect the laid-over structure at the top of the tower that caused the shadow.

C.6 CONNECTION BETWEEN THE SLANT PLANE ANGLES η AND ψ AND THE SLOPE, GRAZING, AND TILT ANGLES

In the main text (see Section 2.5) the complex spotlight-mode SAR image was reconstructed in the ground plane by evaluating the phase history on a slant plane through the origin described by

$$Z = \alpha X + \beta Y \tag{C.36}$$

where $\alpha = \tan\eta$ and $\beta = \tan\psi$ as shown in Figure 2.28.

In terms of our established unit vectors, we can obtain a vector normal to this surface. From analytic geometry, this normal vector is given by

$$\hat{n} = -\alpha\bar{x} - \beta\bar{y} + \bar{z} \; . \tag{C.37}$$

Normalizing this vector forms our slant-plane unit normal

$$\hat{z} = \frac{\hat{n}}{|\hat{n}|} = \frac{-\alpha\bar{x} - \beta\bar{y} + \bar{z}}{\sqrt{\alpha^2 + \beta^2 + 1}} \; . \tag{C.38}$$

We have already established that the dot product between the slant-plane and ground-plane normals yields the cosine of the slope angle; thus

$$\hat{z} \cdot \bar{z} = \cos\varsigma = \gamma = \frac{1}{\sqrt{\alpha^2 + \beta^2 + 1}} \; . \tag{C.39}$$

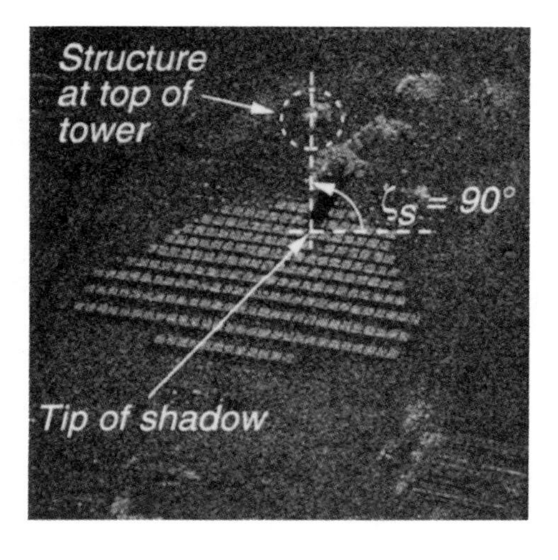

(a)

(c)

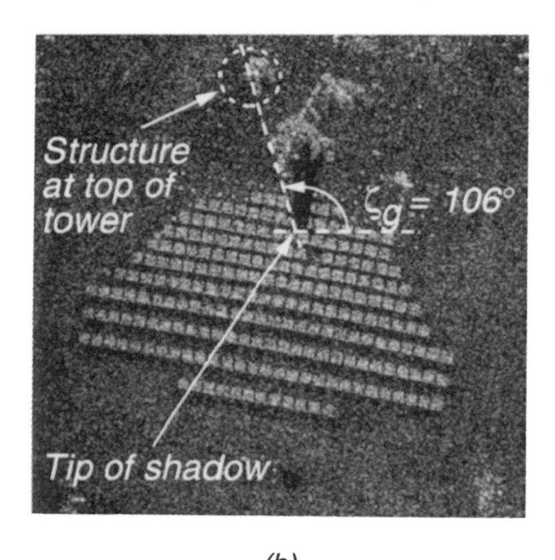

(b)

Figure C.8 A SAR image, gathered with squinted geometry, of the Solar Power Tower at Sandia National Laboratories. The array of heliostats is evident along with the shadow cast by the 60-meter-high tower containing the solar-energy heat-exchange apparatus. The graphic overlay indicates the tip of the shadow cast by the tower. It also shows the shadow angle, which indicates the direction that must be followed to intersect the laid-over structure at the top of the tower that caused the shadow. Other relevant imaging parameters are: (1) Nominal grazing angle $\psi = 45.0918$ degrees, (2) slant-plane squint angle $\theta_s = -20.5087$ degrees, (3) ground-plane squint angle $\theta_g = -29.7532$ degrees, (4) slope angle $\varsigma = 49.1273$ degrees, and (5) tilt angle $\varphi = -22.0403$ degrees. (a) Slant-plane image. Note that the shadow angle, ζ_s, is precisely 90 degrees. (b) Ground-plane image. The computed ground-plane shadow angle $\zeta_g = 106$ degrees. (c) Aerial photo of Power Tower.

Incorporating Equation C.39 into Equation C.38 yields

$$\hat{z} = \gamma(-\alpha\bar{x} - \beta\bar{y} + \bar{z}) \ . \tag{C.40}$$

The slant-plane unit vector $\hat{y}$ (see Figure 2.28) is established as

$$\hat{y} = \cos\psi\bar{y} + \sin\psi\bar{z} \ . \tag{C.41}$$

This now allows us to establish $\hat{x}$ as

$$\hat{x} = \hat{y} \times \hat{z} \ . \tag{C.42}$$

If we incorporate Equations C.40 and C.41 into Equation C.42 and expand the cross product, we obtain

$$\hat{x} = \gamma[(\cos\psi + \beta\sin\psi)\bar{x} - (\alpha\sin\psi)\bar{y} + (\alpha\cos\psi)\bar{z}] \ . \tag{C.43}$$

We have previously defined the dot product between $\hat{x}$ and $\bar{x}$ as the cosine of the tilt angle; thus, (using Equation C.43) we obtain

$$\hat{x} \cdot \bar{x} = \cos\varphi = \gamma(\cos\psi + \beta\sin\psi) \ . \tag{C.44}$$

Applying $\gamma = \cos\varsigma$ and $\beta = \tan\psi$ to Equation C.44 produces the fundamental relationship (given previously in Equation C.9) between the grazing, slope, and tilt angles

$$\cos\varphi = \frac{\cos\varsigma}{\cos\psi} \ . \tag{C.45}$$

If we now manipulate Equation C.39 and use Equation C.45 where required, we obtain three fundamental equations relating the angle η (Figure 2.28) to any two of the fundamental angles of slope, grazing, and tilt

$$\begin{aligned} \tan^2\eta &= \tan^2\varsigma - \tan^2\psi \\ \tan\eta &= \sin\varphi/\cos\varsigma \\ \tan\eta &= \tan\varphi/\cos\psi \ . \end{aligned} \tag{C.46}$$

From Equations C.45 and C.46, we see that if the slope and grazing angles are equal, there is no tilt; therefore, $\eta = 0$ as well.

C.7 COMPUTING THE FRACTIONAL RANGE AND CROSS-RANGE LAYOVER COMPONENTS IN THE GROUND PLANE

In Section C.5, we computed a vector $\bar{p}$ (Equation C.22), which located the ground-plane-projected position of a target elevated above the origin by one unit. Consequently, the components of $\bar{p}$ in the $\bar{x}$ and $\bar{y}$ directions constitute the fractional cross-range and range layovers, respectively. The fractional cross-range layover component is given by

$$\begin{aligned}\frac{\Delta x}{h} &= \bar{p} \cdot \bar{x} \\ &= (\bar{z} - \frac{\hat{z}}{\hat{z} \cdot \bar{z}}) \cdot \bar{x}\end{aligned} \tag{C.47}$$

but, with $\hat{z} \cdot \bar{z} = \gamma = \cos\varsigma$, and $\hat{z}$ given by Equation C.40, we obtain

$$\begin{aligned}\frac{\Delta x}{h} &= \frac{-\hat{z} \cdot \bar{x}}{\gamma} \\ &= \alpha = \tan\eta\ .\end{aligned} \tag{C.48}$$

Similarly, we find the fractional range layover component in the ground plane as

$$\frac{\Delta y}{h} = \bar{p} \cdot \bar{y} = \frac{-\hat{z} \cdot \bar{y}}{\gamma} = \beta = \tan\psi\ . \tag{C.49}$$

Equations C.48 and C.49 agree with those previously presented in Equation 2.73. Figure C.9 shows a ground-plane synthetic image acquired with squinted geometry. The amount of range and cross-range layover for one of the two elevated targets is highlighted.

C.8 COMPUTING THE SLANT-PLANE IMAGE SHEAR ANGLE (ξ)

Another angle that occurs often in image analysis is called the shear angle, denoted by ξ. This angle describes a slant-plane image distortion phenomenon where a target appears offset (sheared) in cross range by an amount proportional to its position in range. Figure C.10 illustrates the apparent shearing of a square array of targets when viewed in the slant plane and acquired with squinted geometry. Transforming a slant-plane image into a ground-plane image requires that the slant-plane image first be warped to remove the shear. The image must then be stretched in the x and y dimensions by the reciprocal of the cosine of the tilt and grazing angles, respectively.

The shear angle can be computed from simple vector operations.[2] The ground-plane unit vector $\bar{y}$ projects orthogonally into the slant plane to form the vector $\hat{w}$ where

$$\hat{w} = \bar{y} - (\bar{y} \cdot \hat{z})\hat{z}\ . \tag{C.50}$$

[2] Again, equal range and cross-range image scale factors are assumed in this analysis.

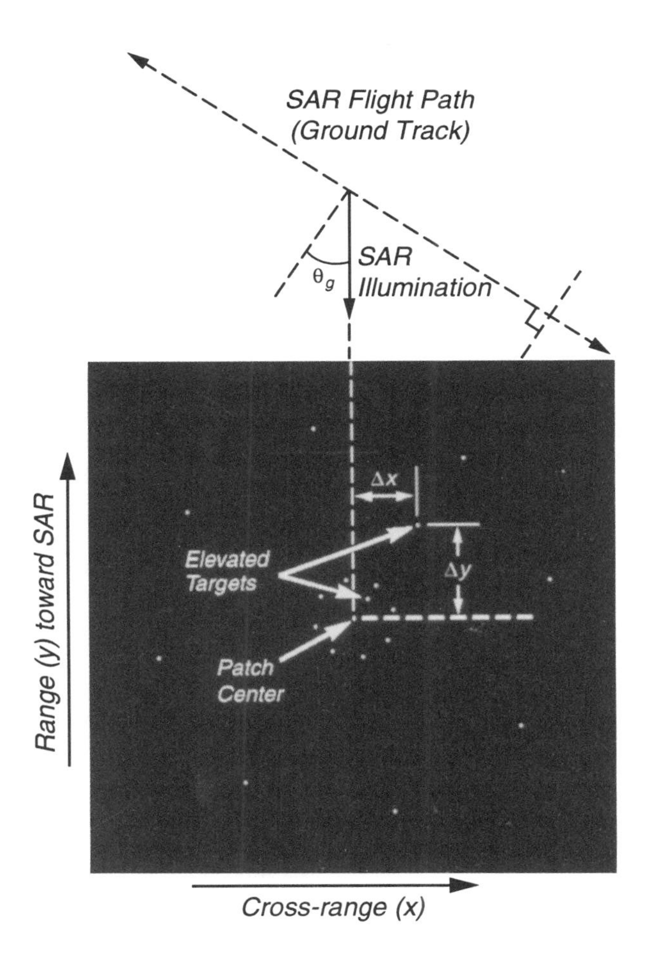

Figure C.9 A synthetically generated ground-plane SAR image containing ground-plane targets and two elevated targets acquired with a slant-plane squint angle $\theta_s = 30$ degrees. The synthetic target parameters are the same as in Figure 3.57. The amount of range (Δy) and cross-range (Δx) layover of one of the two elevated targets are highlighted. For this example: (1) slope angle $\varsigma = 30$ degrees, (2) ground-plane squint $\theta_g = 33.69$ degrees, (3) grazing angle $\psi = 25.66$ degrees, (4) tilt angle $\varphi = 16.1$ degrees, (5) $\Delta y = h \tan \psi = 50 \tan 25.66 = 24.02$ meters, (6) $\Delta x = h \tan \eta = 50 \tan 17.76 = 16.01$ meters.

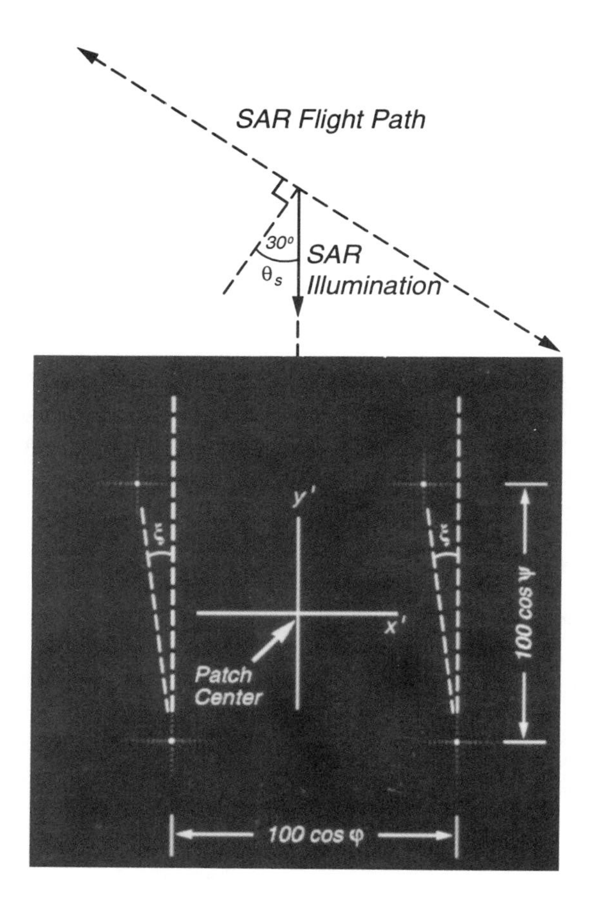

Figure C.10 The apparent shearing of a square array of ground-plane targets when viewed in the slant plane and acquired with squinted geometry. The imaging parameters for this collection are: (1) slant-plane squint angle $\theta_s = 30.0$ degrees, (2) grazing angle $\psi = 25.66$ degrees, (3) slope angle $\varsigma = 30.0$ degrees, and (4) tilt angle $\varphi = 16.1$ degrees. The computed shear angle, $\xi = 7.59$ degrees, is shown on the figure. Transforming a slant-plane image into a ground-plane image requires that the slant-plane image first be warped to remove the shear. The image must then be stretched in the x and y dimensions by the reciprocal of the cosine of the tilt and grazing angles, respectively.

The shear angle in the slant plane is the angle that $\hat{w}$ makes with respect to the slant-range y-axis vector $\hat{y}$. Simple mathematics (with counterclockwise positive as seen from above) yields

$$\begin{aligned} \tan\xi &= -\frac{\hat{w}\cdot\hat{x}}{\hat{w}\cdot\hat{y}} \\ &= -\frac{\bar{y}\cdot\hat{x}}{\bar{y}\cdot\hat{y}} \,. \end{aligned} \tag{C.51}$$

With $\alpha = \tan\eta$, $\gamma = \cos\varsigma = \sin\varphi/\tan\eta$, and $\bar{y}\cdot\hat{x} = -\alpha\gamma\sin\psi$, we obtain the slant-plane shear angle in terms of the grazing angle and tilt angles as

$$\xi = \arctan\left(\tan\psi\sin\varphi\right) . \tag{C.52}$$

It has been demonstrated that formal vector analysis from the established coordinate system unit vectors facilitates the calculation of many useful image and projection quantities. All the previously developed vectors and angles are easily computed during the image-formation process. Subsequently, they can be made available to aid in the interpretation of the image. Equal range and cross-range image scale factors were assumed in all previous analyses. If imagery is processed with unequal range and cross-range scale factors (i.e., non-square pixels), appropriate scale factors must be incorporated into the vector analysis. Any measurements of angles or placement of vectors in the processed imagery must accommodate the scale factors.

Section C.9 summarizes the various angles and vectors implied by the imaging geometry.

C.9 SUMMARY

- Three fundamental angles:

  1. $\hat{x} \cdot \bar{x} = \cos \varphi$, $\tan \varphi = \dfrac{\hat{x} \cdot \bar{y}}{\hat{z} \cdot \bar{y}}$ (tilt) (This angle affects cross-range image scale factor and resolution.)
  2. $\hat{y} \cdot \bar{y} = \cos \psi$ (grazing) (This angle affects range image scale factor and resolution.)
  3. $\hat{z} \cdot \bar{z} = \cos \varsigma$ (slope) (This angle affects elevated target layover magnitude.)

- Relationship between fundamental angles:

  1. $\cos \varphi = \cos \varsigma / \cos \psi$

- Relationship of η to fundamental angles:

  1. $\tan^2 \eta = \tan^2 \varsigma - \tan^2 \psi$
  2. $\tan \eta = \sin \varphi / \cos \varsigma$
  3. $\tan \eta = \tan \varphi / \cos \psi$

- Slant-plane vectors:

  1. $\hat{x} = \gamma[(\cos \psi + \beta \sin \psi)\bar{x} - (\alpha \sin \psi)\bar{y} + (\alpha \cos \psi)\bar{z}]$
  2. $\hat{y} = \cos \psi \bar{y} + \sin \psi \bar{z}$
  3. $\hat{z} = \gamma(-\alpha \bar{x} - \beta \bar{y} + \bar{z})$

- Definition of α, β, γ

  1. $\alpha = \tan \eta$
  2. $\beta = \tan \psi$
  3. $\gamma = \cos \varsigma$

- Line-of-intersection (between slant-plane and ground-plane) unit vector:

  1. $\bar{i} = \dfrac{\bar{z} \times \hat{z}}{|\bar{z} \times \hat{z}|} = \dfrac{\beta \bar{x} - \alpha \bar{y}}{\sqrt{\alpha^2 + \beta^2}}$

- Slant-plane and ground-plane slope vectors:

  1. $\hat{l} = \hat{z} \times \bar{i} = \dfrac{\gamma[\alpha \bar{x} + \beta \bar{y} + (\alpha^2 + \beta^2)\bar{z}]}{\sqrt{\alpha^2 + \beta^2}}$
  2. $\bar{l} = \bar{z} \times \bar{i} = \dfrac{\alpha \bar{x} + \beta \bar{y}}{\sqrt{\alpha^2 + \beta^2}}$

- Generalized slant-plane and ground-plane squint angles:

  1. $\tan\theta_s = -\frac{\hat{y}\cdot\bar{i}}{\hat{y}\cdot\hat{l}}$, or $\sin\theta_s = -(\hat{y}\cdot\bar{i})$
  2. $\tan\theta_g = -\frac{\bar{y}\cdot\bar{i}}{\bar{y}\cdot\bar{l}} = \frac{\tan\eta}{\tan\psi}$, or $\sin\theta_g = -(\bar{y}\cdot\bar{i})$
  3. $\sin\theta_s = \cos\psi\sin\theta_g$

- Slant-plane shear angle:

  1. $\tan\xi = \tan\psi\sin\varphi$

- Slant-plane and ground-plane shadow vectors:

  1. $\hat{s} = \bar{s} - (\bar{s}\cdot\hat{z})\hat{z}$
  2. $\bar{s} = \bar{z} - \frac{\hat{y}}{\bar{z}\cdot\hat{y}}$

- Slant-plane and ground-plane layover vectors:

  1. $\hat{p} = \bar{z} - (\bar{z}\cdot\hat{z})\hat{z}$
  2. $\bar{p} = \bar{z} - \frac{\hat{z}}{\bar{z}\cdot\hat{z}}$

- Slant-plane and ground-plane shadow angle direction unit vectors:

  1. $\hat{v} = \frac{\hat{p}-\hat{s}}{|\hat{p}-\hat{s}|}$
  2. $\bar{v} = \frac{\bar{p}-\bar{s}}{|\bar{p}-\bar{s}|}$

- Slant-plane and ground-plane shadow angles:

  1. $\zeta_s = \arccos(\hat{v}\cdot\hat{x}) = 90$ degrees (always).
  2. $\zeta_g = \arccos(\bar{v}\cdot\bar{x})$

- Ground-plane fractional range and cross-range layover components:

  1. $\frac{\Delta y}{h} = \tan\psi$
  2. $\frac{\Delta x}{h} = \tan\eta$

- Ground-plane fractional layover magnitude:

  1. $\sqrt{(\frac{\Delta y}{h})^2 + (\frac{\Delta x}{h})^2} = \tan\varsigma$

D

DEVELOPMENT OF A SYNTHETIC TARGET GENERATOR

D.1 INTRODUCTION

The synthetic target generator illustrates the importance of having a known dataset that embodies all the important features of actual SAR data from which an image formation algorithm can be tested. Indeed, if one cannot successfully process synthetically generated data, one surely cannot process real data.

D.2 SYNTHETIC TARGET GENERATOR

A useful tool to have in a collection of SAR signal processing codes is a *synthetic target generator*. The target-generator code emulates the motion of the SAR, and mathematically transmits, receives, and demodulates the returns of a number of point reflectors located in a volume of three-dimensional space. It creates a two-dimensional phase history from a collection of point targets that closely matches the signals that an actual SAR would obtain under the same circumstances.

The synthesized dataset can then be processed in various ways to quantify the image properties obtained by a number of image-formation methods. Because the dataset generation is entirely under the operator's control, it removes any unknown properties of the data from the image-formation process. Image-formation code design and code debugging are greatly facilitated by having a known dataset as input. Our experience has verified the utility of such a synthetic target generator. It would have been difficult, if not impossible, to design a robust image-formation code and to analyze image properties (i.e., resolution, scale factors, distortions, defocus, etc.) if the only driving input was data from an actual SAR collection.

In addition, the synthetic target generator allows creation and display of simple and ideal signals that precisely depict those manifestations that ultimately limit image quality and SAR performance. Clearly, this capability is unavailable with an actual SAR collection.[1]

D.2.1 Synthetic Target Generator Algorithm Overview

The heart of the synthetic target generator is the ideal phase from a point scatterer at a specified location in space as seen by the SAR. This instantaneous phase was derived in Equation B.13 and is repeated here for immediate access:

$$\begin{aligned} \phi(t) &= \phi_r(t) - \phi_0(t) \\ &= -\frac{2}{c}(\omega_0 + \dot{\omega} t)(r - r_0) + \frac{2\dot{\omega}}{c^2}(r^2 - r_0^2) \\ &= -\frac{2}{c}\left[\omega_0 + \dot{\omega}\left(t - \frac{2r_0}{c}\right)\right](r - r_0) + \frac{2\dot{\omega}}{c^2}(r - r_0)^2 . \end{aligned} \tag{D.1}$$

Here r_0 is the instantaneous range from the SAR to the imaged patch center; r is the instantaneous range to a point target in the scene; and $\dot{\omega}$ is the radian FM chirp rate: $\dot{\omega} = 2\pi \dot{f} =' 2\alpha$. ($\alpha$ is the chirp-rate parameter used in Chapters 1 and 2.) The time variable, t, is confined to the transmitted pulse duration, $|t| \leq T/2$, where T is the pulse length. This time variable, t, will be called *fast time* because it represents the time along the pulse. During this fast time interval, the SAR is essentially stationary. This precludes the use of fast time to clock the motion of the SAR. A *slow time* variable that allows us to mark the motion of the SAR platform is implicitly embedded in the relation between platform velocity and pulse repetition frequency (PRF).

We now create a constant-amplitude complex video signal from the point target as

$$s(t) = Ae^{j\phi(t)} \tag{D.2}$$

where A is the target reflectivity as seen by the radar. (A is a complex constant independent of instantaneous frequency within the pulse, and independent of viewing angle within the aperture generation interval.)

The signal generated from a collection of individual point targets follows the principle of superposition giving

$$S(t) = \sum_i A_i e^{j\phi_i(t)} \tag{D.3}$$

[1] Appendix E gives examples of generated signals with typical radar parameters selected to highlight important signal properties that directly affect image quality and ultimate SAR performance.

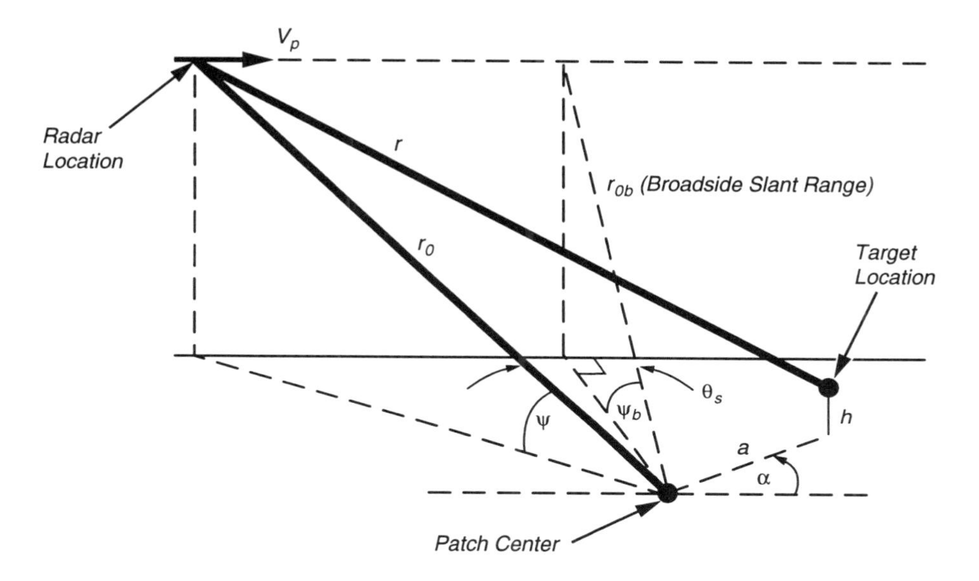

Figure D.1 Geometry for the spotlight synthetic target generator. The SAR platform flight path is specified by sequential (x, y, z) positions in three-dimensional space from which transmission and reception occur. Any number of point targets can be located in a volume of space. This is true whether or not the target signatures can be adequately supported by the chosen sampling parameters.

where the index i refers to the ith target.

Equation D.3 is the basis of synthetic target generation. Now it is necessary to show how actual target locations, scattering amplitudes, radar parameters, and discrete time sampling get embedded in this equation.

Let us begin by defining a collection geometry that can serve as a basis for specifying a realistic SAR imaging scenario. (See Figure D.1.) For simplicity, we have chosen straight and level flight with constant platform velocity; this, of course, is not a requirement. Initial geometry parameters available for operator specification include:

- Radar platform velocity, V_p (m/s).
- Radar platform height above ground plane, h_p (m).
- Depression (or grazing) angle at broadside, ψ_b (deg.) (Also equal to the slope angle ς.)

- Initial squint angle (slant plane), θ_{s_i} (deg.)
- Final squint angle (slant plane), θ_{s_f} (deg.)

Target specification parameters include:

- Number of targets, N_T
- Target amplitude, A_i
- Target distance from patch center, a_i (m)
- Target orientation angle, α_i (deg.)
- Target height above ground plane, h_i (m)

Radar parameters include:

- Radar center frequency, f_0 (Hz)
- Transmitted bandwidth, B (Hz)
- Linear FM chirp rate, $\dot{f}$ (Hz/s)
- Pulse-repetition frequency, PRF (Hz)
- Slant-plane patch diameter, D (m)
- Analog-to-digital (A/D) sample rate, f_s (Hz)

It is necessary to elaborate on the selection of the last three items, namely PRF, D, and f_s. Let us begin by specifying the maximum slant-plane patch diameter D. A spotlight SAR has a physical antenna pattern that actually illuminates the terrain. The amplitude shaping of this antenna pattern defines the uniformity (or lack thereof) of illumination impinging on the ground. Scatterers located within the high field strength region of the antenna return the most signal strength, while scatterers outside the main beam are illuminated with considerably less power. Ideally, the antenna should just illuminate the patch of interest with uniform power, and provide zero power outside the patch. This is, of course, a physically unrealizable situation. Real antenna design is a complicated series of tradeoffs that require considerable analysis and is outside the scope of this book.

The synthetic target generator has no such realizability constraints. Therefore, we can specify the physical region we wish to illuminate with uniform power. By specifying a patch in the slant plane we are defining the region of space in which targets can be placed and are identifying those signals that can be supported by the other two important parameters: the sampling rate f_s, and the pulse repetition frequency PRF.

For example, once the sampling rate and PRF have been selected to support the signals produced by targets whose spatial projection in the slant plane fall within the specified patch, targets outside those dimensions will produce aliased signals that ultimately give rise to image artifacts. Indeed, it is quite simple to create and show the signature of an aliased target exactly like that of a physical target being illuminated by a physical antenna sidelobe that is not sampled rapidly enough by the selected PRF. In summary, the PRF must be high enough to support targets at the extreme edges of the cross-range patch diameter.

Similarly, the A/D (or range) sample rate f_s, must be high enough to support targets that lie at the extreme edges of the slant-range patch diameter. In summary, the sampling rate must be greater than the maximum demodulated video bandwidth; $f_s \geq (2D/c)\dot{f}$.

The synthetic target generator code can compute the minimum sampling rate and PRF required to support all targets whose spatial projections lie within the selected patch given the initial imaging geometry and radar parameters. The operator can, of course, select any f_s and PRF based on knowledge of actual target locations and/or signal effects desired.

The synthetic target generator code also records, in a separate file, the instantaneous (x, y, z) locations (on a pulse-by-pulse basis) of the radar with respect to the imaged patch center. This *pointing vector* file provides pulse-by-pulse information on platform location and its geometrical equivalent in 3-D Fourier space. It is invaluable for coordinate system definition and development, proper polar processing, beam steering (i.e., shifting the processed patch center), out-of-plane and ground-plane projections, aperture overlap computations in interferometry, and a host of other possible compensations required in a SAR collection. Bistatic synthetic target generation results from straightforward modification of the above-mentioned concepts. Target motion is also easily incorporated into the algorithm and allows us to study moving target signatures.

D.2.2 RF and Video Frequency/Time Diagrams

Consider the linear FM chirp frequency/time plot shown in the top half of Figure D.2. The chirp spans an RF bandwidth of B over a pulse interval T. The resulting FM rate is simply $\dot{f} = B/T$. Hypothetical targets lying at the extreme near-range and far-range edge of the imaged patch return linear FM signals that are displaced in time by the round-trip propagation delay and are separated in time by the round-trip patch travel time $2D/c$, as shown in the figure. The round-trip delay to the patch center equals the reference time delay for the demodulation chirp and is given by $\tau_0 = 2r_0/c$.

Demodulation of the returns spanning the patch produce the video frequency/time signal shown in the shaded area of Figure D.2. Targets at the near-range and far-range edges of the patch produce high frequency CW signals, while a hypothetical target *at* the patch center produces a zero frequency (DC) signal. All these signals exist for the pulse time T, but they are linearly skewed relative to one another by the linear propagation delay across the patch.

It is easy to see that the radar receiver should have a time-receive window large enough to encompass signals returning from the extreme range edges of the patch; otherwise, some signals will be cut short with a corresponding loss of image resolution. Mathematically, the receive-time window limits are given by

$$\frac{2r_0 - D}{c} - \frac{T}{2} \leq t_r \leq \frac{2r_0 + D}{c} + \frac{T}{2} \tag{D.4}$$

and spans an effective time, $T_{eff} = T + 2D/c$. The signal processing step required to remove these linear frequency-dependent time delays is called " deskewing" or "deskew correction". The errors and imaging limitations associated with this skew (resulting from the deramp process) are discussed in Chapter 2 of the text and in Appendix B.

If the receive window was selected to encompass the maximum time for which all targets produced signals that exist simultaneously, the effective pulse time is given by $T_{eff} = T - 2D/c$. Thus, the effective receive window must start as soon as the return from the far-range patch edge is received, and must end as soon as the return from the near-range patch edge quits. Mathematically, this truncated receive-time window limits are given by

$$\frac{2r_0 + D}{c} - \frac{T}{2} \leq t_r \leq \frac{2r_0 - D}{c} + \frac{T}{2} \tag{D.5}$$

and spans an effective time, $T_{eff} = T - 2D/c$.

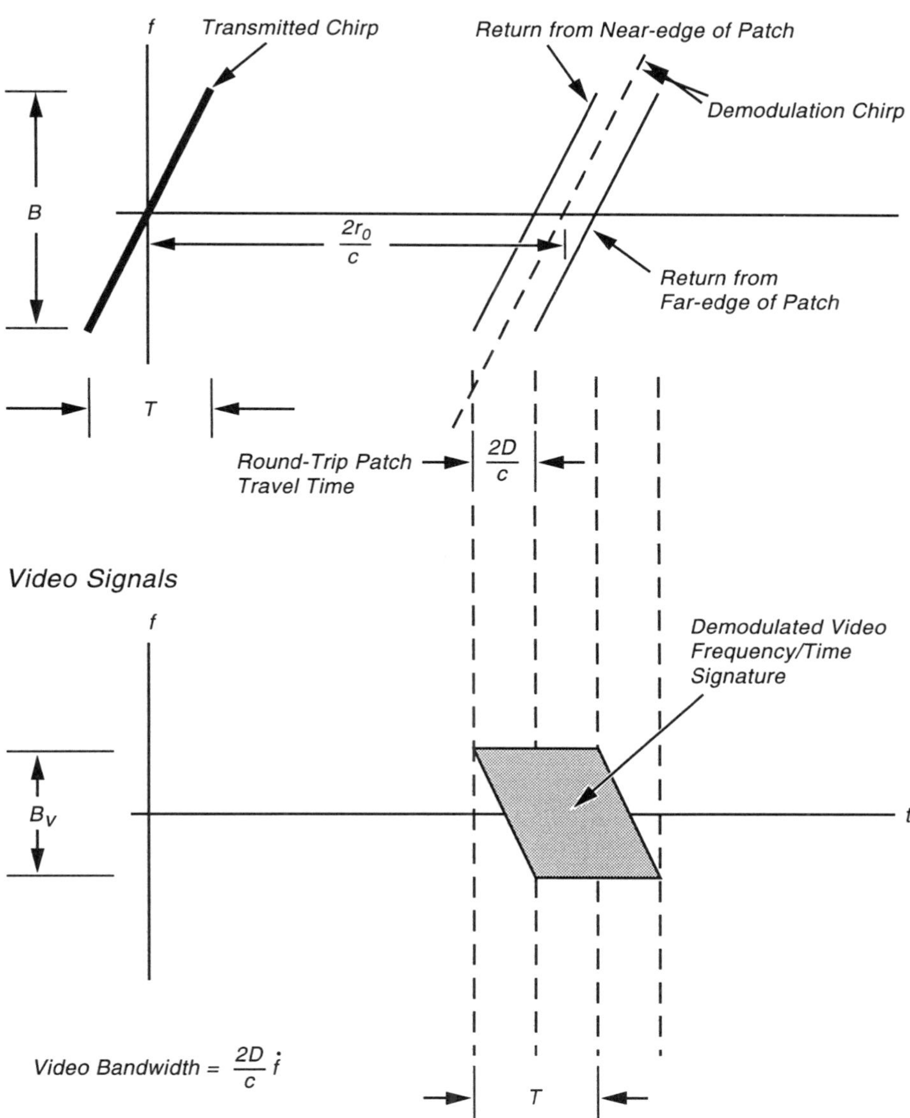

Figure D.2 RF and video frequency/time diagram. Targets at the near-range and far-range edges of the patch produce high frequency CW signals, while a hypothetical target *at* the patch center produces a zero frequency (DC) signal. Each signal exists for the pulse time T. However, the signals are linearly skewed relative to one another by the linear propagation delay across the patch.

Windowing the returns in this manner has the same effect as transmitting a lower bandwidth chirp (with ultimately a loss of range resolution in the final image). This effective bandwidth is given by

$$\begin{aligned} B_{eff} &= \dot{f} T_{eff} \\ &= \dot{f} T - \frac{2D}{c} \dot{f} \\ &= B - B_v \, . \end{aligned} \tag{D.6}$$

The effective bandwidth is simply the original transmitted bandwidth reduced by the video bandwidth. Similarly, the effective system Q factor ($Q = f_0/B$) becomes

$$\begin{aligned} Q_{eff} &= \frac{f_0}{B_{eff}} \\ &= \frac{f_0}{B - B_v} \, . \end{aligned} \tag{D.7}$$

If the transmitted pulse time T is large compared to the round-trip patch travel time $2D/c$, the effective Q is reduced very little from the original system Q and a deskew correction may not be required.

The ultimate performance of the SAR depends on the proper treatment of the transmitted and received signals. Various system constraints may dictate how the demodulated signals are gated and sampled. Knowledge of the associated frequency/time diagram aids understanding of the tradeoffs and resulting consequences.

As a final note, the synthetic-target generator code we use is designed to sample the video signal over the effective pulse time given by Equation D.5 so as to make the time gating simple. Otherwise, without this restriction, the code would have to turn on and turn off target responses depending on their position in the scene and range to the SAR; an unnecessary complication. Target-generator parameters can be easily adjusted to compensate for any loss resulting from this type of time gating and sampling.

E

SOME SIMPLE SYNTHETIC TARGET EXAMPLES

E.1 INTRODUCTION

To help understand the influence of actual radar parameters on the phase history created by simple point targets, it is instructive to generate and view several of these two-dimensional signals. Recall that a spotlight SAR attempts to obtain a slice of the three-dimensional Fourier transform of the target reflectivity. Ideally, a point target produces a three-dimensional complex sinusoid, a planar slice of which is a two-dimensional complex sinusoid (as demodulated in the radar receiver). The synthetically generated data will be displayed on a rectangular array instead of on the polar annulus where they should be placed. It is expected that a point-target phase history thus displayed would be a somewhat distorted version of the true two-dimensional sinusoid when viewed in this rectangular array. The distortion (described in Section 3.4 of the text) indicates the limits to the patch size that can be processed and focused to an image without the need for polar-to-rectangular reformatting.[1]

E.2 EXAMPLES OF SYNTHETICALLY GENERATED PHASE HISTORIES AND IMAGERY

In this section we generate and display several simple phase histories exhibiting various effects. The displayed phase histories will show that fundamental imaging limitations are embedded in the target signals themselves. By viewing these dis-

[1] After proper polar-to-rectangular reformatting, there are still limits to the allowable patch size imposed by the residual errors quantified in Appendix B.

torted signals and the resulting images, we hope to convey some understanding of blurred target signatures. We will also demonstrate the need for polar-to-rectangular reformatting in high-resolution spotlight SAR imaging.

We begin by selecting a set of target-generator parameters for a broadside spotlight collection that provides approximately 0.375 meters resolution in both range and cross range. The patch diameter, A/D sample rate, and PRF will determine a nominal dataset size. The parameters chosen for our particular test case are summarized below.

- Radar platform velocity, $V_p = 500$ (m/s).
- Radar platform height above ground plane, $h_p = 15000$ (m).
- Depression (or grazing) angle at broadside, $\psi_b = \varsigma = 30$ (deg.)
- Initial slant-plane squint angle, $\theta_{s_i} = -1.17$ (deg.)
- Final slant-plane squint angle, $\theta_{s_f} = +1.17$ (deg.)
- Radar center frequency, $f_0 = 1.0 \times 10^{10}$ (Hz)
- Transmitted bandwidth, $B = 4.015 \times 10^8$ (Hz)
- Linear FM chirp rate, $\dot{f} = 1.0 \times 10^{12}$ (Hz/s)
- Pulse repetition frequency, $PRF = 163.5$ (Hz)
- Slant-plane patch diameter, $D = 150$ (m)
- A/D sample rate, $f_s = 1.0 \times 10^6$ (Hz)

Some computed parameters:

- Number of samples per pulse = 400
- Number of pulses = 400
- Maximum valid pulse-time interval, $T_{eff} = 4.005 \times 10^{-4}$ sec.
- Effective transmitted bandwidth, $B_{eff} = 4.005 \times 10^8$ Hz.
- Effective system Q, $Q_{eff} = 24.9688$

Recall from Section 3.4 that the patch size that can be processed by direct 2-D Fourier transformation without polar reformatting is given by

$$D_{max} \leq \frac{4\rho_{x'}\rho_{y'}}{\lambda} \tag{E.1}$$

where $\rho_{x'}$ and $\rho_{y'}$ refer to azimuth (cross-range) and range resolutions, respectively, and λ is the radar wavelength.

We wish to illustrate visually the effect of this limitation on both the phase-history signals and on the resulting image. This can be accomplished by placing point targets at or near the patch diameter limit and by viewing the resultant signals. For our selected parameters, $D_{max} = 18.75$ meters; therefore, targets placed on a radius of 10 meters from the patch center will be slightly beyond this limit and should begin to exhibit some phase errors and defocusing.

Cross-range Target Examples

For our first example, let us place a target 10 meters from the patch center at 0 degrees target angle (see Appendix D, Figure D.1). This target is located entirely in the cross-range dimension of the imaged patch and would ideally produce a 2-D complex sinusoid with spatial frequency *only* in the cross-range dimension. However, because this target is slightly beyond the patch limit, some distortion of the 2-D sinusoid is expected.

We created the phase history corresponding to this target location and placed the sampled returns in a rectangular array on a pulse-by-pulse basis (i.e., without any polar-to-rectangular resampling). Scaling and displaying the real part of such a signal results in the signal shown in Figure E.1(a).

We see the predominant sinusoidal structure in the horizontal (cross-range) dimension, but also notice a compression of the sinusoid toward the top (higher Fourier offset) of the image. A horizontal slice of this 2-D signal would produce a sinusoid whose frequency changes slightly as a function of the vertical position of the slice. This is characteristic of the distortion resulting from direct rectangular display of "cross-range-only" targets. Contrast this figure with Figure E.1(b), which shows the corrected 2-D signal resulting from polar-to-rectangular resampling. The region of support of the 2-D sinusoid in Figure E.1(b) clearly shows the data placed in the proper Fourier annulus, which corrects the distortion.

To exaggerate the distortion of a cross-range target, we simply increase its distance from the patch center. Moving the target out to 50 meters (well beyond the patch

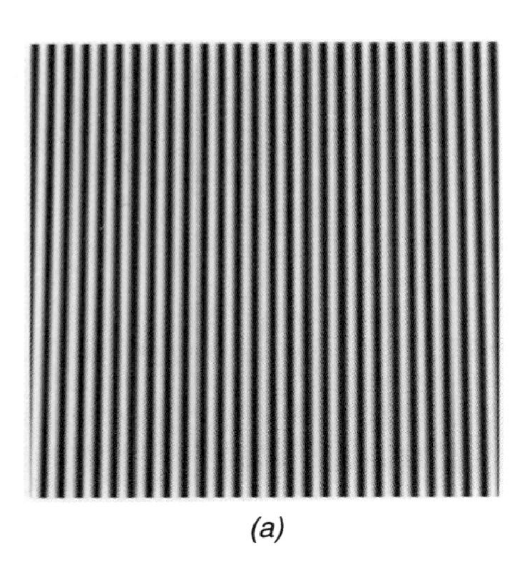

(a)

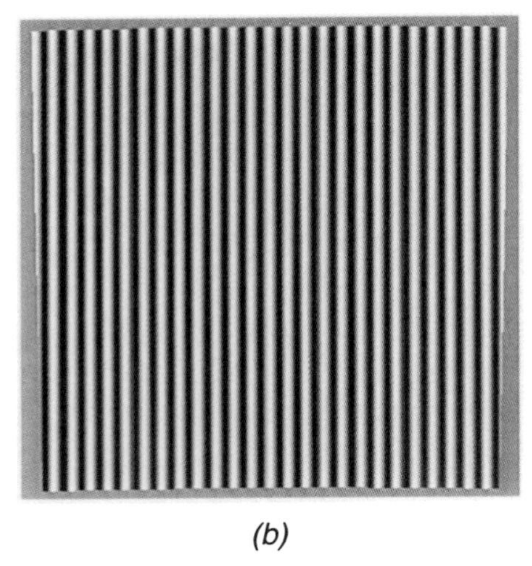

(b)

Figure E.1 Real part of the complex phase history of a single target at 10 meters displacement and 0 degrees target angle (i.e., cross-range target displacement only). (a) Polar reformatting was not performed; instead, the demodulated data were placed on a rectangular array. Notice the slight distortion of the predominant cross-range sinusoidal signal (There is a higher cross-range spatial frequency at the top of the figure than there is at the bottom). (b) Polar processing was applied. The region of support of the 2-D sinusoid clearly shows the data placed in the proper Fourier annulus, which corrects the distortion.

diameter limit) produces a much higher spatial-frequency distorted sinusoid as shown in Figure E.2(a).

In Figure E.2(a) it is difficult to see that the phase history is not a pure sinusoid. Therefore, we have removed the nominal cross-range (horizontal) spatial frequency from Figure E.2(a) and displayed the residual in Figure E.2(b). We find that the residual signal has constant phase contours that are hyperbolas.

It was shown in the fourth term of Equation 3.18 that the phase error resulting from a cross-range target is of the form

$$\phi_\epsilon(\mathcal{X}, \mathcal{Y}) = x_0' \mathcal{X}\mathcal{Y}/k_0 \,. \tag{E.2}$$

Setting ϕ_ϵ to a constant, C, and rearranging yields

$$y = \frac{Ck_0}{x_0'}\left(\frac{1}{\mathcal{X}}\right) . \tag{E.3}$$

This is the equation for a family of hyperbolas. To demonstrate this fact we create the complex signal

$$s(\mathcal{X}, \mathcal{Y}) = e^{j x_0' \mathcal{X}\mathcal{Y}/k_0} \tag{E.4}$$

and display the real part in Figure E.3.

With $\lambda = 0.03$ meters and $x_0 = 50$ meters (from our synthetic target example), $k_0 = 4\pi/\lambda$, and $\rho_{x'} = \rho_{y'} = 0.375$ meters, we find that when Equation E.2 is evaluated at the corners of the 400 by 400 pixel array (which correspond to $\mathcal{X} = \pi/\rho_{x'}$ and $\mathcal{Y} = \pi/\rho_{y'}$), we obtain 8.33 radians of phase error.

A visual comparison of Figures E.2(b) and E.3 shows that the phase error is a hyperbolic function. Figure E.4 shows the magnitude of a simulated slant-plane image, corresponding to the phase history in Figure E.3, obtained by computing the 2-D Fourier transform of Equation E.4. A hypothetical target with a hyperbolic phase error of this form produces the symmetrically defocused image shown. Direct Fourier transformation of the 50-meter cross-range target phase history in Figure E.2(a) produces the slant-plane image shown in Figure E.5. It is apparent that the target is blurred by the same amount as the ideal hyperbolic-phase-error signal in Figure E.4.

Range-displaced Target Examples

We now perform a similar set of experiments with a single target displaced only in the range dimension from the patch center. Figure E.6(a) shows the real part of the

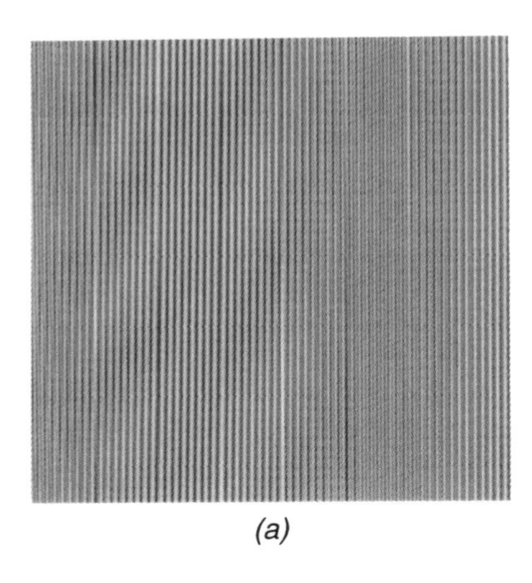

(a)

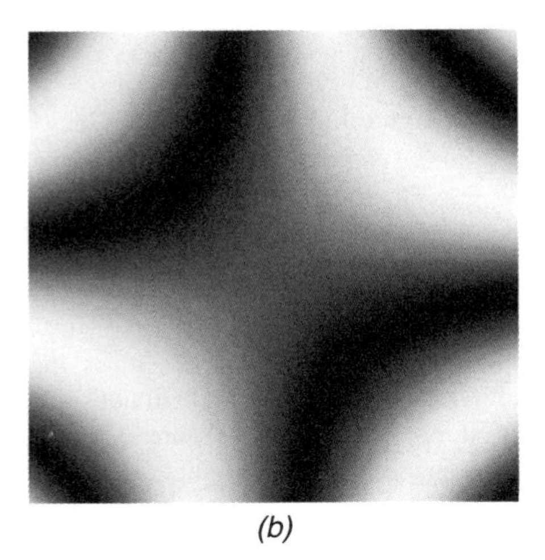

(b)

Figure E.2 Real part of the complex phase history of a single cross-range-displaced target at 50 meters displacement and 0 degrees target angle (no polar processing). (a) The total signal has a high cross-range spatial frequency that makes visualizing the residual distortion difficult. (b) The predominant cross-range spatial frequency is removed to reveal the residual signal. Constant phase contours are hyperbolas that delineate the nature of the residual phase error.

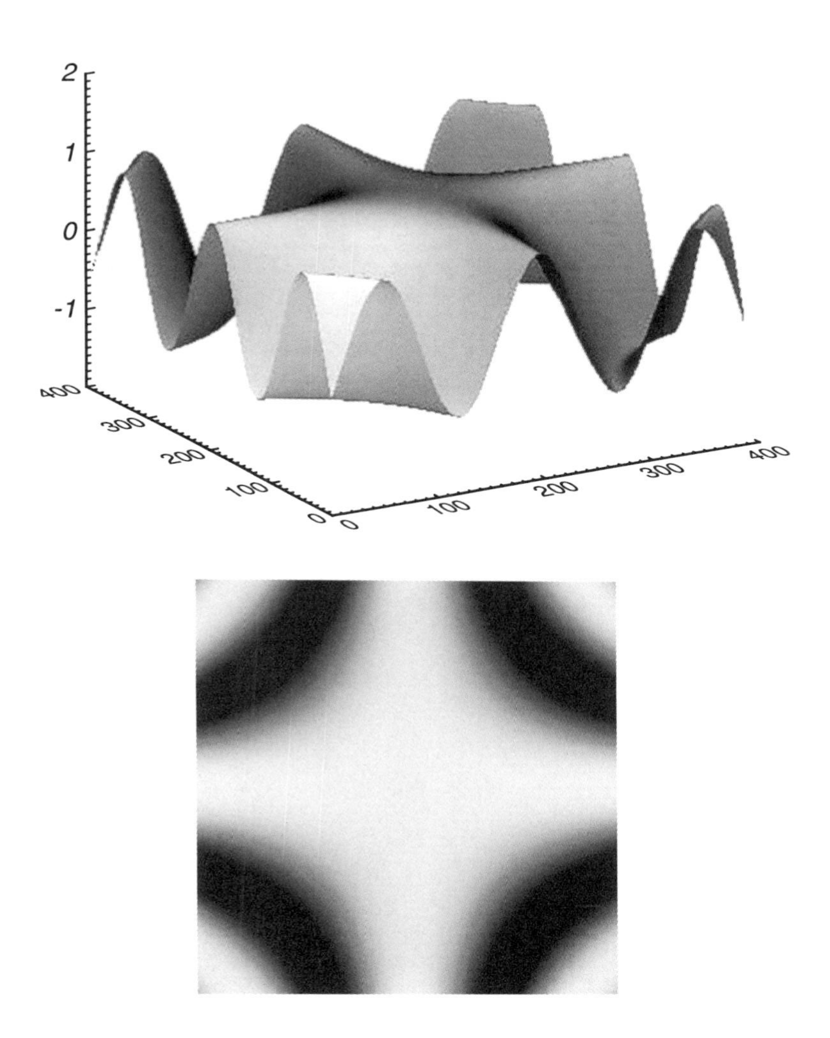

Figure E.3 Real part of the signal with a pure hyperbolic phase residual shown as a shaded surface (top), and as a brightness function (bottom). The phase-error parameters were chosen to agree with the radar and target parameters from the cross-range displaced synthetic target example (with the exception of an inconsequential constant additive phase) to verify the notion that the phase errors are virtually identical. A horizontal or vertical slice through the data produces a sinusoid whose frequency is proportional to the slice position with respect to the image center (i.e., a centered slice produces a zero frequency sinusoid). Compare the bottom part of this figure to Figure E.2(b).

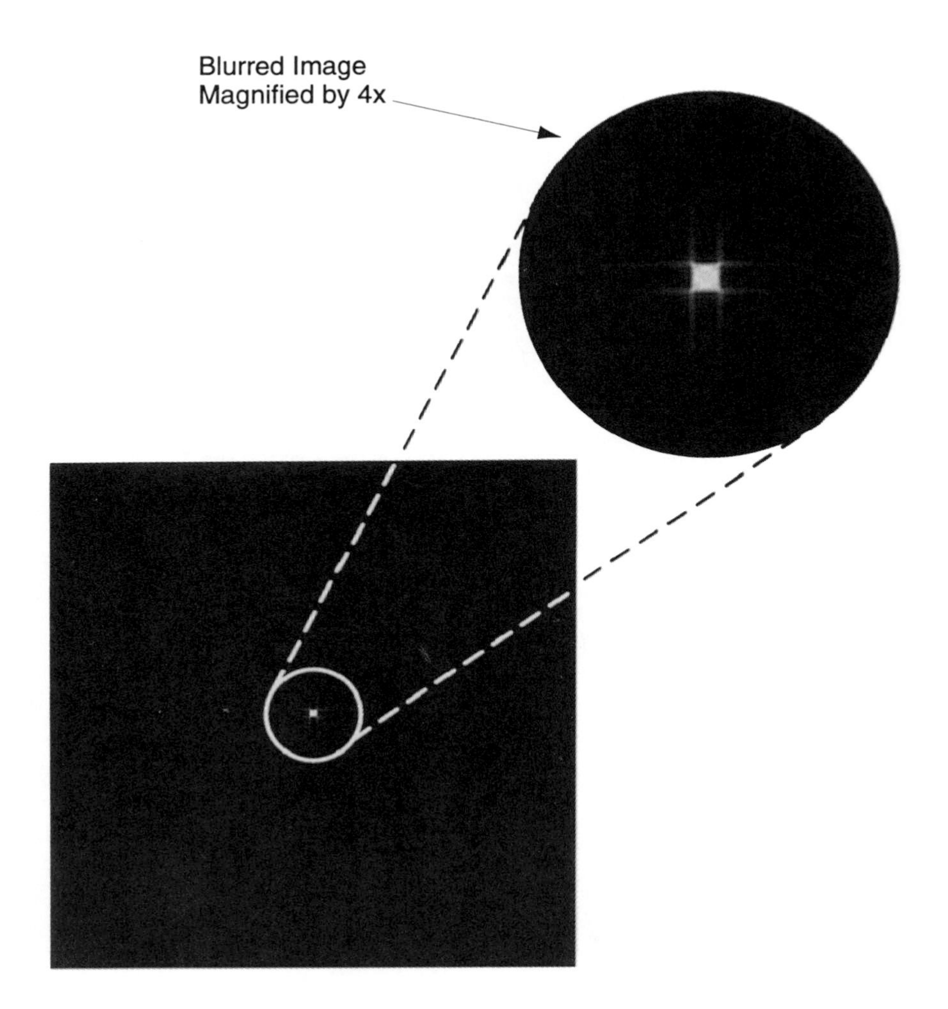

Figure E.4 Magnitude of the image of the signal with a pure hyperbolic phase residual. The blurring is virtually identical to the cross-range-displaced synthetic target image of Figure E.5 because the parameters chosen match the physical parameters used in the synthetic target generation.

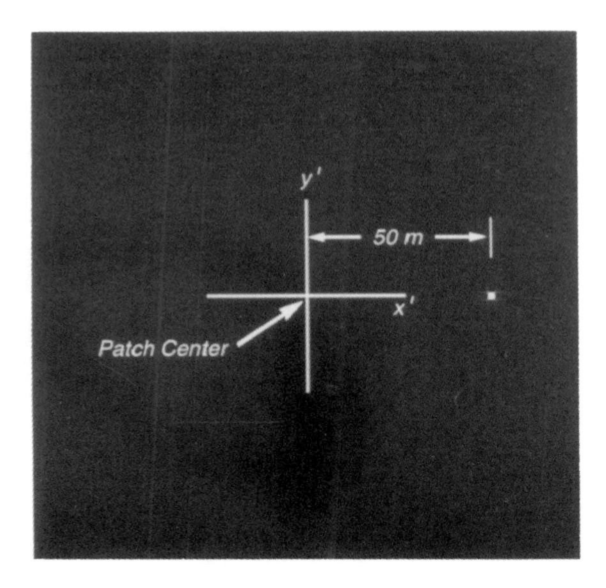

Figure E.5 Magnitude of the slant-plane image of the 50-meter-displaced cross-range target. It is apparent that the target is blurred by the same amount as the ideal hyperbolic-phase-error signal. (See Figure E.4.)

phase history for a single target 10 meters from the patch center at a target-orientation angle of 90 degrees.

Again, we see a predominant sinusoidal signal, but this time it is in the range dimension. The distortion in this signal appears as a definite bow to the constant phase contours. This distortion is such that a vertical slice through the phase history would produce a sinusoidal signal whose frequency does *not* change as a function of the horizontal position of the slice. There is only a phase shift of the sinusoid as a function of horizontal slice position. We can contrast this distortion to the polar-processed phase history shown in Figure E.6(b). The polar annulus region of support is also clearly seen in this figure.

To exaggerate the distortion, we move the target out to 50 meters from the patch center to produce the signal shown in Figure E.7(a). When the dominant sinusoid is removed, we observe the residual signal shown in Figure E.7(b).

This time we see that the constant phase lines run vertically. The residual phase error is strictly one-dimensional in the cross-range direction. A horizontal slice

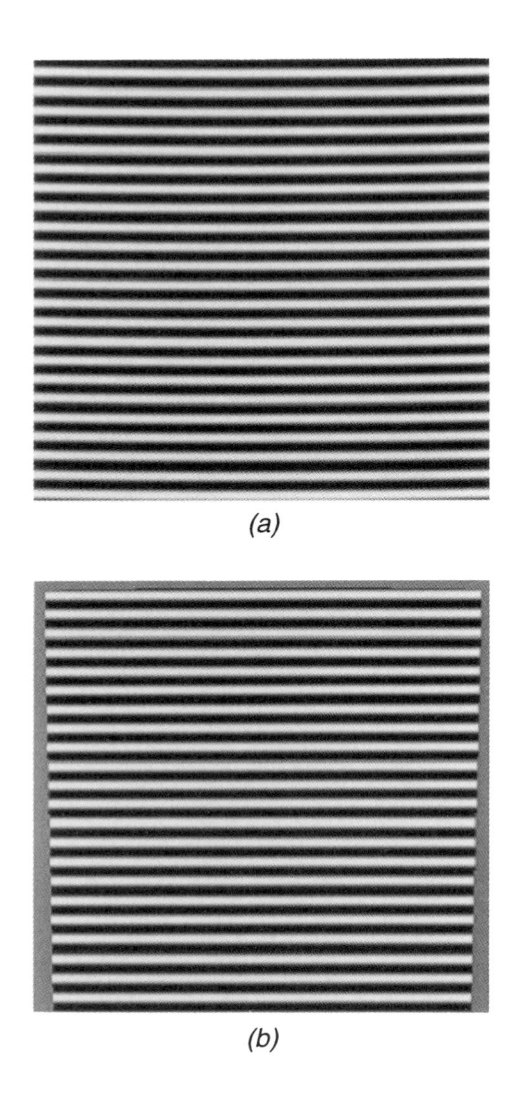

Figure E.6 Real part of the complex phase history of a single target at 10 meters displacement and 90 degrees target angle (i.e., range displacement only). The predominant signal is a sinusoid in the range dimension. (a) Before polar processing is applied, a bow in the constant phase lines is evident. (b) After polar processing, the sinusoidal signal is undistorted and the region of support has the characteristic polar annulus shape.

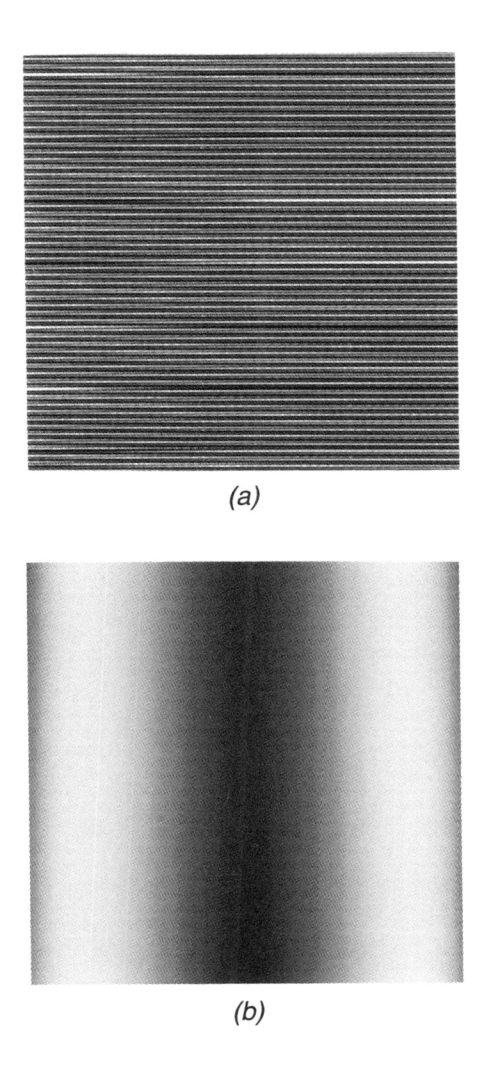

Figure E.7 Real part of the complex phase history of a single target at 50 meters displacement and 90 degrees target angle (i.e., range displacement only). (a) A high spatial frequency in the range dimension makes the distortion more severe but more difficult to visually discern in this example. (b) The predominant range sinusoid has been removed. The residual phase error is quadratic in the *cross-range* direction.

through the residual signal produces a linear frequency chirp. That is, the residual phase error is quadratic in the horizontal dimension and independent of the vertical dimension. Recall from the fifth term in Equation 3.18 that the phase error for a range-displaced target is

$$\phi_\epsilon(\mathcal{X},\mathcal{Y}) = y_0'\mathcal{X}^2/2k_0 \ , \tag{E.5}$$

which is independent of $\mathcal{Y}$.

For comparison, an ideal signal was generated:

$$s(\mathcal{X},\mathcal{Y}) = e^{-jy_0'\mathcal{X}^2/2k_0} \ . \tag{E.6}$$

The parameters of this signal were chosen to match the synthetic target parameters. With $y_0' = 50\cos 30$ (because of the 30 degree grazing angle), we find that 3.6276 radians of phase are produced at the extreme left and right edges of the 400-by-400-pixel phase history. The real part of this signal is shown in Figure E.8. A comparison with Figure E.7(b) indicates that the two phase functions are qualitatively equivalent. The Fourier transform of the phase-history function of Figure E.8 is shown in Figure E.9.

Figure E.10 shows the magnitude of the image formed by direct 2-D Fourier transformation of the synthetic-target phase history in Figure E.7(a). The target is displaced vertically from the patch center by an amount proportional to its distance from the patch center as seen by the SAR in its 30-degree slant plane. Notice that the target is blurred only in the cross-range direction because of the one-dimensional nature of the phase error. Notice also that the target signature is virtually identically to that shown in Figure E.9.

Symmetric Collection of Point Targets Showing Position-Dependent Blurs

One final example shows the phase history and image obtained from a collection of 17 point targets. One target of unity amplitude is placed at the patch center for geometric reference. Eight unit-amplitude targets are placed symmetrically around a circle of 10-meters radius at 45 degree increments. An additional 8 targets (amplitude set to 2.0 to compensate for loss of brightness on display because of blurring) are placed on a circle of 50-meters radius at 45 degree increments. Figure E.11(a) shows the target locations in the ground plane.

The real part of the phase history is shown in Figure E.11(b) and the resultant slant-plane image is shown in Figure E.11(c). Notice that the targets diagonally displaced have blur signatures that are a combination of the hyperbolic and parabolic errors

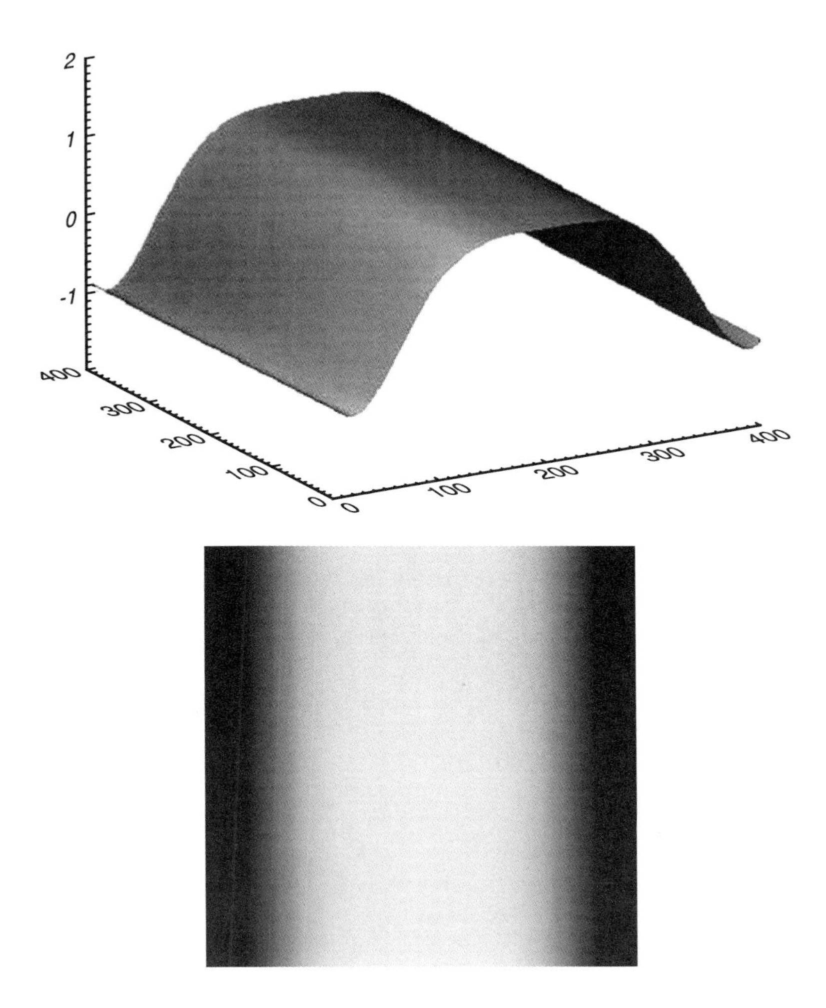

Figure E.8 Real part of the complex phase history of an ideal quadratic phase-error signal shown as a shaded surface (top), and as a brightness function (bottom). The chosen parameters match the range-displaced synthetic target case (with the exception of an inconsequential constant additive phase). A horizontal slice through the data produces a linear FM chirp. A vertical slice is a zero frequency sinusoid (i.e., a constant). Compare to Figure E.7(b).

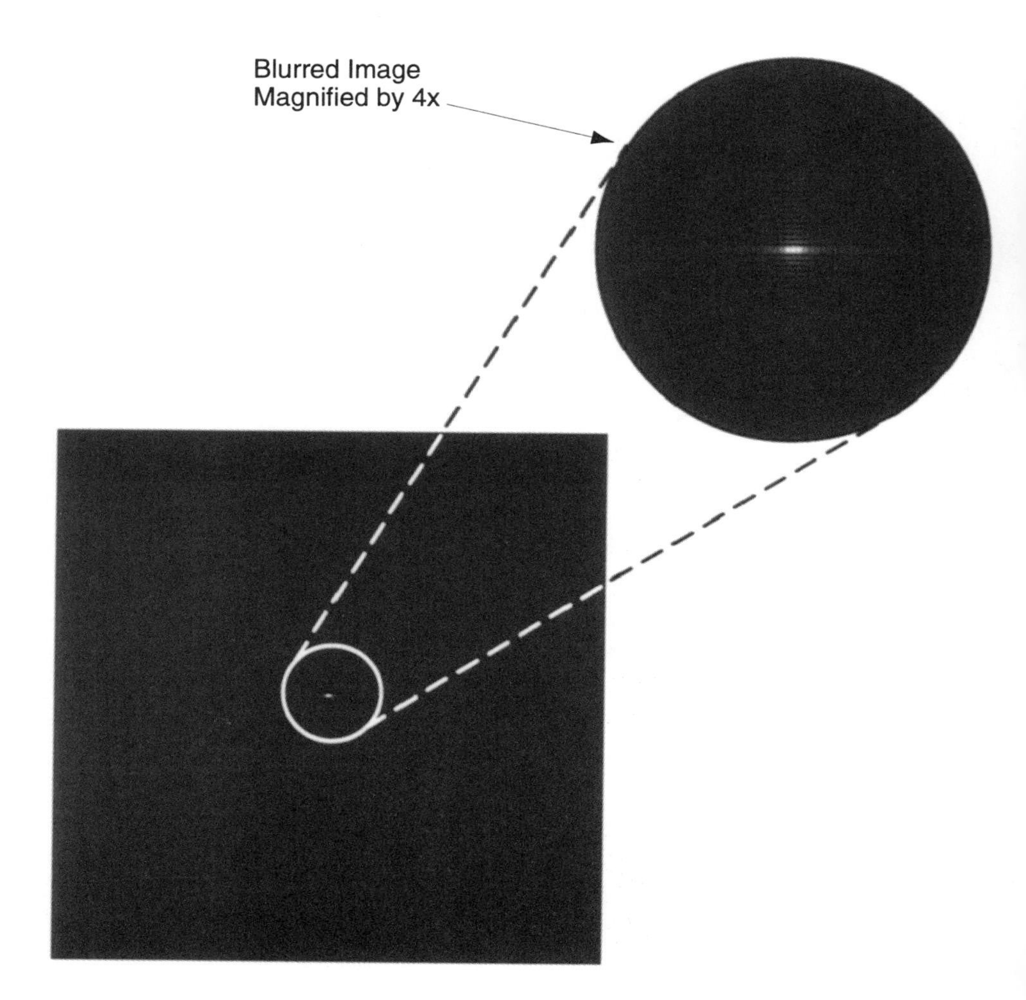

Figure E.9 Magnitude of the image of the ideal quadratic phase-error signal of Equation E.6. The blurring is virtually identical to the range-displaced synthetic target case shown in Figure E.10 because the ideal parameters match the synthetic target parameters.

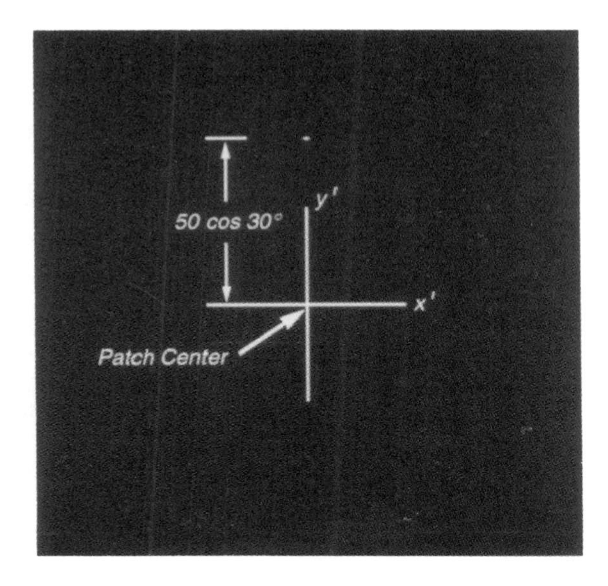

Figure E.10 Magnitude of the slant-plane image of a range displaced target at 50 meters (on ground) displacement and 90 degrees target angle (no polar processing). The blurring is in the cross-range dimension because of the one-dimensional nature of the phase error.

discussed previously. (Compare this image to the non-polar-reformatted SAR image shown in Figure 3.13.) Also, notice that the ideal circular distribution of targets is compressed in the range dimension to an elliptical distribution because of the slant-plane imaging condition. As expected, the ratio of the minor to major ellipse diameter is equal to the cosine of the grazing angle.

We have shown that a high-resolution (e.g., $\rho = 0.375$ meters) SAR image cannot be produced over a usable patch size by direct 2-D Fourier transformation of the raw phase history. The patch size limits of Equation E.1 are real and quite severe for high-resolution, short-wavelength systems. Proper polar-to-rectangular resampling is required to extend the usable patch diameter out to the limits imposed by the second-order effects quantified in Appendix B.

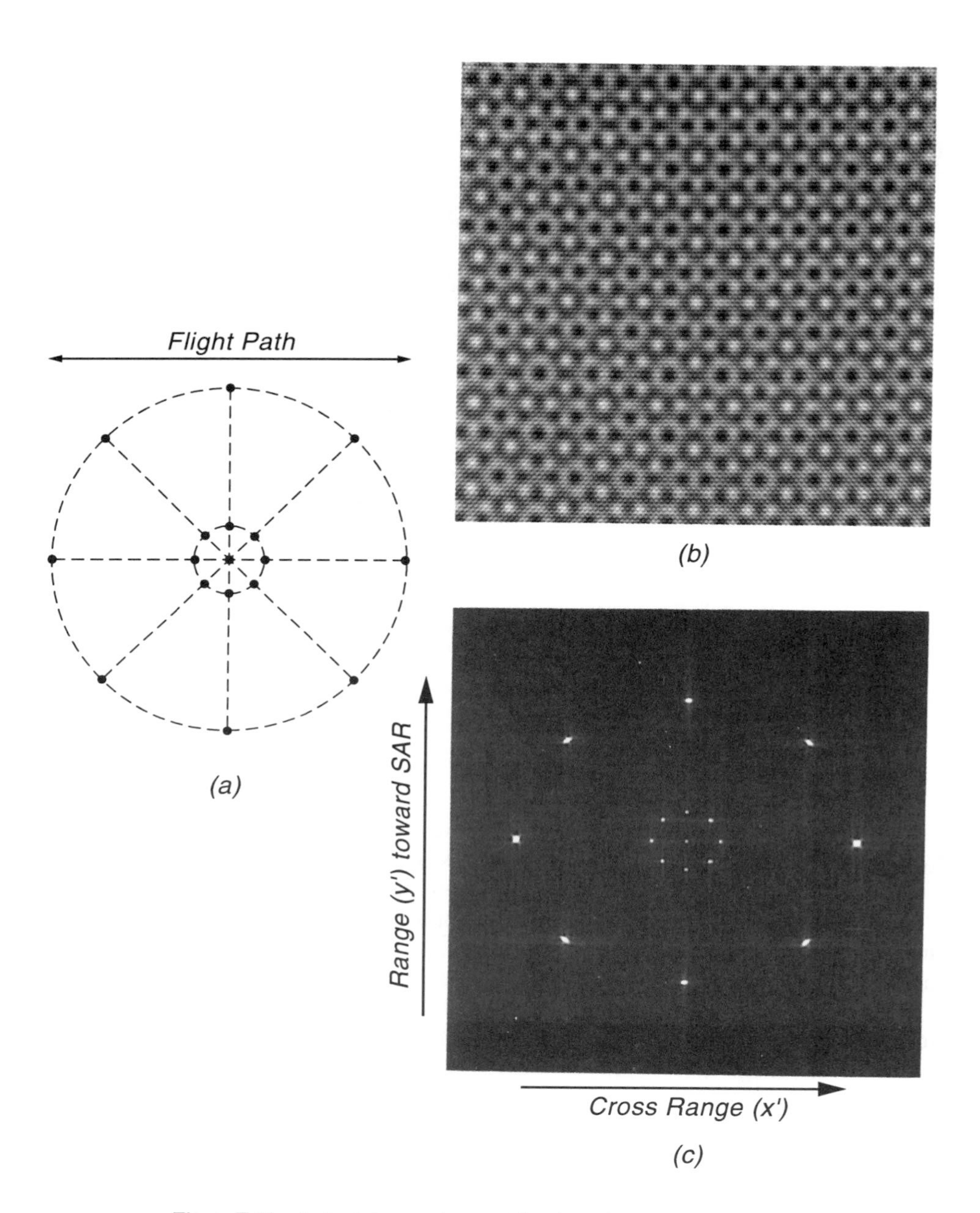

Figure E.11 A simulation to show position-dependent blur. (a) Locations of 17 targets in the ground plane. (b) Real part of the complex phase history without polar processing. (c) Magnitude of the slant-plane image formed without polar reformatting. Notice that the targets diagonally displaced from the center (i.e., offset in range and cross-range) have blur signatures that combine the hyperbolic and parabolic errors discussed in the text. (See also Figure 3.13.)

F

ML PHASE ESTIMATION AND THE CRAMER-RAO LOWER BOUND

This appendix consists of three sections. Section F.1 contains a derivation of a maximum-likelihood estimator for the phase difference of adjacent pulse pairs in the range-compressed domain. Section F.2 gives a discussion of the estimator bias properties. Section F.3 gives a derivation of the Cramer-Rao lower bound (CRLB) which is a theoretical lower bound for any estimation scheme. The performance of the ML estimator is then compared to the CRLB.

F.1 DERIVATION OF THE MAXIMUM-LIKELIHOOD PHASE ESTIMATOR

We assume the SAR image is composed of N range lines and M cross-range columns. After center-shifting and windowing, we treat the real and imaginary (I and Q) components of the complex reflectivity of the point target on each range line as zero-mean Gaussian random variables that are mutually independent and identically distributed (*iid*). The clutter reflectivity I and Q values, which exist at every cell in the image space, are also modeled as *iid* Gaussian random variables. All clutter and target components are mutually independent across all values of range and cross-range. The clutter is thereby modeled as uniform intensity Gaussian white noise.

The model for the data after center-shifting, windowing, and Fourier transforming to the aperture domain is given in Equation 4.38. The model reflects the fact that the image-domain target structure is a single point reflector at the center of each range line. The corresponding target values in the range-compressed domain are simply complex constants across the aperture dimension, different for each range line. The clutter terms in the range-compressed domain remain as uniform intensity white Gaussian noise, as a consequence of the fact that the discrete Fourier transform is a *unitary* linear transformation. Therefore, the I and Q components of the noise components are all mutually independent Gaussian random variables (for all range and cross-range positions). Also, the I and Q components of the target values are mutually independent across the range (k) dimension, and are independent of all clutter values.

We focus our attention on two adjacent pulses m and $m-1$. Using Equation 4.38, we write the model for these pulses as

$$\begin{aligned} \bar{g}(k, m-1) &= a(k)e^{j\phi(m-1)} + \eta(k, m-1) \\ \bar{g}(k, m) &= a(k)e^{j\phi(m)} + \eta(k, m) \,. \end{aligned} \quad \text{(F.1)}$$

For notational convenience, we make the following substitutions:

$$\begin{aligned} \bar{g}_k &= \bar{g}(k, m-1) \\ \bar{h}_k &= \bar{g}(k, m) \\ n_{1k} &= \eta(k, m-1) \\ n_{2k} &= \eta(k, m) \\ \tilde{a}_k &= a(k)e^{j\phi(m-1)} \\ \Delta\phi &= \phi(m) - \phi(m-1) \end{aligned}$$

in Equation F.1. This results in the simplified expression

$$\begin{aligned} \bar{g}_k &= \tilde{a}_k + n_{1k} \\ \bar{h}_k &= \tilde{a}_k \, e^{j\Delta\phi} + n_{2k} \,. \end{aligned} \quad \text{(F.2)}$$

We are interested in estimating the phase difference, $\Delta\phi_m$, between adjacent pulses $m-1$ and m in the range-compressed data as shown Figure F.1. We let the variance of the components of n_{1k} and n_{2k} be $\sigma_n^2/2$, and the variance of the components of $\tilde{a}_k$ be $\sigma_a^2/2$. The *target-to-clutter ratio* for a single range line of this canonical image can therefore be defined as

$$\beta = \frac{\sigma_a^2}{\sigma_n^2} \,. \quad \text{(F.3)}$$

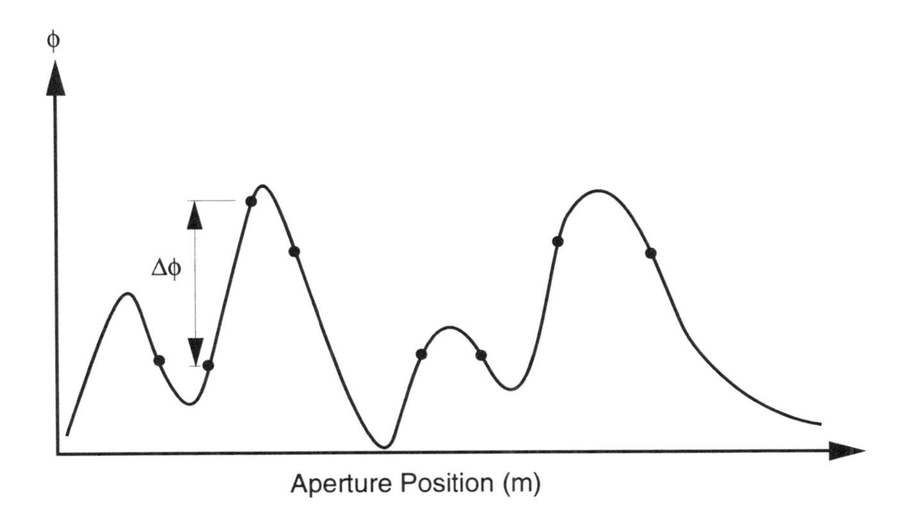

Figure F.1 Phase error to be estimated between two adjacent pulses.

As discussed in section 4.5.1, an important step in the PGA algorithm structure is the *windowing* (filtering) of the image-domain data prior to transformation to the range-compressed space. For low-order phase-error functions, the width of the window is generally estimated from the support (width) of the defocusing (point spread) function. High-order phase errors require a progressive windowing scheme, where the window width is a pre-determined function of the iteration number. The effective value of β, in turn, varies inversely as the chosen window width. This is because σ_n^2 changes in direct proportion to this width. The motivation for windowing is to make β as large as possible, short of rejecting data interior to the defocus width dictated by the degrading phase-error function.

We will now show the major steps in the derivation of the maximum-likelihood estimator for $\Delta\phi$. Let $\mathbf{x}$ be a vector defined as

$$\mathbf{x} = [\ \bar{g}_k \quad \bar{h}_k\ ]^t \tag{F.4}$$

and is composed of the two complex range-compressed values in Equation F.2.

Given the assumptions described above, the conditional probability density function (pdf), given $\Delta\phi$, for this complex vector can be written compactly as[1]

$$p_{\mathbf{x}\,|\Delta\phi}(\mathbf{x}\,|\Delta\phi) = \frac{1}{\pi^2\,|\mathbf{c}|}\exp\{-[\mathbf{x}^{*\mathbf{t}}\ \mathbf{c}^{-1}\ \mathbf{x}]\} \tag{F.5}$$

where $\mathbf{x}^{*\mathbf{t}}$ denotes transpose conjugate and where $|\mathbf{c}|$ denotes the determinant of the covariance matrix, $\mathbf{c}$, which is defined as

$$\mathbf{c} = E\{\mathbf{x}\,\mathbf{x}^{*t}\}\ . \tag{F.6}$$

The calculation of $\mathbf{c}$ is

$$E\left\{\begin{bmatrix} \tilde{a}_k + n_{1k} \\ \tilde{a}_k\ e^{j\Delta\phi} + n_{2k} \end{bmatrix} \begin{bmatrix} \tilde{a}_k^* + n_{1k}^* & \tilde{a}_k^* e^{-j\Delta\phi} + n_{2k}^* \end{bmatrix}\right\} = \begin{bmatrix} \sigma_a^2 + \sigma_n^2 & \sigma_a^2\ e^{-j\Delta\phi} \\ \sigma_a^2\ e^{j\Delta\phi} & \sigma_a^2 + \sigma_n^2 \end{bmatrix}.$$

Equation F.5 can then be rewritten as

$$p_{\mathbf{x}\,|\Delta\phi}(\mathbf{x}\,|\Delta\phi) = \frac{1}{\pi^2|\mathbf{c}|}\ \exp[-Z_k(\Delta\phi)] \tag{F.7}$$

with

$$Z_k(\Delta\phi) \ = \ \frac{\sigma_n^2}{|\mathbf{c}|}\{(\beta+1)(|\bar{g}_k|^2 + |\bar{h}_k|^2) - \beta[2\,\mathbf{Re}(\bar{g}_k\bar{h}_k^* e^{j\Delta\phi})]\} \tag{F.8}$$

and $|\mathbf{c}|$ given by

$$|\mathbf{c}| = \sigma_n^2\ (\sigma_n^2 + 2\sigma_a^2) \tag{F.9}$$

where β is the single range line target-to-clutter ratio defined in Equation F.3.

For N independent range lines of data taken simultaneously, the log of the pdf becomes

$$ln\ p(\mathbf{g},\mathbf{h}\,|\,\Delta\phi) = \alpha_1 - \sum_{k=1}^{N} Z_k(\Delta\phi) \tag{F.10}$$

where $Z_k(\Delta\phi)$ is as defined in Equation F.8 and $\mathbf{g}$ and $\mathbf{h}$ are N-dimensional complex vectors that represent the entire columns (pulses) of range-compressed data for adjacent aperture positions. The term α_1 is a function of the data only and not of $\Delta\phi$.

Combining Equations F.8 and F.10 gives the following:

$$ln\ p(\mathbf{g},\mathbf{h}\,|\,\Delta\phi) = \alpha_2 + \frac{2\sigma_a^2}{|\mathbf{c}|}\ \rho(\mathbf{g},\mathbf{h})\ \left\{\cos\left(\Delta\phi - \angle\left[\sum_{k=1}^{N}\bar{g}_k^*\bar{h}_k\right]\right)\right\} \tag{F.11}$$

[1] N. R. Goodman (reference [14] of Chapter 4) discusses the general conditions under which Equation F.5 is valid.

where

$$\rho(\mathbf{g}, \mathbf{h}) = \left\{ \left\| \left[\sum_{k=1}^{N} \bar{g}_k^* \bar{h}_k \right] \right\| \right\} = ||\mathbf{g}^{*t}\ \mathbf{h}|| \tag{F.12}$$

and α_2 is not a function of $\Delta\phi$.

The ML estimator of $\Delta\phi$ is now calculated by finding that value of $\Delta\phi$ which maximizes the log-likelihood function of Equation F.11.[2] Inspection of this equation gives the solution as

$$\widehat{\Delta\phi}_{ML} = \angle \left[\sum_{k=1}^{N} \bar{g}_k^* \bar{h}_k \right] \tag{F.13}$$

where $\angle$ denotes the value of the angle of the complex quantity, computed on the interval $[-\pi, \pi]$. (We use the notation $\hat{\phi}$ to denote an estimate for ϕ.)

Equation F.13 has the desirable property that the target-to-clutter ratio, β, is not an input. This is a consequence of the assumption that the clutter variance is constant over all range lines. The ultimate performance of the estimator, of course, *is* a function of β, as will be seen in section F.3.

F.2 ESTIMATOR BIAS

In this section we will demonstrate that the ML estimator for $\Delta\phi$ derived above is unbiased even for reasonably small values of the target-to-clutter ratio, β. To show that the ML estimator is unbiased, we must demonstrate that

$$E\{\widehat{\Delta\phi}_{ML}\} = \Delta\phi\ . \tag{F.14}$$

Using equations F.2 and F.13, we have

$$\begin{aligned} E\{\widehat{\Delta\phi}_{ML}\} &= E\left\{ \angle \left[\sum_{k=1}^{N} \bar{g}_k^* \bar{h}_k \right] \right\} = E\left\{ \angle \sum_{k=1}^{N} (\tilde{a}_k^* + n_{1k}^*)\, (\tilde{a}_k\, e^{j\Delta\phi} + n_{2k}) \right\} \\ &= E\left\{ \angle \left[\sum_{k=1}^{N} |\tilde{a}_k|^2\, e^{j\Delta\phi} + \sum_{k=1}^{N} (n_{1k}^* \tilde{a}_k\, e^{j\Delta\phi} + \tilde{a}_k^* n_{2k} + n_{1k}^* n_{2k}) \right] \right\} \\ &= E\left\{ \angle \left[S + N \right] \right\} \end{aligned} \tag{F.15}$$

[2] Because the logarithm is a monotonic function, this will result in maximization of the likelihood function itself

where

$$\begin{aligned} S &= \sum_{k=1}^{N} |\tilde{a}_k|^2 \, e^{j\Delta\phi} \\ N &= \sum_{k=1}^{N} (n_{1k}^* \tilde{a}_k \, e^{j\Delta\phi} + \tilde{a}_k^* n_{2k} + n_{1k}^* n_{2k}) \, . \end{aligned} \quad \text{(F.16)}$$

Because the expectation operator cannot simply be "passed through" the $\angle$ operator, an expression for the expected value in Equation F.15 is difficult to obtain. However, a heuristic argument can be made as follows. The two summations in the last line of Equation F.15 can be interpreted as two vectors. It is required to compute the expected value of the angle of the sum of these two vectors. The first sum can be represented as a vector with an angle equal to $\Delta\phi$, (shown as vector **A** in Figure F.2), because every vector in the sum has an angle equal to $\Delta\phi$. Its magnitude is the sum of the squared magnitudes of the central point targets on all range lines. The second sum can be thought of as a small random vector (shown by vector **B** in Figure F.2), centered on the tip of the first vector and uniformly distributed in angle (see reference [9] of Chapter 4). For moderate to high values of target-to-clutter ratio, the length of vector **B** is, with very high probability, much smaller than the length of vector **A**. The average angle of the sum vector, **C**, is then equal to $\Delta\phi$, so that the estimate is unbiased. The curve of Figure F.3 shows the expected value of $\widehat{\Delta\phi}_{ML}$ vs β for the case of 512 range lines. The true value of $\Delta\phi$ was 90°. Note the avalanche effect on estimator bias that occurs when the target-to-clutter ratio falls below that threshold for which the length of vector **B** is no longer very small compared to the length of vector **A**. The estimator angle then becomes totally random, i.e., uniformly distributed on $[-\pi, \pi]$, with an expected value of 0. Note that β need only be on the order of -10 dB for the estimator to remain unbiased for the case of 512 range lines.

F.3 THE CRAMER-RAO LOWER BOUND

We are now in a position to compute the Cramer-Rao lower bound (CRLB) on estimation error variance and to compare the performance of the ML estimator with that bound. One expression for the CRLB (see p. 66, reference [15] of Chapter 4) is given by

$$Var[\widehat{\Delta\phi} - \Delta\phi] \geq \left\{ -E\left[\frac{\partial^2}{\partial \Delta\phi^2} \{ln \; p(\mathbf{g}, \mathbf{h} \mid \Delta\phi)\} \right] \right\}^{-1} . \quad \text{(F.17)}$$

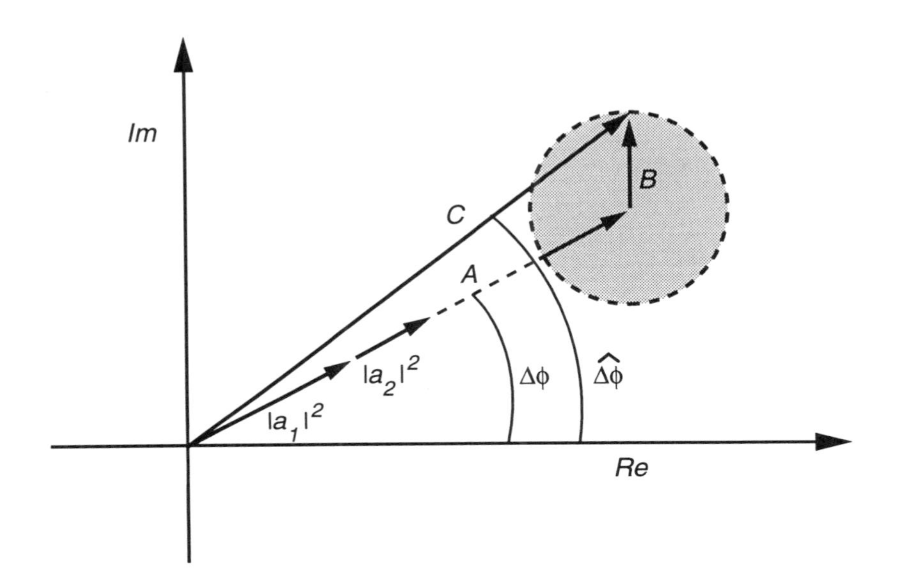

Figure F.2 Vector interpretation of phase difference measurement. The vertical direction is the imaginary part and the horizontal direction is the real part.

Using Equations F.8 and F.10, we have

$$\frac{\partial^2}{\partial \Delta\phi^2}\{ln\ p(\mathbf{g},\mathbf{h} \mid \Delta\phi)\} = \frac{-\sigma_a^2}{|\mathbf{c}|}\left[\sum_{k=1}^{N} \bar{g}_k^* \bar{h}_k e^{-j\Delta\phi} + \sum_{k=1}^{N} \bar{g}_k \bar{h}_k^* e^{j\Delta\phi}\right] . \quad \text{(F.18)}$$

The negative inverse of the expected value of the above expression is

$$\frac{\sigma_n^2(\sigma_n^2 + 2\sigma_a^2)}{2N\sigma_a^4} . \quad \text{(F.19)}$$

For large values of β, this expression is approximated as

$$CRLB \approx \frac{1}{N\,\beta} \quad \text{(F.20)}$$

while for small values of β, this expression becomes

$$CRLB \approx \frac{1}{2\,N\,\beta^2} . \quad \text{(F.21)}$$

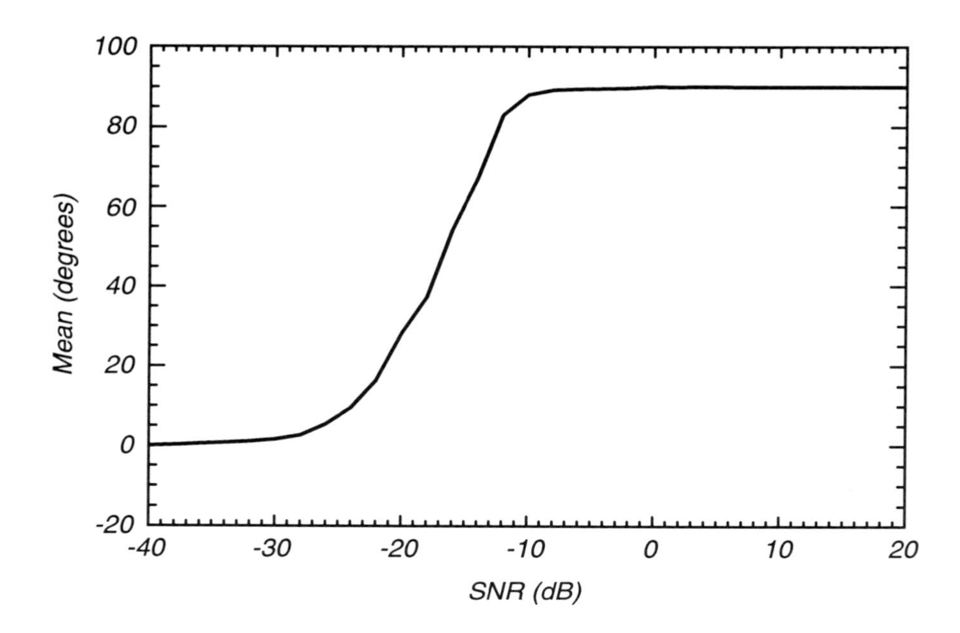

Figure F.3 Mean ML estimator derived via a Monte-Carlo simulation.

The question that next arises is: "Over what range of target-to-clutter ratios does the variance of the ML estimator for $\Delta\phi$ meet the Cramer-Rao lower bound?" All ML estimators have the property that in the limit as the estimation error becomes very small (e.g., the number of observations becomes very large and/or the signal-to-noise ratio becomes very large), the estimator is *efficient*, i.e., its variance achieves the value dictated by the Cramer-Rao lower bound. The performance of the ML estimator for moderate to low signal-to-noise ratios, however, may or may not achieve the CRLB. A necessary and sufficient condition for the estimator to be efficient is that the first partial of the log-likelihood function can be expressible in the form

$$\frac{\partial}{\partial \Delta\phi}\{ln\ p(\mathbf{g},\mathbf{h} \mid \Delta\phi)\} = [\widehat{\Delta\phi}(\mathbf{g},\mathbf{h}) - \Delta\phi]\ k(\Delta\phi) \qquad \text{(F.22)}$$

for all $\mathbf{g},\mathbf{h}$, and $\Delta\phi$ (see p. 73, reference [15] of Chapter 4).

From Equation F.11, we obtain this expression for the first partial of the log-likelihood function:

$$\frac{\partial}{\partial \Delta\phi}\{ln\ p(\mathbf{g},\mathbf{h} \mid \Delta\phi)\} = \frac{-2\sigma_a^2}{|\mathbf{c}|}\rho(\mathbf{g},\mathbf{h}) \left\{ \sin\left(\Delta\phi - \angle\left[\sum_{k=1}^{N} \bar{g}_k^* \bar{h}_k\right]\right)\right\}. \qquad \text{(F.23)}$$

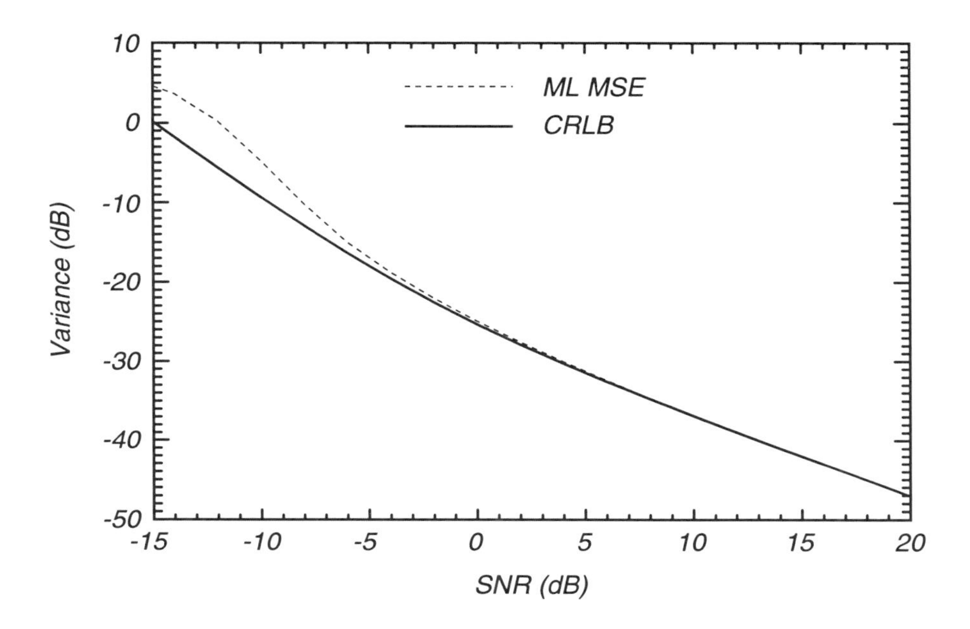

Figure F.4 MSE of ML estimator vs. the CRLB. The simulation was performed using 512 range lines.

The above expression does not meet the form required by Equation F.22. In the limit as the number of observations (range lines) becomes large, however, it is clear that the required form is in fact attained, because $\sin(\Delta\theta) \approx \Delta\theta$ for small $\Delta\theta$, and $\Delta\theta$ is the estimation error, which becomes small as the number of observations grows. Also, the term $\rho(\mathbf{g}, \mathbf{h})$ will be essentially constant (equal to $N\sigma_a^2$) as N becomes large.

Figure F.4 shows a plot of the mean-square error (MSE) of the ML estimator vs. target-to-clutter ratio, as determined via a Monte Carlo simulation. The simulation started with a pair of columns, g and $\mathbf{h}$, from the canonical image of Equation F.2, with $N = 512$. Also plotted here is the Cramer-Rao lower bound of Equation F.19. Note that the ML estimator essentially achieves the CRLB down to a β value of -5 dB. Values for β found in real SAR imagery are often on the order of -5 dB to -10 dB.

INDEX